ALGEBRA 1 INTERACTIVE STUDENT GUIDE

FOR THE COMMON CORE

mheonline.com

Send all inquiries to:
McGraw-Hill Education
8787 Orion Place
Columbus, OH 43240

ISBN: 978-0-02-143921-8
MHID: 0-02-143921-4

Printed in the United States of America.

4 5 6 7 8 9 QTN 19 18 17

Contents

Chapter 4 Systems of Linear Equations and Inequalities

Chapter 5 Linear Functions

Chapter 6 Equations of Linear Functions

Chapter 7 Exponential Functions

Chapter 8 Polynomials

Chapter 9 Quadratic Equations

Chapter 10 Quadratic Functions

Chapter 11 Statistics

How to Use This Program

Each chapter of the *Common Core Interactive Student Guide* has features to help you succeed.

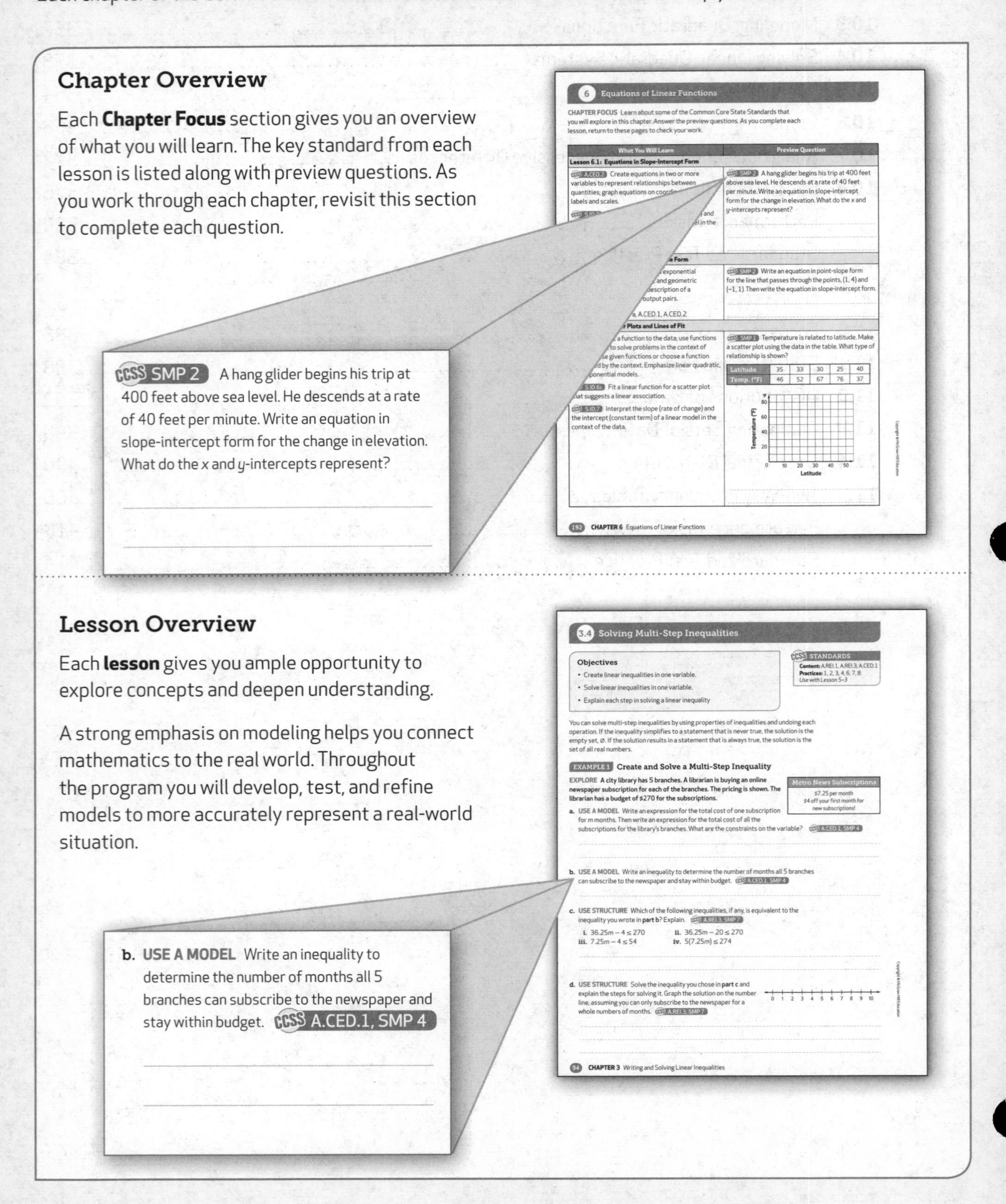

Chapter Overview

Each **Chapter Focus** section gives you an overview of what you will learn. The key standard from each lesson is listed along with preview questions. As you work through each chapter, revisit this section to complete each question.

Lesson Overview

Each **lesson** gives you ample opportunity to explore concepts and deepen understanding.

A strong emphasis on modeling helps you connect mathematics to the real world. Throughout the program you will develop, test, and refine models to more accurately represent a real-world situation.

Assessment Practice

Performance Tasks at the end of each chapter allow you to practice abstract reasoning, perseverance, and problem solving on tasks similar to those found on Common Core assessments.

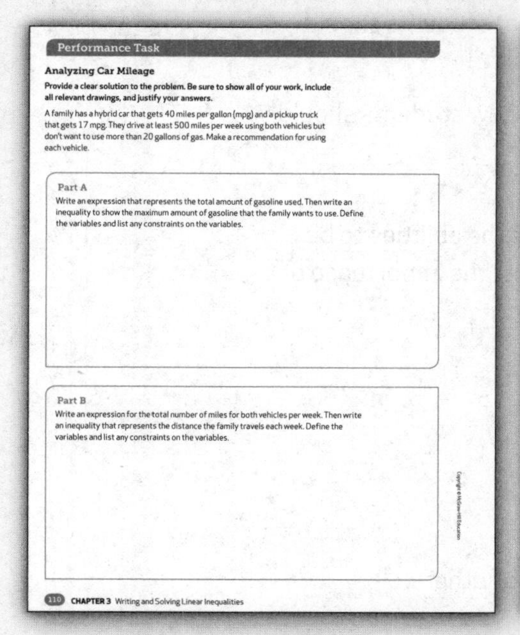

Performance Task

Analyzing Car Mileage

Provide a clear solution to the problem. Be sure to show all of your work, include all relevant drawings, and justify your answers.

A family has a hybrid car that gets 40 miles per gallon (mpg) and a pickup truck that gets 17 mpg. They drive at least 500 miles per week using both vehicles but don't want to use more than 20 gallons of gas. Make a recommendation for using each vehicle.

Part A

Write an expression that represents the total amount of gasoline used. Then write an inequality to show the maximum amount of gasoline that the family wants to use. Define the variables and list any constraints on the variables.

Part B

Write an expression for the total number of miles for both vehicles per week. Then write an inequality that represents the distance the family travels each week. Define the variables and list any constraints on the variables.

Copyright © McGraw-Hill Education

110 CHAPTER 3 Writing and Solving Linear Inequalities

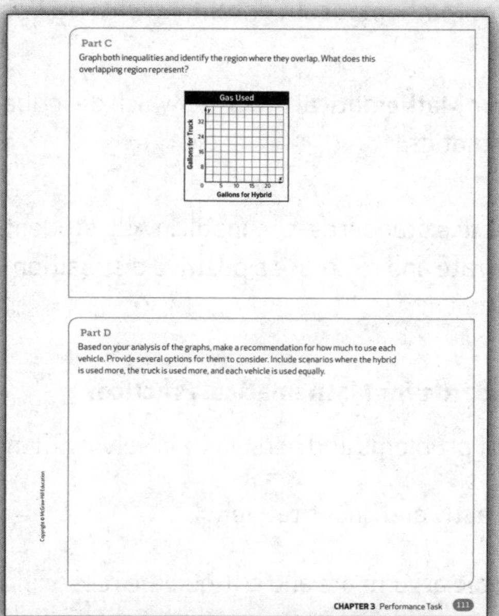

Part C

Graph both inequalities and identify the region where they overlap. What does this overlapping region represent?

Part D

Based on your analysis of the graphs, make a recommendation for how much to use each vehicle. Provide several options for them to consider. Include scenarios where the hybrid is used more, the truck is used more, and each vehicle is used equally.

Copyright © McGraw-Hill Education

CHAPTER 3 Performance Task 111

At the end of each chapter, there is a **Test Practice** section that is formatted like the Common Core assessments. Use the test practice to familiarize yourself with new kinds of assessment questions.

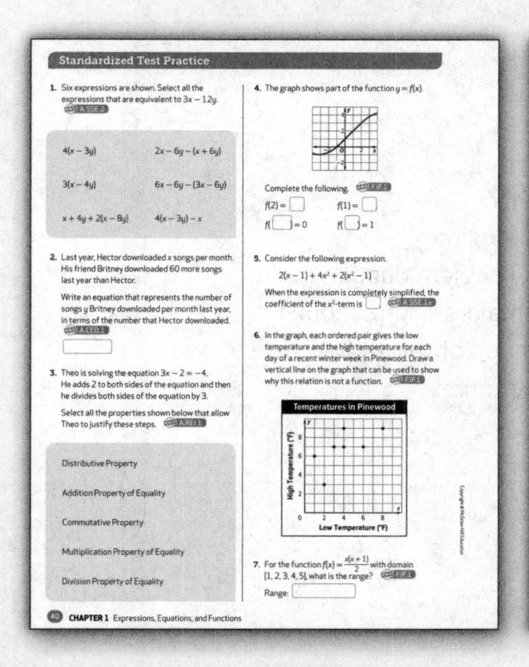

Standardized Test Practice

1. Six expressions are shown. Select all the expressions that are equivalent to $3x - 12y$. CCSS A.SSE.2

$4(x - 3y)$	$2x - 6y - (x + 6y)$
$3(x - 4y)$	$6x - 6y - (3x - 6y)$
$x + 4y + 2(x - 8y)$	$4(x - 3y) - x$

2. Last year, Hector downloaded x songs per month. His friend Britney downloaded 60 more songs last year than Hector.

   Write an equation that represents the number of songs y Britney downloaded per month last year, in terms of the number that Hector downloaded. CCSS A.CED.1

3. Theo is solving the equation $3x - 2 = -4$. He adds 2 to both sides of the equation and then he divides both sides of the equation by 3.

   Select all the properties shown below that allow Theo to justify these steps. CCSS A.REI.1

   - Distributive Property
   - Addition Property of Equality
   - Commutative Property
   - Multiplication Property of Equality
   - Division Property of Equality

4. The graph shows part of the function $y = f(x)$.

   Complete the following. CCSS F.IF.1

   $f(2) = \square$ $f(1) = \square$

   $f(\square) = 0$ $f(\square) = 1$

5. Consider the following expression.

   $2(x - 1) + 4x^2 + 2(x^2 - 1)$

   When the expression is completely simplified, the coefficient of the x^2-term is $\square$. CCSS A.SSE.1a

6. In the graph, each ordered pair gives the low temperature and the high temperature for each day of a recent winter week in Pinewood. Draw a vertical line on the graph that can be used to show why this relation is not a function. CCSS F.IF.1

7. For the function $f(x) = \frac{x(x+1)}{2}$ with domain [1, 2, 3, 4, 5], what is the range? CCSS F.IF.1

   Range:

Copyright © McGraw-Hill Education

40 CHAPTER 1 Expressions, Equations, and Functions

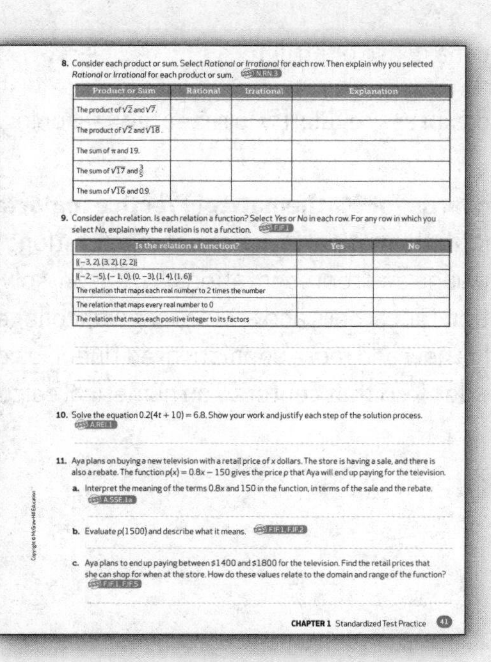

8. Consider each product or sum. Select *Rational* or *Irrational* for each row. Then explain why you selected *Rational* or *Irrational* for each product or sum. CCSS N.RN.3

Product or Sum	Rational	Irrational	Explanation
The product of $\sqrt{2}$ and $\sqrt{7}$.			
The product of $\sqrt{2}$ and $\sqrt{18}$.			
The sum of π and 19.			
The sum of $\sqrt{17}$ and $\frac{3}{5}$.			
The sum of $\sqrt{16}$ and 0.9.			

9. Consider each relation. Is each relation a function? Select *Yes* or *No* in each row. For any row in which you select *No*, explain why the relation is not a function. CCSS F.IF.1

Is the relation a function?	Yes	No
{(−3, 2), (3, 2), (2, 2)}		
{(−2, −5), (−1, 0), (0, −3), (1, 4), (1, 6)}		
The relation that maps each real number to 2 times the number		
The relation that maps every real number to 0		
The relation that maps each positive integer to its factors		

10. Solve the equation $0.2(4t + 10) = 6.8$. Show your work and justify each step of the solution process. CCSS A.REI.1

11. Aya plans on buying a new television with a retail price of x dollars. The store is having a sale, and there is also a rebate. The function $p(x) = 0.8x - 150$ gives the price p that Aya will end up paying for the television.

    a. Interpret the meaning of the terms 0.8x and 150 in the function, in terms of the sale and the rebate. CCSS A.SSE.1a

    b. Evaluate $p(1500)$ and describe what it means. CCSS F.IF.1, F.IF.2

    c. Aya plans to end up paying between \$1400 and \$1800 for the television. Find the retail prices that she can shop for when at the store. How do these values relate to the domain and range of the function? CCSS F.IF.1, F.IF.5

Copyright © McGraw-Hill Education

CHAPTER 1 Standardized Test Practice 41

Guide to Developing the Standards for Mathematical Practice
This guide provides a standard-by-standard analysis of the approach taken to each Mathematical Practice, including its meaning and the types of questions students can use to enhance mathematical development.

The Common Core State Standards are made up of:

- the **Standards for Mathematical Content**, which detail what students should learn.
- the **Standards for Mathematical Practice**, which describe how students should approach mathematics.

The goal of the practice standards is to instill in ALL students the abilities to be mathematically literate and to create a positive disposition for the importance of using math effectively.

What are the Standards for Mathematical Practice?

1. Make sense of problems and persevere in solving them.
2. Reason abstractly and quantitatively.
3. Construct viable arguments and critique the reasoning of others.
4. Model with mathematics.
5. Use appropriate tools strategically.
6. Attend to precision.
7. Look for and make use of structure.
8. Look for and express regularity in repeated reasoning.

Why are the Standards for Mathematical Practice important?
The Standards for Mathematical Practice set expectations for using mathematical language and representations to reason, solve problems, and model in preparation for careers and a wide range of college majors. High school mathematics builds new and more sophisticated fluencies on top of the earlier fluencies from grades K–8 that centered on numerical calculation.

1 Make sense of problems and persevere in solving them.

Mathematically proficient students start by explaining to themselves the meaning of a problem and looking for entry points to its solution. They analyze givens, constraints, relationships, and goals. They make conjectures about the form and meaning of the solution and plan a solution pathway rather than simply jumping into a solution attempt. They consider analogous problems, and try special cases and simpler forms of the original problem in order to gain insight into its solution. They monitor and evaluate their progress and change course if necessary. Older students might, depending on the context of the problem, transform algebraic expressions or change the viewing window on their graphing calculator to get the information they need. Mathematically proficient students can explain correspondences between equations, verbal descriptions, tables, and graphs or draw diagrams of important features and relationships, graph data, and search for regularity or trends. Younger students might rely on using concrete objects or pictures to help conceptualize and solve a problem. Mathematically proficient students check their answers to problems using a different method, and they continually ask themselves, "Does this make sense?" They can understand the approaches of others to solving complex problems and identify correspondences between different approaches.

What does it mean?	What questions do I ask?
Solving a mathematical problem takes time. Use a logical process to make sense of problems, understand that there may be more than one way to solve a problem, and alter the process if needed.	• What am I being asked to do or find? What do I know? • How does the given information relate to each other? Does a graph or diagram help? • Is this problem similar to any others I have solved? • What is my plan for solving the problem? • What should I do if I get "stuck"? • Does the answer make sense? • Is there another way to solve the problem? • Now that I've solved the problem, what did I do well? How would I approach a similar problem next time?

2 Reason abstractly and quantitatively.

Mathematically proficient students make sense of quantities and their relationships in problem situations. They bring two complementary abilities to bear on problems involving quantitative relationships: the ability to decontextualize—to abstract a given situation and represent it symbolically and manipulate the representing symbols as if they have a life of their own, without necessarily attending to their referents—and the ability to contextualize, to pause as needed during the manipulation process in order to probe into the referents for the symbols involved. Quantitative reasoning entails habits of creating a coherent representation of the problem at hand; considering the units involved; attending to the meaning of quantities, not just how to compute them; and knowing and flexibly using different properties of operations and objects.

What does it mean?	What questions do I ask?
You can start with a concrete or real-world context and then represent it with abstract numbers or symbols (decontextualize), find a solution, then refer back to the context to check that the solution makes sense (contextualize).	• What do the numbers represent? What are the variables, and how are they related to each other and to the numbers? • How can the relationships be represented mathematically? Is there more than one way? • How did I choose my method? • Does my answer make sense in this problem? • Does my answer fit the facts given in the problem? If not, why not?

3 Construct viable arguments and critique the reasoning of others.

Mathematically proficient students understand and use stated assumptions, definitions, and previously established results in constructing arguments. They make conjectures and build a logical progression of statements to explore the truth of their conjectures. They are able to analyze situations by breaking them into cases and can recognize and use counterexamples. They justify their conclusions, communicate them to others, and respond to the arguments of others. They reason inductively about data, making plausible arguments that take into account the context from which the data arose. Mathematically proficient students are also able to compare the effectiveness of two plausible arguments, distinguish correct logic or reasoning from that which is flawed, and—if there is a flaw in an argument—explain what it is. Elementary students can construct arguments using concrete referents such as objects, drawings, diagrams, and actions. Such arguments can make sense and be correct, even though they are not generalized or made formal until later grades. Later, students learn to determine domains to which an argument applies. Students at all grades can listen or read the arguments of others, decide whether they make sense, and ask useful questions to clarify or improve the arguments.

What does it mean?	What questions do I ask?
Sound mathematical arguments require a logical progression of statements and reasons. You can clearly communicate their thoughts and defend them.	• How did I get that answer? • Is that always true? • Why does that work? What mathematical evidence supports my answer? • Can I use objects in the classroom to show that my answer is correct? • Can I give a "nonexample" or a counterexample? • What conclusion can I draw? What conjecture can I make? • Is there anything wrong with that argument?

4 Model with mathematics.

Mathematically proficient students can apply the mathematics they know to solve problems arising in everyday life, society, and the workplace. In early grades, this might be as simple as writing an addition equation to describe a situation. In middle grades, a student might apply proportional reasoning to plan a school event or analyze a problem in the community. By high school, a student might use geometry to solve a design problem or use a function to describe how one quantity of interest depends on another. Mathematically proficient students who can apply what they know are comfortable making assumptions and approximations to simplify a complicated situation, realizing that these may need revision later. They are able to identify important quantities in a practical situation and map their relationships using such tools as diagrams, two-way tables, graphs, flowcharts and formulas. They can analyze those relationships mathematically to draw conclusions. They routinely interpret their mathematical results in the context of the situation and reflect on whether the results make sense, possibly improving the model if it has not served its purpose.

What does it mean?	What questions do I ask?
Modeling links classroom mathematics and statistics to everyday life, work, and decision-making. High school students at this level are expected to apply key takeaways from earlier grades to high school level problems.	• How might I represent the situation mathematically? • How does my equation or diagram model the situation? • What assumptions can I make? Should I make them? • What is the best way to organize the information? What other information is needed? • Can I make a good estimate of the answer? • Does my answer make sense?

5 Use appropriate tools strategically.

Mathematically proficient students consider the available tools when solving a mathematical problem. These tools might include pencil and paper, concrete models, a ruler, a protractor, a calculator, a spreadsheet, a computer algebra system, a statistical package, or dynamic geometry software. Proficient students are sufficiently familiar with tools appropriate for their grade or course to make sound decisions about when each of these tools might be helpful, recognizing both the insight to be gained and their limitations. For example, mathematically proficient high school students analyze graphs of functions and solutions generated using a graphing calculator. They detect possible errors by strategically using estimation and other mathematical knowledge. When making mathematical models, they know that technology can enable them to visualize the results of varying assumptions, explore consequences, and compare predictions with data. Mathematically proficient students at various grade levels are able to identify relevant external mathematical resources, such as digital content located on a website, and use them to pose or solve problems. They are able to use technological tools to explore and deepen their understanding of concepts.

What does it mean?	What questions do I ask?
Certain tools, including estimation and virtual tools, are more appropriate than others. You should understand the benefits and limitations of each tool.	• What tools would help to visualize the situation? • What are the limitations of using this tool? • Is an exact answer needed? • How can I use estimation as a tool? • Can I find additional information on the Internet? • Can I solve this problem using another tool?

6 Attend to precision.

Mathematically proficient students try to communicate precisely to others. They try to use clear definitions in discussion with others and in their own reasoning. They state the meaning of the symbols they choose, including using the equal sign consistently and appropriately. They are careful about specifying units of measure and labeling axes to clarify the correspondence with quantities in a problem. They calculate accurately and efficiently, express numerical answers with a degree of precision appropriate for the problem context. In the elementary grades, students give carefully formulated explanations to each other. By the time they reach high school, they have learned to examine claims and make explicit use of definitions.

What does it mean?	What questions do I ask?
Precision in mathematics is more than accurate calculations. It is also the ability to communicate with the language of mathematics. In high school mathematics, precise language makes for effective communication and serves as a tool for understanding and solving problems.	• How can the everyday meaning of a math term help me remember the math meaning? • Can I give some examples and nonexamples of that term? • Is this term similar to something I already know? • What does the math symbol mean? How do I know? • How do the terms in the problem help to solve it? • What does the variable represent, and in what units? • Does the question require a precise answer or is an estimate sufficient? If the answer needs to be precise, how precise? • Have I checked my answer for the correct labels?

Look for and make use of structure.

Mathematically proficient students look closely to discern a pattern or structure. Young students, for example, might notice that three and seven more is the same amount as seven and three more, or they may sort a collection of shapes according to how many sides the shapes have. Later, students will see 7×8 equals the well—remembered $7 \times 5 + 7 \times 3$, in preparation for learning about the distributive property. In the expression $x^2 + 9x + 14$, older students can see the 14 as 2×7 and the 9 as $2 + 7$. They recognize the significance of an existing line in a geometric figure and can use the strategy of drawing an auxiliary line for solving problems. They also can step back for an overview and shift perspective. They can see complicated things, such as some algebraic expressions, as single objects or as being composed of several objects. For example, they can see $5 - 3(x - y)^2$ as 5 minus a positive number times a square and use that to realize that its value cannot be more than 5 for any real numbers x and y.

What does it mean?	What questions do I ask?
Mathematics is based on a well-defined structure. Mathematically proficient students look for that structure to find easier ways to solve problems.	• Can I think of an easier way to find the solution? • How can using what I already know help solve this problem? • How are numerical expressions and algebraic expressions the same? How are they different? • Can the terms of this expression be grouped in a way that would allow it to be simplified or give us more information? • How can what I know about integers help with polynomials? • What shapes do I see in the figure? Could a line or segment be added to the figure that would give us more information?

8 Look for and express regularity in repeated reasoning.

Mathematically proficient students notice if calculations are repeated and look for general methods and for shortcuts. Upper elementary students might notice when dividing 25 by 11 that they are repeating the same calculations over and over again and conclude they have a repeating decimal. By paying attention to the calculation of slope as they repeatedly check whether points are on the line through (1, 2) with slope 3, middle school students might abstract the equation $\frac{y-2}{x-1} = 3$. Noticing the regularity in the way terms cancel when expanding $(x - 1)(x + 1)$, $(x - 1)(x^2 + x + 1)$, and $(x - 1)(x^3 + x^2 + x + 1)$ might lead them to the general formula for the sum of a geometric series. As they work to solve a problem, mathematically proficient students maintain oversight of the process, while attending to the details. They continually evaluate the reasonableness of their intermediate results.

What does it mean?	What questions do I ask?
Mathematics has been described as the study of patterns. Recognizing a pattern can lead to results more quickly and efficiently.	• Is there a pattern? • Is this pattern like one I've seen before? How is it different? • What does this problem remind me of? • Is this problem similar to something already known? • What would happen if I...? • How would I prove that? • How would this work with other numbers? Does it work all the time? How do I know? • Would technology help model this situation? How?

Number and Quantity

The Real Number System N-RN

Extend the properties of exponents to rational exponents.

1. Explain how the definition of the meaning of rational exponents follows from extending the properties of integer exponents to those values, allowing for a notation for radicals in terms of rational exponents.
2. Rewrite expressions involving radicals and rational exponents using the properties of exponents.

Use properties of rational and irrational numbers.

3. Explain why the sum or product of two rational numbers is rational; that the sum of a rational number and an irrational number is irrational; and that the product of a nonzero rational number and an irrational number is irrational.

Quantities* N-Q

Reason quantitatively and use units to solve problems.

1. Use units as a way to understand problems and to guide the solution of multi-step problems; choose and interpret units consistently in formulas; choose and interpret the scale and the origin in graphs and data displays.
2. Define appropriate quantities for the purpose of descriptive modeling.
3. Choose a level of accuracy appropriate to limitations on measurement when reporting quantities.

Algebra

Seeing Structure in Expressions A-SSE

Interpret the structure of expressions

1. Interpret expressions that represent a quantity in terms of its context.*
   a. Interpret parts of an expression, such as terms, factors, and coefficients.
   b. Interpret complicated expressions by viewing one or more of their parts as a single entity.
2. Use the structure of an expression to identify ways to rewrite it.

Write expressions in equivalent forms to solve problems.

3. Choose and produce an equivalent form of an expression to reveal and explain properties of the quantity represented by the expression.*
   a. Factor a quadratic expression to reveal the zeros of the function it defines.
   b. Complete the square in a quadratic expression to reveal the maximum or minimum value of the function it defines.
   c. Use the properties of exponents to transform expressions for exponential functions.

Arithmetic with Polynomials and Rational Expressions A-APR

Perform arithmetic operations on polynomials.

1. Understand that polynomials form a system analogous to the integers, namely, they are closed under the operations of addition, subtraction, and multiplication; add, subtract, and multiply polynomials.

Creating Equations* A-CED

Create equations that describe numbers or relationships.

1. Create equations and inequalities in one variable and use them to solve problems.
2. Create equations in two or more variables to represent relationships between quantities; graph equations on coordinate axes with labels and scales.

*Mathematical Modeling Standards

3. Represent constraints by equations or inequalities, and by systems of equations and/or inequalities, and interpret solutions as viable or nonviable options in a modeling context.

4. Rearrange formulas to highlight a quantity of interest, using the same reasoning as in solving equations.

Reasoning with Equations and Inequalities A-REI

Understand solving equations as a process of reasoning and explain the reasoning.

1. Explain each step in solving a simple equation as following from the equality of numbers asserted at the previous step, starting from the assumption that the original equation has a solution. Construct a viable argument to justify a solution method.

Solve equations and inequalities in one variable.

3. Solve linear equations and inequalities in one variable, including equations with coefficients represented by letters.

4. Solve quadratic equations in one variable.
   a. Use the method of completing the square to transform any quadratic equation in x into an equation of the form $(x - p)^2 = q$ that has the same solutions. Derive the quadratic formula from this form.
   b. Solve quadratic equations by inspection (e.g., for $x^2 = 49$), taking square roots, completing the square, the quadratic formula and factoring, as appropriate to the initial form of the equation. Recognize when the quadratic formula gives complex solutions and write them as $a \pm bi$ for real numbers a and b.

Solve systems of equations.

5. Prove that, given a system of two equations in two variables, replacing one equation by the sum of that equation and a multiple of the other produces a system with the same solutions.

6. Solve systems of linear equations exactly and approximately (e.g., with graphs), focusing on pairs of linear equations in two variables.

7. Solve a simple system consisting of a linear equation and a quadratic equation in two variables algebraically and graphically.

Represent and solve equations and inequalities graphically.

10. Understand that the graph of an equation in two variables is the set of all its solutions plotted in the coordinate plane, often forming a curve (which could be a line).

11. Explain why the x-coordinates of the points where the graphs of the equations $y = f(x)$ and $y = g(x)$ intersect are the solutions of the equation $f(x) = g(x)$; find the solutions approximately, e.g., using technology to graph the functions, make tables of values, or find successive approximations. Include cases where $f(x)$ and/or $g(x)$ are linear, polynomial, rational, absolute value, exponential, and logarithmic functions.*

12. Graph the solutions to a linear inequality in two variables as a halfplane (excluding the boundary in the case of a strict inequality), and graph the solution set to a system of linear inequalities in two variables as the intersection of the corresponding half-planes.

Functions

Interpreting Functions F-IF

Understand the concept of a function and use function notation.

1. Understand that a function from one set (called the domain) to another set (called the range) assigns to each element of the domain exactly one element of the range. If f is a function and x is an element of its domain, then $f(x)$ denotes the output of f corresponding to the input x. The graph of f is the graph of the equation $y = f(x)$.

2. Use function notation, evaluate functions for inputs in their domains, and interpret statements that use function notation in terms of a context.

3. Recognize that sequences are functions, sometimes defined recursively, whose domain is a subset of the integers.

Interpret functions that arise in applications in terms of the context.

4. For a function that models a relationship between two quantities, interpret key features of graphs and tables in terms of the quantities, and sketch graphs showing key features given a verbal description of the relationship.*

5. Relate the domain of a function to its graph and, where applicable, to the quantitative relationship it describes.

6. Calculate and interpret the average rate of change of a function (presented symbolically or as a table) over a specified interval. Estimate the rate of change from a graph.*

Analyze functions using different representations.

7. Graph functions expressed symbolically and show key features of the graph, by hand in simple cases and using technology for more complicated cases.*
   a. Graph linear and quadratic functions and show intercepts, maxima, and minima.
   b. Graph square root, cube root, and piecewise-defined functions, including step functions and absolute value functions.
   e. Graph exponential and logarithmic functions, showing intercepts and end behavior, and trigonometric functions, showing period, midline, and amplitude.

8. Write a function defined by an expression in different but equivalent forms to reveal and explain different properties of the function.
   a. Use the process of factoring and completing the square in a quadratic function to show zeros, extreme values, and symmetry of the graph, and interpret these in terms of a context.
   b. Use the properties of exponents to interpret expressions for exponential functions.

9. Compare properties of two functions each represented in a different way (algebraically, graphically, numerically in tables, or by verbal descriptions).

Building Functions F-BF

Build a function that models a relationship between two quantities.

1. Write a function that describes a relationship between two quantities.*
   a. Determine an explicit expression, a recursive process, or steps for calculation from a context.
   b. Combine standard function types using arithmetic operations.

2. Write arithmetic and geometric sequences both recursively and with an explicit formula, use them to model situations, and translate between the two forms.*

Build new functions from existing functions.

3. Identify the effect on the graph of replacing $f(x)$ by $f(x) + k$, $k\,f(x)$, $f(kx)$, and $f(x + k)$ for specific values of k (both positive and negative); find the value of k given the graphs. Experiment with cases and illustrate an explanation of the effects on the graph using technology.

4. Find inverse functions.
   a. Solve an equation of the form $f(x) = c$ for a simple function f that has an inverse and write an expression for the inverse.

Linear, Quadratic, and Exponential Models F-LE

Construct and compare linear, quadratic, and exponential models and solve problems.

1. Distinguish between situations that can be modeled with linear functions and with exponential functions.
   a. Prove that linear functions grow by equal differences over equal intervals, and that exponential functions grow by equal factors over equal intervals.

b. Recognize situations in which one quantity changes at a constant rate per unit interval relative to another.

c. Recognize situations in which a quantity grows or decays by a constant percent rate per unit interval relative to another.

2. Construct linear and exponential functions, including arithmetic and geometric sequences, given a graph, a description of a relationship, or two input-output pairs (include reading these from a table).

3. Observe using graphs and tables that a quantity increasing exponentially eventually exceeds a quantity increasing linearly, quadratically, or (more generally) as a polynomial function.

Interpret expressions for functions in terms of the situation they model.

5. Interpret the parameters in a linear or exponential function in terms of a context.

Statistics and Probability

Interpreting Categorical and Quantitative Data S-ID

Summarize, represent, and interpret data on a single count or measurement variable.

1. Represent data with plots on the real number line (dot plots, histograms, and box plots).

2. Use statistics appropriate to the shape of the data distribution to compare center (median, mean) and spread (interquartile range, standard deviation) of two or more different data sets.

3. Interpret differences in shape, center, and spread in the context of the data sets, accounting for possible effects of extreme data points (outliers).

Summarize, represent, and interpret data on two categorical and quantitative variables.

5. Summarize categorical data for two categories in two-way frequency tables. Interpret relative frequencies in the context of the data (including joint, marginal, and conditional relative frequencies). Recognize possible associations and trends in the data.

6. Represent data on two quantitative variables on a scatter plot, and describe how the variables are related.
   a. Fit a function to the data; use functions fitted to data to solve problems in the context of the data. Use given functions or choose a function suggested by the context. Emphasize linear, quadratic, and exponential models.

   b. Informally assess the fit of a function by plotting and analyzing residuals.

   c. Fit a linear function for a scatter plot that suggests a linear association.

Interpret linear models.

7. Interpret the slope (rate of change) and the intercept (constant term) of a linear model in the context of the data.

8. Compute (using technology) and interpret the correlation coefficient of a linear fit.

9. Distinguish between correlation and causation.

Expressions, Equations, and Functions

CHAPTER FOCUS Learn about some of the Common Core State Standards that you will explore in this chapter. Answer the preview questions. As you complete each lesson, return to these pages to check your work.

What You Will Learn	Preview Question
Lesson 1.1: Rational and Irrational Numbers	
CCSS N.RN.3 Explain why the sum or product of two rational numbers is rational; that the sum of a rational number and an irrational number is irrational; and that the product of a nonzero rational number and an irrational number is irrational.	CCSS SMP 3 When you multiply two rational numbers, is the result always another rational number? Why or why not? CCSS SMP 7 Give an example of a product of two irrational numbers that is also irrational and a product of two irrational numbers that is rational.
Lesson 1.2: Variables and Expressions	
CCSS A.SSE.1a Interpret parts of an expression, such as terms, factors, and coefficients. CCSS A.SSE.2 Use the structure of an expression to identify ways to rewrite it.	CCSS SMP 2 The expression $14x + 5$ represents the cost of ordering x copies of a book online, including a flat fee for shipping. Which term in the expression represents the shipping fee? How do you know?
Lesson 1.3: Order of Operations	
CCSS A.SSE.1b Interpret complicated expressions by viewing one or more of their parts as a single entity. CCSS A.SSE.2 Use the structure of an expression to identify ways to rewrite it.	CCSS SMP 6 How do you evaluate the expression $5^2 - 18 \div 6 + 3$? Do you get a different result if the expression is written as $5^2 - 18 \div (6 + 3)$? Why or why not?

What You Will Learn	Preview Question
Lesson 1.4: Properties of Numbers	
CCSS A.SSE.2 Use the structure of an expression to identify ways to rewrite it. CCSS A.SSE.1b Interpret complicated expressions by viewing one or more of their parts as a single entity.	CCSS SMP 7 How can you rewrite the expression $8 \cdot 89$ so it can be simplified using mental math?
Lesson 1.5: Equations	
CCSS A.CED.1 Create equations and inequalities in one variable and use them to solve problems. CCSS A.REI.3 Solve linear equations and inequalities in one variable, including equations with coefficients represented by letters. CCSS A.REI.1 Explain each step in solving a simple equation as following from the equality of numbers asserted at the previous step, starting from the assumption that the original equation has a solution. Construct a viable argument to justify a solution method.	CCSS SMP 4 Jonas spent \$7.75, excluding tax, at the store when he bought a notebook that cost \$2.50 and 3 pens. Write and solve an equation to find the price of each pen.
Lesson 1.6: Relations	
CCSS A.REI.10 Understand that the graph of an equation in two variables is the set of all its solutions plotted in the coordinate plane, often forming a curve (which could be a line). **Also addresses:** F.IF.1	CCSS SMP 1 Use set notation to write the domain and range of the relation $\{(-2, 3), (4, 4), (4, 5), (5, -8)\}$.
Lesson 1.7: Functions	
CCSS F.IF.1 Understand that a function from one set (called the domain) to another set (called the range) assigns to each element of the domain exactly one element of the range. If f is a function and x is an element of its domain, then $f(x)$ denotes the output of f corresponding to the input x. The graph of f is the graph of the equation $y = f(x)$. CCSS F.IF.2 Use function notation, evaluate functions for inputs in their domains, and interpret statements that use function notation in terms of a context. CCSS F.IF.5 Relate the domain of a function to its graph and, where applicable, to the quantitative relationship it describes.	CCSS SMP 3 A student said that the relation $\{(-2, 3), (4, 4), (4, 5), (5, -8)\}$ is a function. Do you agree? Why or why not? CCSS SMP 2 The amount of money in Trina's bank account, in dollars, n months after she opens the account is given by the function $b(n) = 50n + 100$. What are the meanings of the values 50 and 100? What is the balance in Trina's account after 10 months?

1.1 Rational and Irrational Numbers

Objectives

- Explain why the sum or product of two rational numbers is rational.
- Explain why the sum of a rational number and an irrational number is irrational.
- Explain why the product of a nonzero rational number and an irrational number is irrational.

CCSS STANDARDS

Content: N.RN.3
Practices: 2, 3, 4, 5, 6, 7
Use with Extend 10–2

A **rational number** is a number that can be written in the form $\frac{a}{b}$, where a and b are integers and $b \neq 0$. The decimal form of a rational number is a repeating decimal or a terminating decimal.
An **irrational number** is a number that cannot be expressed in the form $\frac{a}{b}$, where a and b are integers and $b \neq 0$. The decimal form of an irrational number is neither repeating nor terminating.

EXAMPLE 1 Sums of Rational and Irrational Numbers CCSS N.RN.3

EXPLORE The table shows examples of rational and irrational numbers.

Rational Numbers
$\frac{1}{3}$ 0.4 -7 $-\frac{3}{5}$

Irrational Number
$\sqrt{2}$ π $-\sqrt{5}$

a. CALCULATE ACCURATELY Choose two of the rational numbers from the table and find their sum. Is the sum rational or irrational? Explain how you know. CCSS SMP 6

b. MAKE A CONJECTURE Compare your result for **part a** with those of other students. Then make a conjecture about the sum of two rational numbers. CCSS SMP 3

c. USE TOOLS Choose a rational number and an irrational number from the table and find their sum using a calculator. Does the sum appear to be rational or irrational? Why? CCSS SMP 5

d. MAKE A CONJECTURE Compare your result for **part c** with those of other students. Then make a conjecture about the sum of a rational number and an irrational number. CCSS SMP 3

You can use reasoning and the definitions of rational and irrational numbers to construct logical arguments to show that the conjectures you made are true.

EXAMPLE 2 Justify a Conjecture CCSS N.RN.3

Complete the following argument to explain why the sum of two rational numbers is rational.

a. USE STRUCTURE Suppose x and y are both rational numbers. You must show that $x + y$ is also a rational number. Since x and y are rational numbers, you can write $x = \frac{a}{b}$ and $y = \frac{c}{d}$. What must be true about a, b, c, and d? CCSS SMP 7

b. USE STRUCTURE Write the sum of x and y as a single fraction in terms of a, b, c, and d. CCSS SMP 7

c. CONSTRUCT ARGUMENTS Explain why this shows that $x + y$ is a rational number. CCSS SMP 3

EXAMPLE 3 Justify a Conjecture Using a Contradiction CCSS N.RN.3

Complete the following argument to explain why the sum of a rational number and an irrational number is irrational.

a. USE STRUCTURE Suppose x is a rational number and y is an irrational number. You must show that $x + y$ is irrational. In this argument, you will assume that the sum is a rational number and show that this leads to a contradiction. Suppose $x + y = z$ and z is a rational number. Then you can write $x = \frac{a}{b}$ and $z = \frac{c}{d}$. What must be true about a, b, c, and d? CCSS SMP 7

b. USE STRUCTURE Solving for y shows that $y = z - x$. Use this to write y as a single fraction in terms of a, b, c, and d. CCSS SMP 7

c. USE STRUCTURE What does this say about y? Explain why this is a contradiction. CCSS SMP 7

d. CONSTRUCT ARGUMENTS What does the contradiction allow you to conclude? Why? CCSS SMP 3

PRACTICE

1. **CONSTRUCT ARGUMENTS** Explain why the product of two rational numbers is rational. CCSS N.RN.3, SMP 3

2. The product of a nonzero rational number and an irrational number is irrational.

   a. **CONSTRUCT ARGUMENTS** Explain why this statement is true. Use an argument that leads to a contradiction. CCSS N.RN.3, SMP 3

   b. **COMMUNICATE PRECISELY** Why is the word *nonzero* important in the statement? How do you use this fact in your argument? CCSS N.RN.3, SMP 6

3. **CRITIQUE REASONING** Alyssa said the square of an irrational number must also be irrational. Do you agree or disagree? Justify your answer. CCSS N.RN.3, SMP 3

4. **CONSTRUCT ARGUMENTS** A set is *closed* under an operation if for any numbers in the set, the result of the operation is also in the set. For example, the set of integers is closed under addition because the sum of two integers is another integer. Is the set of irrational numbers closed under multiplication? If so, explain why. If not, give a counterexample. CCSS N.RN.3, SMP 3

COMMUNICATE PRECISELY **The table shows how Ms. Rodriguez assigns scores in her biology class. Determine whether each statement is *always*, *sometimes*, or *never* true. Explain.** CCSS N.RN.3, SMP 6

Category	Type of Score
Midterm Exam	Whole number from 0 to 100
Final Exam	Whole number from 0 to 100
Quizzes	Rational number from 0 to 10
Homework	$0, \frac{1}{4}, \frac{1}{2}, \frac{3}{4}, 1$

5. The average of a student's midterm exam score and final exam score is a whole number.

6. The average of a student's quiz scores is a rational number.

7. The average of a student's homework scores is an irrational number.

8. **USE A MODEL** Determine irrational values for x and y so that the area in square feet of the rectangular carpet is a rational number greater than 100 but less than 200. Justify your answer. CCSS N.RN.3, SMP 4

9. **REASON QUANTITATIVELY** Without performing any calculations, determine if the expression $5.323232\ldots + 6\frac{2}{3}$ represents a rational number. Justify your answer. CCSS N.RN.3, SMP 2

10. **CRITIQUE REASONING** Amanda claims that the product $\sqrt{3} \cdot (\sqrt{3} - 7)$ is irrational. Her argument is shown at the right. Do you agree with Amanda's argument? What about her conclusion? Explain. CCSS N.RN.3, SMP 3

The difference $\sqrt{3} - 7$ can be written as sum of an irrational and (nonzero) rational number, as $\sqrt{3} + (-7)$. This represents an irrational number, because the sum of an irrational and rational number is irrational. The number $\sqrt{3}$ is also irrational, so the product of two irrational numbers is also irrational.

1.2 Variables and Expressions

Objectives

- Write algebraic expressions.
- Use the structure of an expression to identify ways to rewrite it.
- Interpret parts of an expression.

Content: A.SSE.1a, A.SSE.2
Practices: 2, 3, 4, 7, 8
Use with Lesson 1-1

An **algebraic expression** consists of sums and/or products of numbers and variables. A **term** of an expression may be a number, a variable, or a product or quotient of numbers and variables. For example, in the expression $24m + 5n + 0.1$, there are three terms: $24m$, $5n$, and 0.1.

EXAMPLE 1 Write an Algebraic Expression CCSS A.SSE.2

EXPLORE The Watkins family is designing a new path for their garden. They use black and white square tiles to make a pattern that they will use to build the path.

Stage 1 Stage 2 Stage 3

a. **FIND A PATTERN** Complete the table. CCSS SMP 8

Stage	1	2	3	4	5
Number of White Tiles					
Number of Black Tiles					
Total Tiles					

b. **FIND A PATTERN** Suppose the Watkins family wants to make stage *n* of the pattern. Write an expression for the number of white tiles they will need. Explain how you wrote the expression. CCSS SMP 8

c. **FIND A PATTERN** Write an expression for the number of black tiles they will need to make stage *n* of the pattern. Explain how you wrote the expression. CCSS SMP 8

d. **USE STRUCTURE** Write an expression for the total number of tiles they will need to make stage *n* of the pattern. Is there more than one way to write the expression? Explain. CCSS SMP 7

e. USE STRUCTURE Evaluate your expression from **part d** for $n = 1, 2, 3, 4, 5$ to verify that it produces the same values as in the bottom row of the table in **part a.** CCSS SMP 7

A verbal expression like "3 more than the number of white tiles" may be written as the algebraic expression $n + 3$ because the words "more than" correspond to addition.

KEY CONCEPT

Complete the table by writing the operation that corresponds to each set of verbal phrases.

Verbal Phrases	Operation
more than, sum, plus, increased by, added to	
less than, subtracted from, difference, decreased by, minus	
product of, multiplied by, times, of	
quotient of, divided by	

EXAMPLE 2 Interpret Parts of an Expression CCSS A.SSE.1a, SMP 7

The expression $8x + 12.5y + 6$ gives the total cost in dollars of ordering x T-shirts and y caps from a Web site. The cost includes a fee for shipping that is the same no matter how many shirts or caps you order.

a. USE STRUCTURE What does the coefficient 8 represent in the expression? What does the term $8x$ represent? Explain.

b. USE STRUCTURE What does the coefficient 12.5 represent in the expression? What does the term $12.5y$ represent? Explain.

c. USE STRUCTURE What is the fee for shipping? Explain how you know.

d. USE STRUCTURE How would the expression for the total cost be different if the company decides to increase the price of each shirt and cap by $1.50? Explain.

PRACTICE

1. Diego used gray and white counters to make the pattern shown here. CCSS A.SSE.2

Stage 1 Stage 2 Stage 3

a. **FIND A PATTERN** Write an expression for the number of gray counters at stage n and an expression for the number of white counters at stage n. What do the expressions tell you about the number of counters of each color at stage n? CCSS SMP 7

b. **USE STRUCTURE** Write two different expressions for the total number of counters at stage n. Explain how you know the two expressions are both correct. CCSS SMP 7

2. **USE STRUCTURE** The expression $13.25d + 6.5$ gives the total cost in dollars of renting a bicycle and helmet for d days. The fee for the helmet does not depend upon the number of days. CCSS A.SSE.1a, SMP 7

a. What is the coefficient of d? What does it represent?

b. How would the expression be different if the cost of the helmet were doubled?

3. **REASON ABSTRACTLY** Gabrielle makes a pattern using pennies. The pattern grows in a predictable way at each stage of the pattern. The expression $3n + 1$ gives the total number of pennies at each stage of the pattern. Make a sketch to show what Gabrielle's pattern might look like. Draw stages 1, 2, and 3. CCSS A.SSE.1a, SMP2

The table shows the prices of several items at an office supply store. Use the table for Exercises 4–6.

Item	Price
Stapler	\$5.99
Tape dispenser	\$3.50
Sticky notes	\$3.25
Gel pen	\$1.75

4. USE STRUCTURE Jemma buys s staplers and g gel pens. She has a coupon for \$2 off the total cost of her purchase. Write two different expressions that can be used to find her final cost before tax. CCSS A.SSE.2, SMP 7

5. CRITIQUE REASONING DeMarco buys x tape dispensers and x packs of sticky notes. He says he can use the expression $6.75x$ to find the total cost of the items before tax. Do you agree? Why or why not? CCSS A.SSE.2, SMP 3

6. USE STRUCTURE Tyler buys s packs of sticky notes and one gel pen. He uses the expression $1.08(3.25s + 1.75)$ to find the final cost. What do you think the 1.08 in the expression represents? Explain. CCSS A.SSE.1a, SMP 7

7. USE A MODEL The figure shows a floor plan for a two-room apartment. Write an expression for the area of the apartment, in square feet, by first finding the area of each room and then adding. Then describe how you can write the expression in a different way. CCSS A.SSE.2, SMP 4

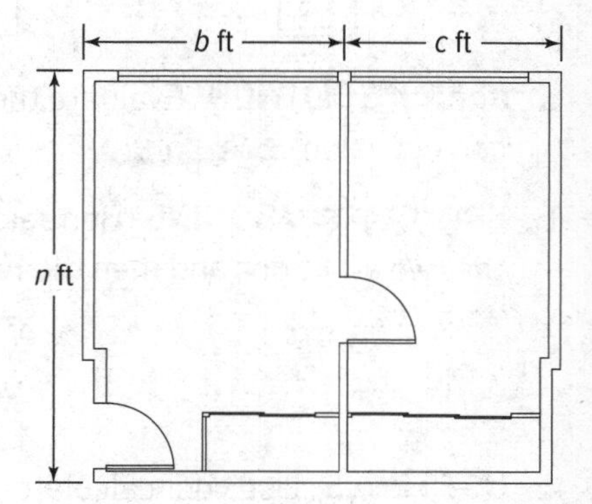

8. USE STRUCTURE Describe a situation that could be represented by each expression. CCSS A.SSE.1a, SMP 7

a. $9.95 + 0.75b$

b. $15x - 5x$

c. $59c - 25c - 30$

1.3 Order of Operations

Content: A.SSE.1b, A.SSE.2
Practices: 1, 2, 3, 4, 5, 6, 7
Use with Lesson 1–2

Objectives

- Evaluate expressions by using the order of operations.
- Use the structure of an expression to identify ways to rewrite it.
- Interpret parts of an expression.

To **evaluate** an expression means to find its value. When you evaluate a complicated expression containing multiple operations, you must perform the operations in the correct order to get the correct value for the expression.

EXAMPLE 1 Evaluate Expressions CCSS A.SSE.2

EXPLORE Rima created a game. She wrote expressions on slips of paper that represent different sums of money. The player has 10 seconds to match each expression to its sum.

$26 - 15 + 4 - 1 + 14$	$48 \div 2 + 4 \times 2$	$5[10 - (2 + 5)]$	$(1 + 6)^2 - (3 + 7)$
\$15	\$32	\$39	\$28

a. PLAN A SOLUTION Evaluate the expressions. Draw a line from each expression to the correct value. CCSS SMP 1

b. CRITIQUE REASONING Ben said that to evaluate $26 - 15 + 4 - 1 + 14$ you perform the two additions and then the two subtractions. Do you agree? Explain. CCSS SMP 3

c. USE TOOLS Use your calculator to evaluate $48 \div 2 + 4 \times 2$. Does it give the correct value? If so, in what order does it perform the operations? If not, in what order do you think it performs the operations? CCSS SMP 5

d. COMMUNICATE PRECISELY What rules can you state about grouping symbols, such as parentheses, and exponents in order to get the correct values for $5[10 - (2 + 5)]$ and $(1 + 6)^2 - (3 + 7)$? CCSS SMP 6

The rule that describes the sequence in which you should perform operations is called the **order of operations.**

KEY CONCEPT

Complete the table by writing the order of operations.

Operations	Order of Operations
Multiply and/or divide from left to right.	Step 1:
Add and/or subtract from left to right.	Step 2:
Evaluate expressions inside grouping symbols.	Step 3:
Evaluate all powers.	Step 4:

EXAMPLE 2 **Write and Evaluate an Expression**

Jared is buying carpet for a square room with sides that are *s* feet long. The table shows the price of the carpet and the price of the metal strip that holds down the edge of the carpet.

Item	Price
Carpet	$2.65 per square foot
Metal Strip	$0.20 per foot

a. USE A MODEL The metal strip holds down the carpet around the entire perimeter of the room, except at the doorway, which is 3 feet wide. Write an expression for the total length of the strip that Jared will need. Explain. CCSS A.SSE.1b, SMP 4

b. USE A MODEL Write an expression that Jared can use to calculate the total cost of the carpet and the metal strip for a room with sides *s* feet long. Explain what each term of your expression represents. CCSS A.SSE.1b, SMP 4

c. USE STRUCTURE Explain how you can write the expression in a different way. CCSS A.SSE.2, SMP 7

d. COMMUNICATE PRECISELY Explain how you can use the expression from **part b** to find the total cost of the carpet and metal strip for a room with sides 16 feet long. CCSS A.SSE.2, SMP 6

e. USE STRUCTURE How would the expression from **part b** be different if Jared had a coupon for 10% off the total cost? (*Hint:* 10% off the total cost means Jared pays 90% of the original cost). How much would Jared pay in this case? CCSS A.SSE.1b, SMP 7

PRACTICE

CALCULATE ACCURATELY **Evaluate each expression.** CCSS A.SSE.2, SMP 6

1. $10^2 - 20 + 4^3$

2. $32 - (7 - 1)^2$

3. $64 \div (2 + 6) - 14$

CALCULATE ACCURATELY **Evaluate each expression if $x = 2$, $y = 7$, and $z = -1$.** CCSS A.SSE.2, SMP 6

4. $z(2x - y)$

5. $y^2 - 4y \div 2$

6. $x(z + 5y) - 2x^2$

7. The table shows how scores are calculated at diving competitions. Each of the five judges scores each dive from 1 to 10 in 0.5-point increments.

Calculating a Diving Score	
Step 1	Drop the highest and lowest of the five judges' scores.
Step 2	Add the remaining scores to find the raw score.
Step 3	Multiply the raw score by the degree of difficulty.

a. **CALCULATE ACCURATELY** Roberto performs a dive with a degree of difficulty of 2.5. His scores from the judges are 8.0, 7.5, 6.5, 7.5, and 7.0. Write and evaluate an expression to find his score for the dive. CCSS A.SSE.2, SMP 6

b. **CONSTRUCT ARGUMENTS** Jennifer performs a dive and uses the expression $(7.5 + 8.5 + 8.0) \times 3.2$ to find her score. What is her score for the dive? What can you conclude about the highest score she received from the five judges? Explain. CCSS A.SSE.1b, SMP 3

c. **REASON QUANTITATIVELY** Mai performs a dive with a degree of difficulty of 3.5. Her score for the dive is 70. What scores could Mai have received from the five judges? Explain. CCSS A.SSE.1b, SMP 3

d. **REASON QUANTITATIVELY** Eli also performs a dive with a degree of difficulty of 3.5. His score is 75.25. What scores could Eli have received from the five judges? Explain. CCSS A.SSE.1b, SMP 2

e. **REASON QUANTITATIVELY** Skylar does a dive with a degree of difficulty of 3.3. Four of his scores are 7.5, 7.0, 6.5, and 6.5. Toby does a dive with a degree of difficulty of 3.1 and receives scores of 7.0, 7.5, 7.5, 6.0, and 8.0. If Skylar's final score was greater than Toby's, what can you say about Skylar's fifth score? Explain. CCSS A.SSE.1b, SMP 2

8. CRITIQUE REASONING A student was asked to evaluate an expression. The student's work is shown at right. Critique the student's work. If there are any errors, describe them and find the correct value of the expression. CCSS A.SSE.2, SMP 3

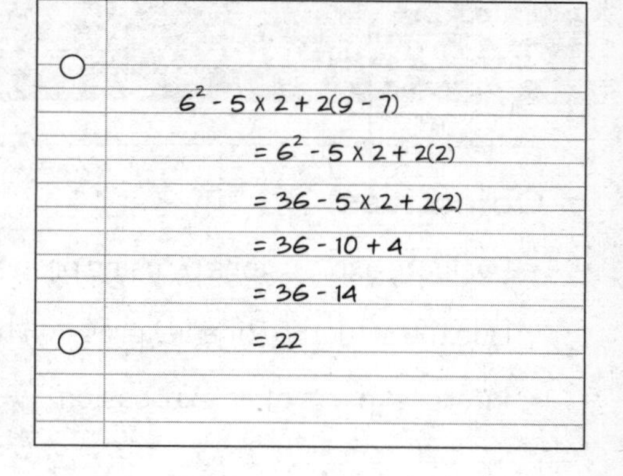

9. USE TOOLS Kelly buys 3 video games that cost \$18.95 each. She also buys 2 pairs of earbuds that cost \$11.50 each. She has a coupon for \$2 off the price of each video game. Kelly uses a calculator, as shown, to find that the total cost of the items is \$77.85. The cashier tells her that the total cost is \$73.85. Who is correct? Explain. CCSS A.SSE.1b, SMP 5

```
3*18.95-2+2*11.5
                77.85
```

REASON QUANTITATIVELY Determine whether each statement about the expression $a^2 - 5(b + 2)$ is *always, sometimes,* or *never* true. Explain. CCSS A.SSE.1b, SMP 2

10. If $a = 10$, then the value of the expression is less than 100.

11. If $b = 1$, then the expression is equivalent to $a^2 - 15$.

12. The value of the expression is 0.

13. USE A MODEL The side panel of a skateboard ramp is a trapezoid, as shown in the figure. The expression $\frac{h}{2}(b_1 + b_2)$ can be used to find the area of a trapezoid. Write and evaluate an expression to find the amount of wood needed to build the two side panels of a skateboard ramp where $h = 24$ inches, $b_1 = 30$ inches, and $b_2 = 50$ inches. CCSS A.SSE.1b, SMP 1

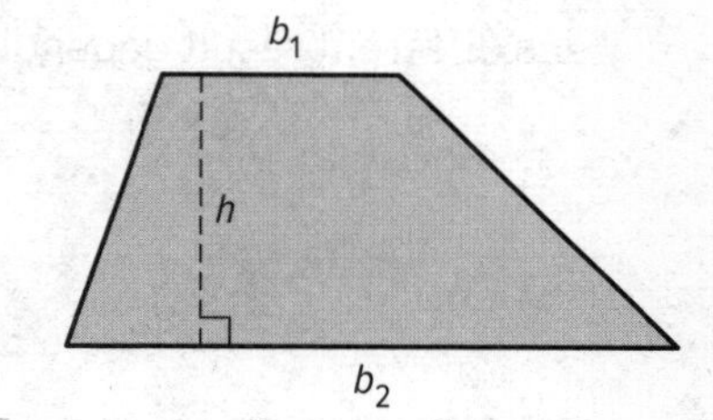

14. REASON QUANTITATIVELY Write an expression that includes the numbers 2, 4, and 5, and has a value of 50. Your expression should include one set of parentheses. CCSS A.SSE.2, SMP 2

15. REASON QUANTITATIVELY Isabel wrote the expression $6 + 3 \times 5 - 6 + 8 \div 2$ and asked Tamara to evaluate it. When Tamara evaluated it, she got a value of 19. Isabel told Tamara that her value was incorrect and said that the value should have been 38. With whom do you agree? Explain. CCSS A.SSE.2, SMP 2

1.4 Properties of Numbers

Objectives

- Evaluate expressions by using properties of numbers.
- Use the structure of an expression to identify ways to rewrite it.
- Interpret parts of an expression.

CCSS STANDARDS

Content: A.SSE.1b, A.SSE.2
Practices: 2, 3, 4, 6, 7, 8
Use with Lessons 1-3, 1-4

EXAMPLE 1 Explore Properties of Numbers CCSS SMP 7

EXPLORE Rectangle *ABCD* represents Arletta's garden. She plants part of the garden with vegetables and part of the garden with flowers, as shown.

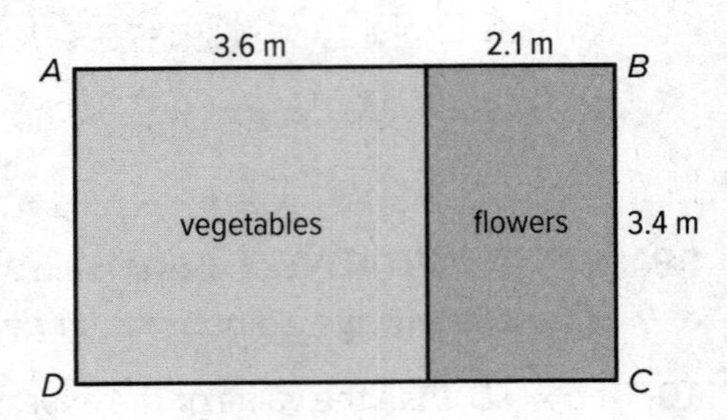

a. USE STRUCTURE Arletta wants to put a straight path along the garden from *D* to *C*. Write two different expressions she can use to find the length of the path. Explain why it makes sense that the two expressions give the same length. CCSS A.SSE.2

b. USE STRUCTURE Arletta also wants to put a fence along the border of the garden from *A* to *B* to *C*. Write an expression she can use to find the length of the fence. Then show two different ways she can group a pair of numbers in the expression. Does she get a different result depending on the grouping? Explain. CCSS A.SSE.2

c. USE STRUCTURE Arletta wants to find the area of the garden. She finds the area of the vegetable plot and the area of the flower plot then the sum of these areas. Write and evaluate an expression to show how she finds the area. CCSS A.SSE.1b

d. USE STRUCTURE Arletta's friend, Troy, finds the area of the garden by adding to find the distance from *A* to *B* and then multiplying by the distance from *B* to *C*. Write and evaluate an expression to show how Troy finds the area. Does he get the same result as Arletta? Explain why this makes sense. CCSS A.SSE.1b

In the previous exploration, you may have discovered the following properties of numbers.

KEY CONCEPT

Complete the table by using numbers to write examples for each property.

Property	Examples
Commutative Property For any numbers a and b, $a + b = b + c$ and $a \cdot b = b \cdot a$.	
Associative Property For any numbers a, b, and c, $(a + b) + c = a + (b + c)$ and $(ab)c = a(bc)$.	
Distributive Property For any numbers a, b, and c, $a(b + c) = ab + ac$ and $a(b - c) = ab - ac$.	

EXAMPLE 2 Use Mental Math

Giovanni is buying equipment for his soccer team. The table shows the price of some of the items he is buying.

Item	Price
Soccer Balls	$22 each
Portable Goal	$93 per pair

a. USE STRUCTURE Giovanni is buying 7 soccer balls. Write an expression for the total cost of the soccer balls. Then explain how he can use the Distributive Property to rewrite the expression to find the cost using mental math. CCSS A.SSE.1b, SMP 7

b. USE STRUCTURE Giovanni is buying 6 pairs of portable goals. Show two different ways he can find the cost of the goals using the Distributive Property. CCSS A.SSE.2, SMP 7

c. USE STRUCTURE Fred's business donated 3 soccer balls and 3 portable goals to Giovanni's team. Write an expression using the Distributive Property for the cost of the donated items. CCSS SMP 8

d. DESCRIBE A METHOD Describe a general method for using the Distributive Property to find a product by mental math. CCSS A.SSE.1b, SMP 8

Like terms are terms that contain the same variables, with corresponding variables having the same exponent. In the expression $5x^2 + 4x + 7x^2 + 2y^2$, the terms $5x^2$ and $7x^2$ are like terms. You can use properties of numbers to simplify expressions by combining like terms.

EXAMPLE 3 Combine Like Terms

Follow these steps to simplify the expression $-4m^3 + 4m + 6n^3 + 2m$.

a. **USE STRUCTURE** What are the like terms in the expression? Explain how you know. CCSS A.SSE.2, SMP 7

b. **COMMUNICATE PRECISELY** Which property allows you to rewrite the expression as $-4m^3 + 4m + 2m + 6n^3$? Why? CCSS A.SSE.2, SMP 6

c. **USE STRUCTURE** Show how to use the Distributive Property to rewrite the middle two terms ($4m$ and $2m$) as a single term. What is the simplified expression? CCSS A.SSE.1b, SMP 7

d. **DESCRIBE A METHOD** Describe how you can simplify an expression by combining like terms. CCSS A.SSE.2, SMP 8

PRACTICE

USE STRUCTURE Which property is illustrated by each equation? CCSS A.SSE.2, SMP 7

1. $9y + 3x + 6y = 3x + 9y + 6y$

2. $c(5 + d) = 5c + cd$

3. $4(5m) = (4 \cdot 5)m$

USE STRUCTURE The table shows the prices of tickets to a theme park. Use the table for Exercises 4–6. For each situation, explain how to use the Distributive Property and mental math to find the total cost of the tickets. CCSS A.SSE.1b, SMP 7

Type of Ticket	Price
Adults	\$29
Students	\$21
Seniors	\$18

4. 7 tickets for adults

5. 9 tickets for students

6. 12 tickets for seniors

USE STRUCTURE Simplify each expression. CCSS A.SSE.2, SMP 7

7. $-2k + 2k^2 - 3k + k^3$

8. $2n^2 + 5n^4 - 6n^2 - 3n^4$

9. $-b + 3a + b^2 - 2b + 6a - 9a$

10. **CRITIQUE REASONING** Angela is shipping 8 bags of granola to a customer. Each bag weighs 22 ounces and the maximum weight she can ship in one box is 10 pounds 5 ounces. She makes the calculation at right and decides that she can ship the bags in one box. Do you agree? Explain. CCSS A.SSE.1b, SMP 3

$$
\begin{aligned}
8(22) &= 8(20 + 2)\\
&= 8(20) + 2\\
&= 160 + 2\\
&= 162 \text{ ounces}
\end{aligned}
$$

11. A theater has m seats per row on the left side of the aisle and n seats per row on the right side of the aisle. There are r rows of seats. CCSS A.SSE.1b

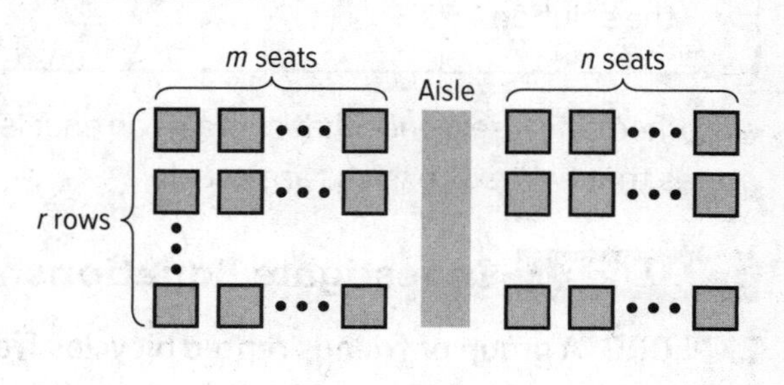

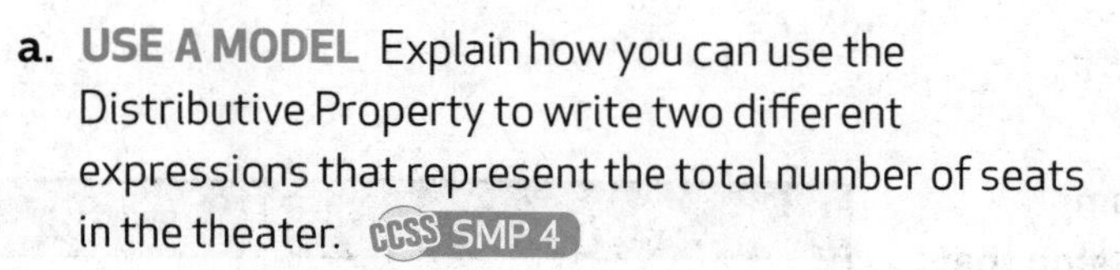

a. **USE A MODEL** Explain how you can use the Distributive Property to write two different expressions that represent the total number of seats in the theater. CCSS SMP 4

b. **CONSTRUCT ARGUMENTS** Suppose you double the number of seats in each row on the left side of the aisle. Does this double the number of seats in the theater? Use one of the expressions you wrote in **part a** to justify your answer. CCSS SMP 3

12. **CONSTRUCT ARGUMENTS** Is there a Commutative Property or Associative Property for subtraction? Explain why or why not for each property. CCSS A.SSE.2, SMP 3

13. **REASON QUANTITATIVELY** Provide a counterexample to show that there is no Commutative Property or Associative Property for division. What is the relationship between the results when the order of division of two numbers is switched? CCSS A.SSE.2, SMP 2

1.5 Equations

Objectives

- Write equations in one variable and use them to solve problems.
- Solve linear equations in one variable and explain the steps of the solution.

STANDARDS

Content: A.CED.1, A.REI.1, A.REI.3
Practices: 1, 2, 3, 4, 7, 8
Use with Lesson 1-5

A mathematical sentence that contains an equals sign (=) is an **equation**. An equation states that two expressions are equal.

EXAMPLE 1 Investigate Equations

EXPLORE A group of friends rented bicycles from different bike shops. The table shows the expression that each shop uses to calculate the cost of renting one of their bikes for *h* hours.

Shop	Cost for *h* Hours ($)
Real Wheels	$5(h+1)$
Easy Bike	$3.5h+8.5$
Pedal Power	$2(3h-1)$

a. USE A MODEL Kaden rented his bike from Easy Bike and he paid $26 for the rental. Write an equation that relates the expression the bike shop uses to calculate the cost and the amount Kaden paid. Do you think Kaden rented his bike for 6 hours? Justify your answer using the equation you wrote. CCSS A.CED.1, SMP 4

b. CONSTRUCT ARGUMENTS Do you think Kaden rented his bike for 5 hours? Justify your answer using the equation you wrote. CCSS A.REI.3, SMP 3

c. CRITIQUE REASONING Megan rented her bike from Real Wheels and she paid $25. She claims that she rented the bike for 5 hours. Do you agree? Use an equation to explain why or why not. CCSS A.CED.1, SMP 3

d. USE REASONING Kim-Ly rented her bike from Pedal Power and she paid $22. Did she rent the bike for 3, 4, or 5 hours? Use an equation to explain your answer. CCSS A.CED.1, SMP 2

A **solution** of an equation is a value of the variable that makes the equation true. A set of numbers from which replacements for a variable may be chosen is called a **replacement set**. A **solution set** is the set of all solutions in the replacement set.

EXAMPLE 2 Solve an Equation

Complete these steps to solve the equation $(8^2 \div 4 - 11)p - 2p = 12$.

a. USE STRUCTURE Use the order of operations to simplify the expression in parentheses and write the resulting equivalent equation. Explain your steps. 

b. USE STRUCTURE Explain how to simplify the resulting equation. What property justifies this step of the process? CCSS A.REI.1, SMP 7

c. USE STRUCTURE What is the solution of the equation? How do you know? CCSS A.REI.3, SMP 7

d. INTERPRET PROBLEMS Explain how you can check that your solution is correct. CCSS A.REI.3, SMP 1

e. CRITIQUE REASONING The set $\{3 < p < 5\}$ is given as the replacement set for the equation $(8^2 \div 4 - 11)p - 2p = 15$. This set is shown on the number line. Martina says that this changes the solution of the equation. Do you agree with Martina? Explain your answer. CCSS A.REI.3, SMP 3

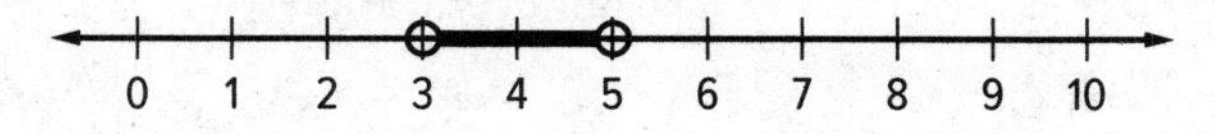

f. DESCRIBE A METHOD If no replacement set is given for an equation, describe how you could go about solving the equation. How does having a replacement set given change your approach? CCSS A.REI.1, SMP 8

Some equations have no solution. Other equations have more than one solution. An equation that is true for every value of the variable is called an **identity**. For example, $x + 3 = 3 + x$ is an identity.

EXAMPLE 3 Solve an Equation

Complete these steps to solve the equation $5x + 4x + 3 = 2x + 2 + 7x$.

a. USE STRUCTURE Use one or more properties to justify each step of the solution process shown below. CCSS A.REI.1, SMP 7

$5x + 4x + 3 = 2x + 2 + 7x$ Original equation

$9x + 3 = 2x + 2 + 7x$ ____________________

$9x + 3 = 2x + 7x + 2$ ____________________

$9x + 3 = 9x + 2$ ____________________

b. CONSTRUCT ARGUMENTS What is the solution to the original equation? Justify your answer. CCSS A.REI.3, SMP 3

____

____

c. REASON QUANTITATIVELY Change one term in the original equation so that the equation has exactly one solution, when $x = 1$. How do you know that you are correct? CCSS A.REI.1, SMP 2

____

____

EXAMPLE 4 Solve an Equation

Complete these steps to solve the equation $-3b + 9b + 17 = 5b + 15 + b + 2$.

a. USE STRUCTURE Use one or more properties to justify each step of the solution process shown below. CCSS A.REI.1, SMP 7

$-3b + 9b + 17 = 5b + 15 + b + 2$ Original equation

$-3b + 9b + 17 = 5b + b + 15 + 2$ ____________________

$-3b + 9b + 17 = 6b + 17$ ____________________

$6b + 17 = 6b + 17$ ____________________

b. CONSTRUCT ARGUMENTS What is the solution to the original equation? Justify your answer. CCSS A.REI.3, SMP 3

____

____

c. REASON QUANTITATIVELY Suppose the 17 in the original equation was an 18. Without solving another equation, what would the solution to this equation be? Explain how you know. CCSS A.REI.1, SMP 2

____

____

EXAMPLE 5 Write and Solve an Equation

The table shows the fees charged for overdue items at the Cedarville Library.

Item	Late Fees
Book	\$1.10 plus \$0.25 per day
CD	\$1.50 plus \$0.75 per day
DVD	\$3.00 plus \$1.25 per day

a. USE A MODEL Let d be the number of days that a book is overdue. Let F be the total late fee that is charged for the book. Write an equation of the form $F =$ [expression] that shows how to calculate the late fee. Explain how you wrote the equation. CCSS A.CED.1, SMP 4

b. USE A MODEL Jamar has a book that is 5 days overdue. Show how to solve an equation to find the late fee for the book. Use the spaces provided to show the steps of your solution and write an explanation of each step in the spaces at the right. CCSS A.REI.3, SMP 4

_____ = _____ Original equation from **part a**

_____ = _____ _____

_____ = _____ _____

_____ = _____ _____

c. REASON QUANTITATIVELY Write a sentence explaining what the last line of your solution tells you. CCSS A.REI.3, SMP 2

d. USE A MODEL Write an equation that shows how to calculate the total late fee F for a CD that is d days overdue. Then show how to use the equation to find the total fee for a CD that is 7 days overdue. CCSS A.CED.1, SMP 4

e. USE A MODEL Write an equation that shows how to calculate the total late fee F for a DVD that is d days overdue. Hailey has a DVD with a late fee of \$10.50. Use your equation to determine whether Hailey's DVD is 9 days overdue. Explain your answer. CCSS A.CED.1, SMP 4

f. REASON QUANTITATIVELY Liza returns a CD to the library and his charged a late fee of \$6.00. How many days overdue was Liza's CD? Explain your reasoning. CCSS A.REI.1, SMP 2

PRACTICE

Three Web sites sell food for dogs that require a special diet. The table shows expressions that give the total cost of ordering the dog food from each Web site. Use the table for Exercises 1–3.

Web Site	Cost for p pounds ($)
Canine Kitchen	$1.75(p-1)$
Pet Zone	$2.50(p+2)$
Super Chow	$1.50p+6$

1. **USE A MODEL** Natasha bought dog food from Super Chow and paid a total of $16.50. Write an equation that relates the expression the Web site uses to calculate the cost and the amount Natasha paid. Then use the equation to explain whether you think Natasha bought 7 pounds of dog food. CCSS A.CED.1, SMP 4

2. **REASON QUANTITATIVELY** Isaac bought dog food from Pet Zone and paid a total of $32.50. Did he buy 10, 11, or 12 pounds of dog food? Use an equation to justify your answer. CCSS A.CED.1, SMP 2

3. **CONSTRUCT ARGUMENTS** The equation $1.75(p-1)=42$ can be used to find the amount of dog food the Cheng family ordered from Canine Kitchen. Did the Chengs order more or less than 20 pounds of dog food? Justify your response. CCSS A.CED.1, SMP 3

4. **USE STRUCTURE** Describe the steps you use to solve the equation $16=x+4+(2^4-6)$. Then explain how you know your solution is correct. CCSS A.REI.1, SMP 7

5. **USE STRUCTURE** The values on the number line are the replacement set for the equation $m^2+1=5$. Which of the values, if any, are solutions of the equation? Explain. CCSS A.REI.3, SMP 7

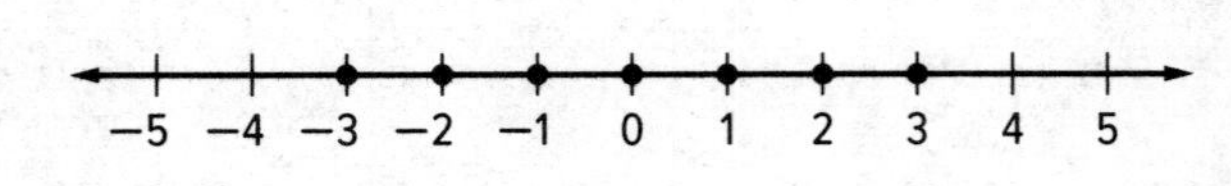

USE STRUCTURE Solve each equation. First simplify the equation and use a property to justify each step. Then write the solution and explain how you found it. CCSS A.REI.1, SMP 7

6. $c + 12 + 6c = 10c - 3c + 11 + 1$ Original equation

Solution:

7. $2n + 2 + 2n = 6n - 2n - 2$ Original equation

Solution:

8. A Web site offers its members a special rate for online movies, as shown in the advertisement. Let m be the number of movies you watch and let C be the total cost to watch the movies. CCSS A.CED.1

Movie Mania

One-time sign-up fee: \$6.85
Then watch as many movies as you like for just \$2.99 per movie!

a. USE A MODEL Write an equation that relates the total cost to the number of movies you watch. CCSS SMP 4

b. USE A MODEL Jeffrey watches 16 movies this month. Explain how to use the equation to find his total cost. CCSS SMP 4

c. CRITIQUE REASONING Madison said she joined the site and paid exactly \$19 to watch some movies. Her sister said this is impossible. Who is correct? Explain. CCSS SMP 3

d. REASON QUANTITATIVELY SuperFlix has no sign-up fee, just a flat rate per movie. If renting 13 movies at MovieMania costs the same as renting 9 movies at SuperFlix, what does SuperFlix charge per movie? CCSS SMP 2

1.6 Relations

Objectives

- Represent a relation in multiple ways.
- Identify the domain and range of a relation.
- Interpret the graph of a relation.

CCSS STANDARDS

Content: A.REI.10, F.IF.1
Practices: 1, 2, 3, 4, 6
Use with Lesson 1-6

A **relation** is a set of ordered pairs. The set of the first numbers in the ordered pairs is the **domain**. The set of the second numbers in the ordered pairs is the **range**.

EXAMPLE 1 Represent a Relation

EXPLORE A newspaper reporter asked five teenagers about the last time they babysat. Each teenager gave the number of hours he or she babysat and the amount he or she earned. The graph shows the results.

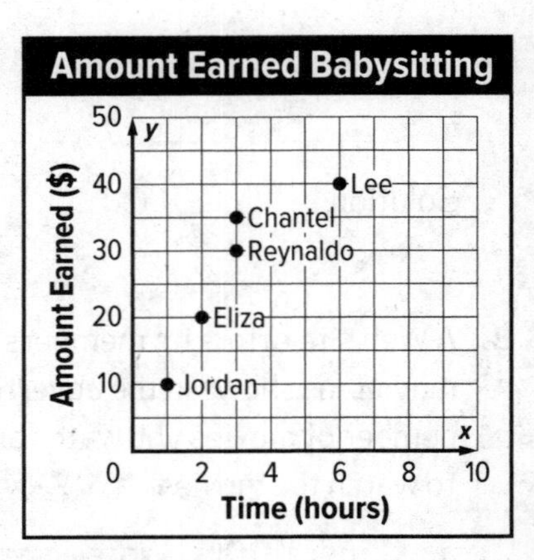

a. USE A MODEL Write the set of ordered pairs shown in the graph. What do the numbers in each ordered pair represent? CCSS A.REI.10, SMP 4

b. USE A MODEL You can use a table or a mapping to represent a relation. A **mapping** is a diagram that shows how each element in the domain is paired with an element in the range. Complete the table and mapping shown below. CCSS F.IF.1, SMP 4

x	y
1	10

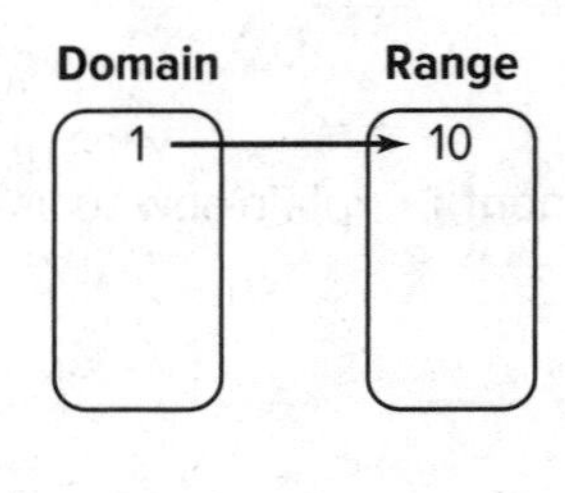

c. INTERPRET PROBLEMS Write the domain and range for the relation. Write each as a set within brackets, { }. CCSS F.IF.1, SMP 1

d. USE A MODEL How can you tell from the graph which of the teenagers babysat for the same number of hours? CCSS A.REI.10, SMP 4

e. REASON QUANTITATIVELY Which teenager was paid the highest hourly rate? Justify your answer. CCSS F.IF.1, SMP 2

In a relation, the value that determines the output is the **independent variable**. The variable with a value that is dependent on the value of the independent variable is the **dependent variable**.

EXAMPLE 2 Interpret a Graph

Michelle started at her house and went for a long walk. The graph represents her distance from home since the walk began.

Michelle's Walk

a. USE A MODEL What are the independent variable and dependent variable? Explain. CCSS F.IF.1, SMP 4

b. COMMUNICATE PRECISELY Describe what happens in the graph. CCSS A.REI.10, SMP 6

c. USE A MODEL How would the graph be different if Michelle decided to stop at her aunt's house on the way home and spend the night there? CCSS A.REI.10, SMP 4

d. COMMUNICATE PRECISELY The graph shows another walk that Michelle took. What does this graph show? Compare the starting and ending points to those in the first graph. CCSS A.REI.10, SMP 6

Michelle's Walk

PRACTICE

1. The following ordered pairs give the length in feet and the weight in pounds of five snakes at the reptile house of a zoo: {(5.5, 4.5), (3, 0.5), (3, 2), (8, 4.5), (2, 0.5)}

   a. **INTERPRET PROBLEMS** Write the domain and range for the relation. CCSS F.IF.1, SMP 1

   b. **USE A MODEL** What are the independent and dependent variables? CCSS F.IF.1, SMP 4

   c. **USE A MODEL** Complete the table, the mapping, and the graph below. CCSS F.IF.1, SMP 4

x	y

Domain Range

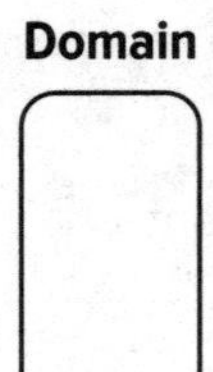

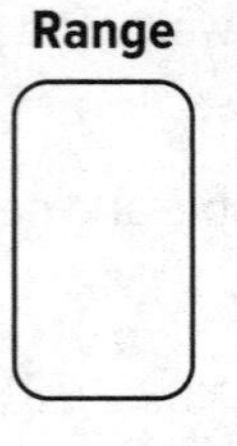

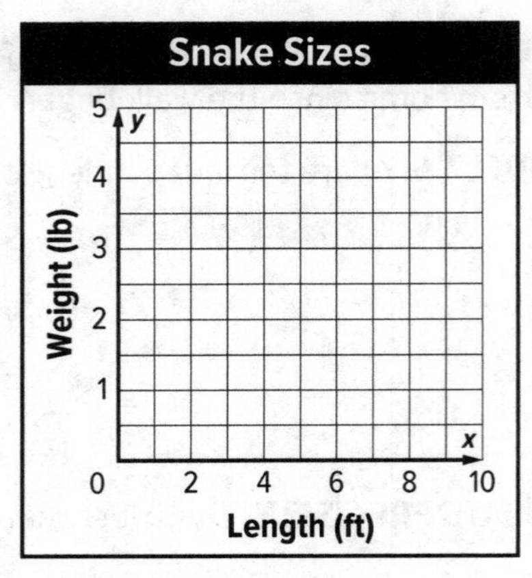

   d. **USE A MODEL** How can you tell from the graph which snakes have the same weight? CCSS A.REI.10, SMP 4

   e. **USE A MODEL** What does it mean if there are two points on the graph that lie on a vertical line? CCSS A.REI.10, SMP 4

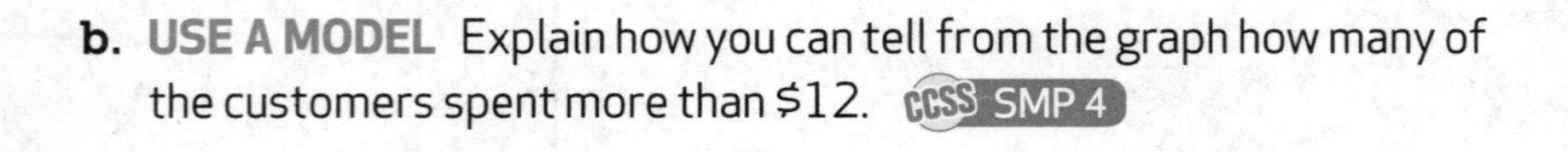

2. The graph shows the number of items that eight customers bought at a supermarket and the total cost of the items. CCSS F.IF.1

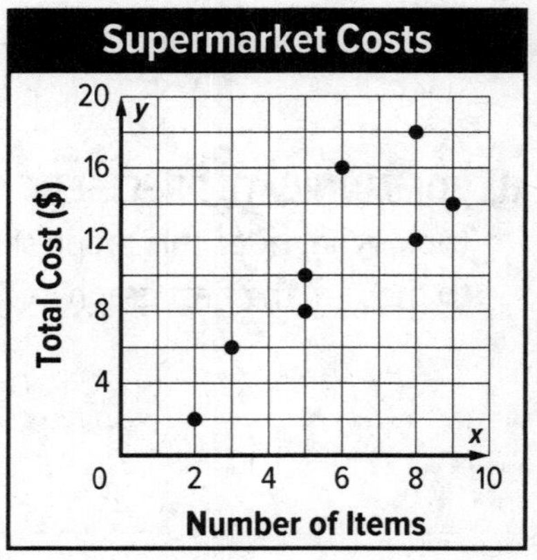

   a. **INTERPRET PROBLEMS** Write the domain and range for the relation. CCSS SMP 1

   b. **USE A MODEL** Explain how you can tell from the graph how many of the customers spent more than $12. CCSS SMP 4

   c. **CRITIQUE REASONING** A student said you can add the values in the domain to find the total number of items these customers bought. Do you agree? Explain. CCSS SMP 3

Tim and Lauren use their cars to deliver pizzas. The graph represents their distance from the pizzeria starting at 6 PM. Use the graph for Exercises 3–6.  CCSS A.REI.10

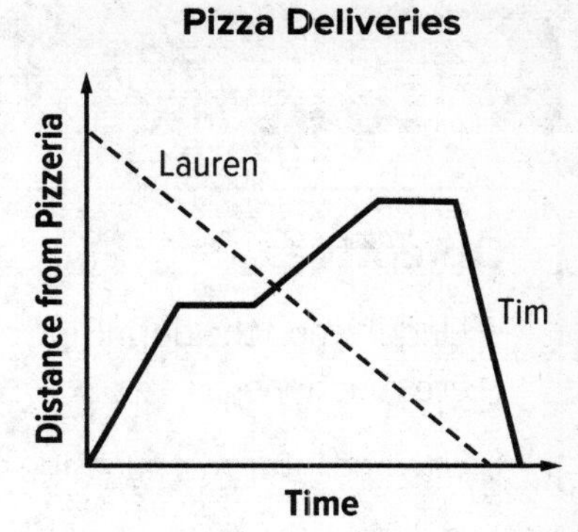

3. **COMMUNICATE PRECISELY** Describe what happens in Tim's graph. CCSS SMP 6

4. **COMMUNICATE PRECISELY** Describe what happens in Lauren's graph. CCSS SMP 6

5. **CRITIQUE REASONING** A student said that Tim's and Lauren's graphs intersect, so their cars must have crashed at some time after 6 PM. Do you agree or disagree? Explain. CCSS SMP 3

6. **USE A MODEL** After 6 PM, which delivery person was the first to return to the pizzeria? How do you know? CCSS SMP 4

7. **CRITIQUE REASONING** Cameron said that for any relation, the number of elements in the domain must be greater than or equal to the number of elements in the range. Do you agree? If so, explain why. If not, give a counterexample. CCSS F.IF.1, SMP 3

8. **USE A MODEL** The graph shows the height of an elevator above the ground. Describe the domain and range for this relation in words and by using inequalities. Then give three ordered pairs in the relation. CCSS F.IF.1, SMP 2

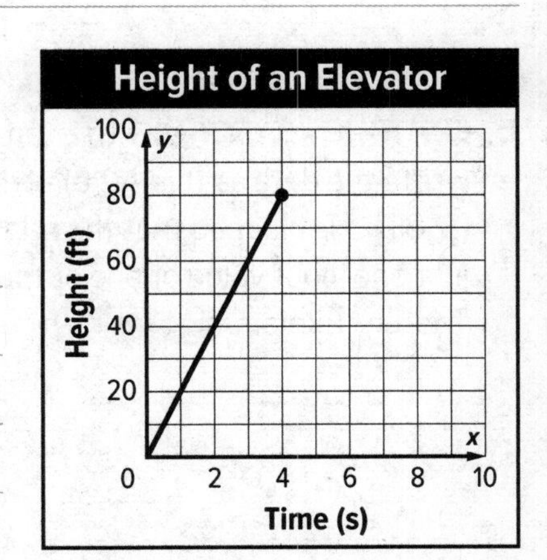

1.7 Functions

Objectives

- Understand the definition of a function and identify relations that are functions.
- Use and interpret function notation.
- Relate the domain of a function to its graph.

Content: F.IF.1, F.IF.2, F.IF.5
Practices: 1, 2, 3, 4, 6
Use with Lesson 1-7

A **function** is a relation in which there is exactly one output for each input. In other words, each element of the domain is assigned to exactly one element of the range.

EXAMPLE 1 Identify Functions CCSS F.IF.1

EXPLORE Tristan surveyed students at some local high schools. At each school, he asked six students how long they studied for their last exam and the score they received on the exam. His data is shown in the table, mapping, and graph below.

Central High Score	
Time (h)	**Score**
0.5	81
1	81
3	92
1.5	75
2	90
1.5	94

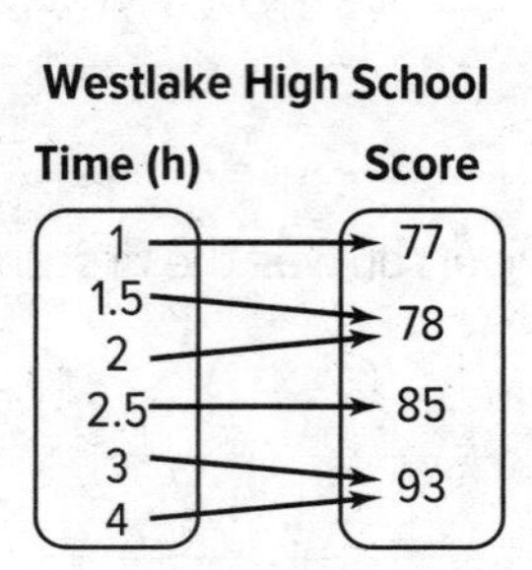

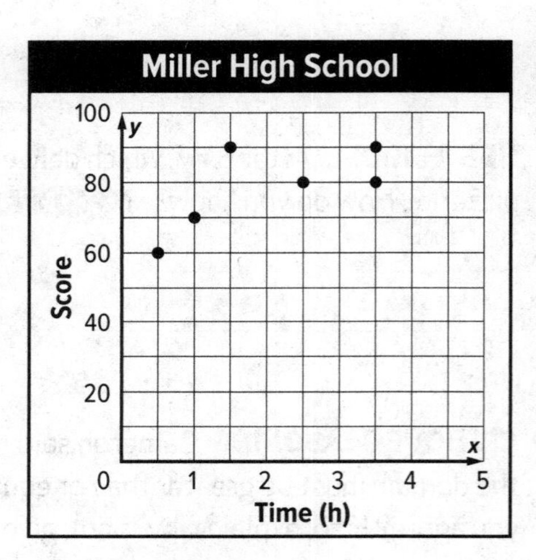

a. USE A MODEL For each school, is the relation a function? Why or why not? CCSS SMP 4

b. CRITIQUE REASONING Tristan surveyed six students at Chavez High School, and wrote the data as this set of ordered pairs: {(3, 87), (4, 98), (2.5, 70), (1.5, 70), (0.5, 67), (3, 81)}. He claimed that the relation is not a function, but he said that he could change just one input value or one output value and make it a function. Do you agree with Tristan? Explain. CCSS SMP 3

A graph that consists of points that are not connected is a **discrete function**. A function with a graph that is a line or a smooth curve is a **continuous function**.

EXAMPLE 2 Graph a Function

Tickets to the county fair cost $5 each.

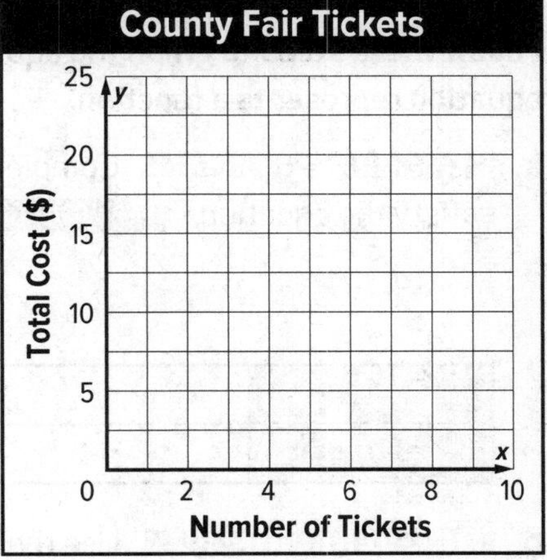

a. **USE A MODEL** Make a graph that shows the relationship between the number of tickets you buy and the total cost of the tickets. CCSS F.IF.1, SMP 4

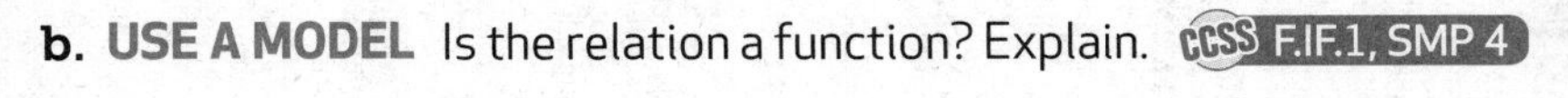

b. **USE A MODEL** Is the relation a function? Explain. CCSS F.IF.1, SMP 4

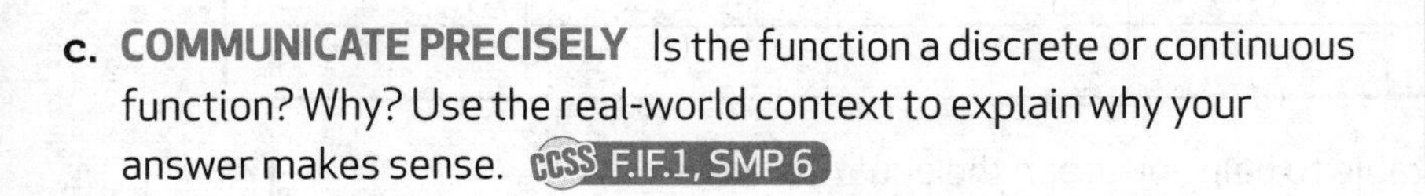

c. **COMMUNICATE PRECISELY** Is the function a discrete or continuous function? Why? Use the real-world context to explain why your answer makes sense. CCSS F.IF.1, SMP 6

d. **REASON QUANTITATIVELY** What is the domain of the function? How is the domain related to the real-world context? CCSS F.IF.5, SMP 2

e. **REASON QUANTITATIVELY** What is the range of the function? How is the range related to the real-world context? CCSS F.IF.1, SMP 2

f. **CRITIQUE REASONING** A student said that if you write the function as a set of ordered pairs, the ordered pair (25, 100) will be an element of the set. Do you agree or disagree? Explain. CCSS F.IF.1, SMP 3

g. **COMMUNICATE PRECISELY** Describe a situation that could be modeled by a function that includes the same ordered pairs as the function for the cost of county fair tickets, but is of the type that you did not select as your answer to **part c**. Explain why this situation leads to a function of the other type. CCSS F.IF.1, SMP 6

The **vertical line test** is a way to check whether a graph represents a function. If there is a vertical line that intersects the graph in more than one point, then the graph is not a function. Otherwise, the relation is a function.

EXAMPLE 3 Graph a Function

Follow these steps to graph the equation $y - 2x = 4$ and determine whether the equation represents a function.

a. **INTERPRET PROBLEMS** Complete the table below by finding five ordered pairs that satisfy the equation. CCSS F.IF.1, SMP 1

x					
y					

b. **INTERPRET PROBLEMS** Use the table to help you graph the equation on the coordinate plane provided. CCSS F.IF.1, SMP 1

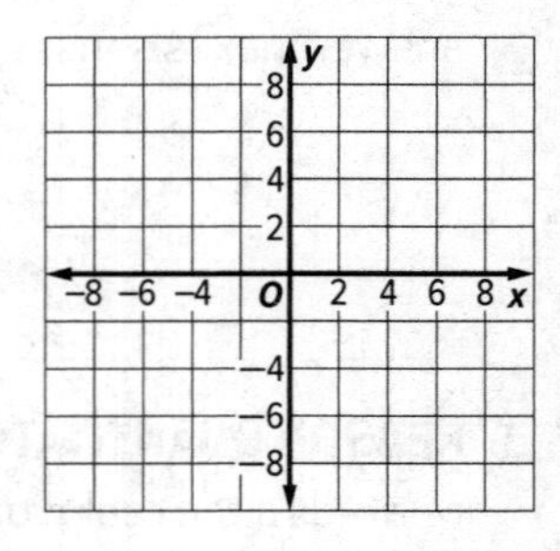

c. **CONSTRUCT ARGUMENTS** Is the relation a function? Use your graph to justify your answer. CCSS F.IF.1, SMP 3

d. **REASON ABSTRACTLY** What are the domain and range of the function? Explain how these are related to the graph of the function. CCSS F.IF.5, SMP 2

e. **COMMUNICATE PRECISELY** Describe how you can use the graph of the function to find the output value that corresponds to the input value −6. CCSS F.IF.1, SMP 6

f. **COMMUNICATE PRECISELY** How did you know to draw the graph as a continuous function? Use the given equation in your justification. How can you tell if a relation is discrete? CCSS F.IF.5, SMP 6

Function notation is a way to use an equation to write a rule for a function. For example, in function notation, the equation $y = x + 7$ is written as $f(x) = x + 7$. The graph of the function $f(x)$ is the graph of the equation $y = f(x)$.

In function notation, $f(x)$ denotes the element of the range corresponding to the element x of the domain. For example, for the function $f(x) = x + 7$, $f(9)$ represents the output value that corresponds to the input value $x = 9$. Therefore, $f(9) = 9 + 7 = 16$.

EXAMPLE 4 Use and Interpret Function Notation

Candace runs a company that installs fences. She calculates the total cost C of installing a fence using the function rule $C(x) = 5x + 25$, where x is the length of the fence in feet.

a. REASON QUANTITATIVELY What is the value of $C(17.5)$? What does $C(17.5)$ represent? CCSS F.IF.2, SMP 2

b. USE A MODEL Explain how Candace can use the function rule to find the cost of the installing a fence that is 11 yards long. CCSS F.IF.2, SMP 4

c. USE A MODEL Graph $C(x)$ on the coordinate plane at the right. CCSS F.IF.1, SMP 4

d. REASON ABSTRACTLY What are the domain and range of the function? Explain how these are related to the graph of the function. CCSS F.IF.5, SMP 2

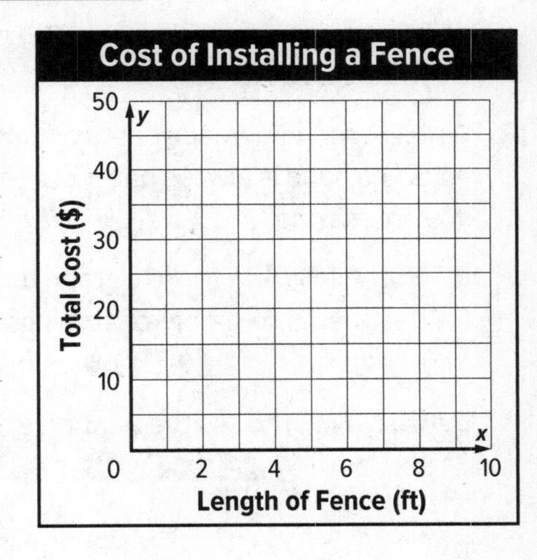

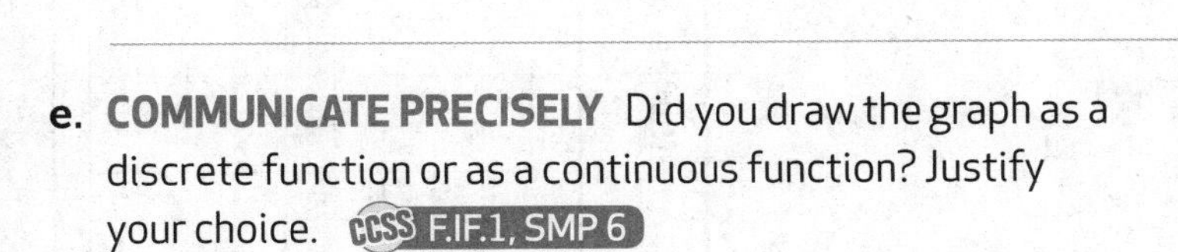

e. COMMUNICATE PRECISELY Did you draw the graph as a discrete function or as a continuous function? Justify your choice. CCSS F.IF.1, SMP 6

f. COMMUNICATE PRECISELY Would your answer to **part e** change if the fencing material were only sold in one-foot increments? If someone needed to fence a length that was not a whole number, what should they do? Explain. CCSS F.IF.5, SMP 6

PRACTICE

1. **USE A MODEL** Mario collected data about some of the players on a women's basketball team. The data is shown in the table, mapping, and graph. Is each relation a function? Why or why not? CCSS F.IF.1, SMP 4

Team History	
Years on Team	Games Played
1	24
2	45
3	82
3	88
5	120

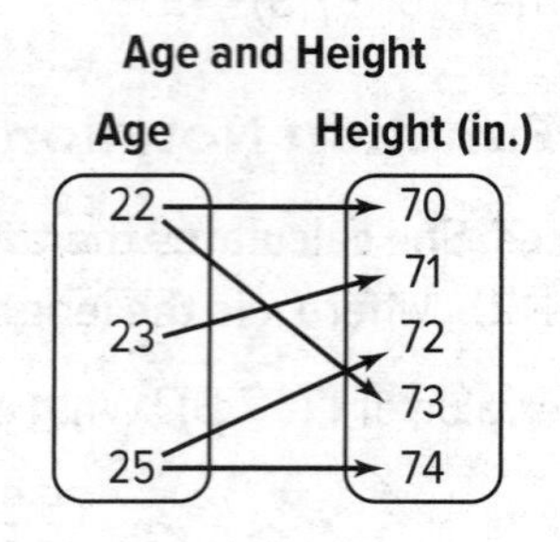

2. A recipe for homemade pasta dough says that the number of eggs you need is always one more than the number of servings you are making.

   a. **USE A MODEL** Make a graph that shows the relationship between the number of servings and the number of eggs. CCSS F.IF.1, SMP 4

   b. **USE A MODEL** Is the relation a function? Explain. CCSS F.IF.1, SMP 4

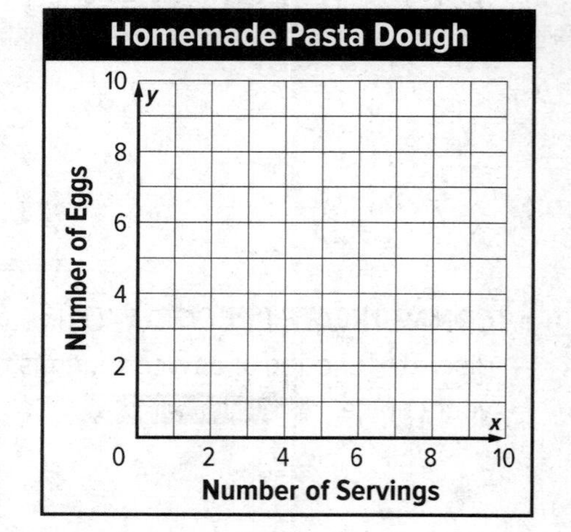

   c. **REASON QUANTITATIVELY** What is the domain of the function? How is the domain related to the real-world context? CCSS F.IF.5, SMP 2

INTERPRET PROBLEMS Graph each equation. Then explain whether or not the equation represents a function. CCSS F.IF.1, SMP 1

3. $2x + y = 6$

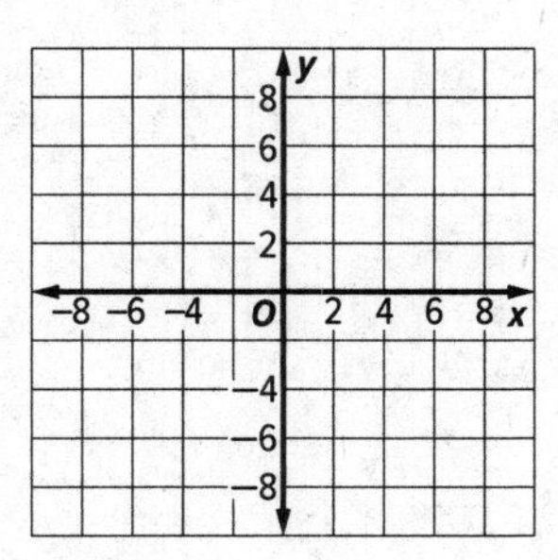

4. $y = x^2$

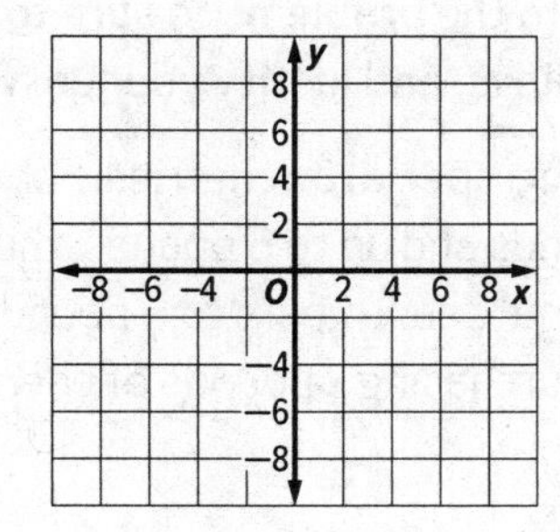

5. REASON QUANTITATIVELY The height h of a balloon, in feet, t seconds after it is released is given by the function $h(t) = 2t + 6$. CCSS SMP 5

a. What is the value of $h(20)$, and what does it tell you? CCSS F.IF.2

b. Explain how to use the function to find the height of the balloon 2 minutes after it is released. CCSS F.IF.2

c. What is the height of the balloon just before it is released? How do you know? CCSS F.IF.2

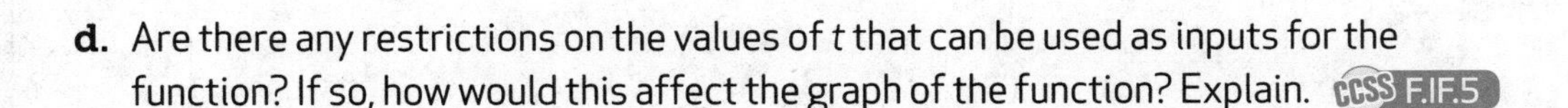

d. Are there any restrictions on the values of t that can be used as inputs for the function? If so, how would this affect the graph of the function? Explain. CCSS F.IF.5

6. CONSTRUCT ARGUMENTS The following set of ordered pairs represents a function, but one of the values is missing and has been replaced by a question mark: $\{(-4, -1), (-3, -1), (3, 2), (5, 2), (?, 2)\}$. What conclusions can you make about the missing value? Explain. CCSS F.IF.1, SMP 3

Performance Task

Finding a Sale Price

Provide a clear solution to the problem. Be sure to show all of your work, include all relevant drawings, and justify your answers.

Sander's Market is having a special on cherries. For every pound of cherries purchased beyond 3 pounds and up to 6 pounds, the price per pound is discounted by $1. Sander's Market limits customers to 6 pounds of cherries. The graph shows the cost y in dollars for purchasing x pounds of cherries.

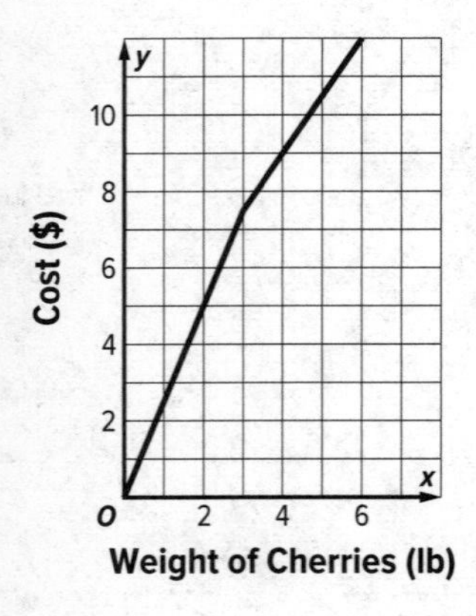

Part A

Find the domain and range. Describe their meaning in the context of this situation.

Part B

Use the graph to determine the cost per pound for buying 3 or fewer pounds of cherries. Write and solve an equation to find the total cost for purchasing 2.5 pounds of cherries. Describe how the graph can be used to check your answer.

Part C

Explain whether the graph represents a relation and/or a function.

Part D

Write a function $f(x)$ that gives the cost for purchasing x pounds of cherries, where $3 < x \leq 6$. Explain how you arrived at your answer.

Performance Task

At the Box Office

Provide a clear solution to the problem. Be sure to show all of your work, include all relevant drawings, and justify your answers.

Toby and his friends want to go see a play in the new theater downtown. He has the following options to save money on a purchase of several tickets. Each ticket agent sells the tickets at the same full-price before any discounts are applied.

Ticket-Time	SUPERSTUB	Bill's Box Office
For every two full-price tickets purchased, receive one free ticket!	20% off any order of 4 or more tickets!	Get $10 off your total order price.

Part A

Define a variable and write expressions to represent the cost of 10 tickets if purchased at Ticket-Time, Superstub, or Bill's Box Office. Simplify each expression. Based on your expressions, is it possible to determine whether the total price for 10 tickets is less at one ticket agent than at another? Justify your answer.

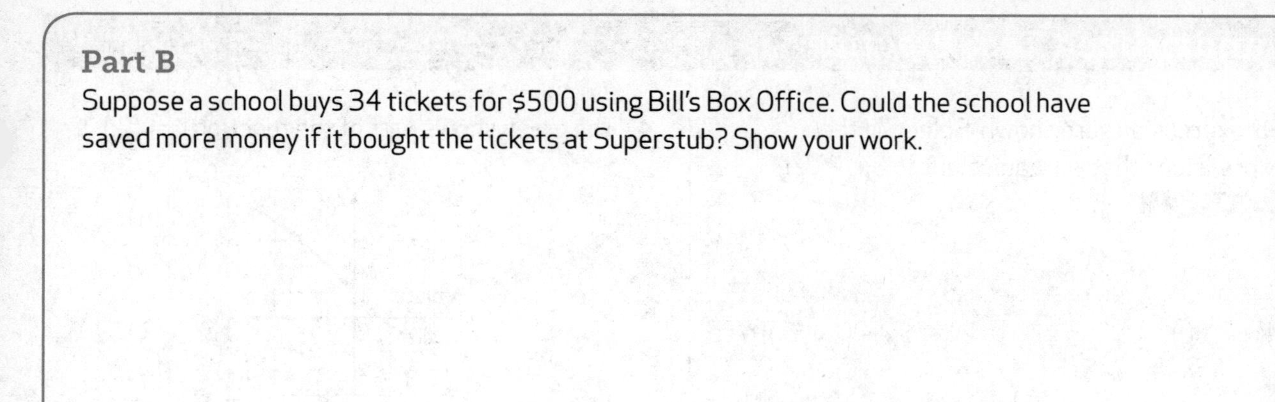

Part B

Suppose a school buys 34 tickets for $500 using Bill's Box Office. Could the school have saved more money if it bought the tickets at Superstub? Show your work.

Part C

Suppose a full-price ticket costs $15. A customer buys n tickets at Ticket-Time, where n is a multiple of 3. Write a function $C(n)$ that gives the cost of the tickets with the discount. Evaluate $C(6)$ and describe its meaning.

Part D

Explain which ticket agent gives the worst or the best deal, as the number of tickets purchased increases.

Standardized Test Practice

1. Six expressions are shown. Select all the expressions that are equivalent to $3x - 12y$.

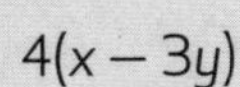

$4(x - 3y)$	$2x - 6y - (x + 6y)$
$3(x - 4y)$	$6x - 6y - (3x - 6y)$
$x + 4y + 2(x - 8y)$	$4(x - 3y) - x$

2. Last year, Hector downloaded x songs per month. His friend Britney downloaded 60 more songs last year than Hector.

Write an equation that represents the number of songs y Britney downloaded per month last year, in terms of the number that Hector downloaded.

☐

3. Theo is solving the equation $3x - 2 = -4$. He adds 2 to both sides of the equation and then he divides both sides of the equation by 3.

Select all the properties shown below that allow Theo to justify these steps. CCSS A.REI.1

- Distributive Property
- Addition Property of Equality
- Commutative Property
- Multiplication Property of Equality
- Division Property of Equality

4. The graph shows part of the function $y = f(x)$.

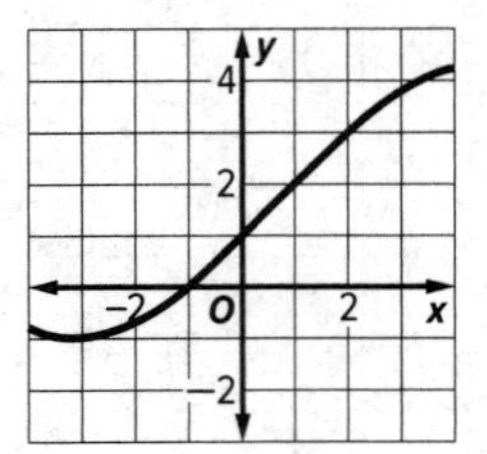

Complete the following.

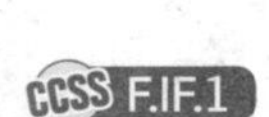

$f(2) =$ ☐. $f(1) =$ ☐.

$f($☐$) = 0$ $f($☐$) = 1$

5. Consider the following expression.

$$2(x - 1) + 4x^2 + 2(x^2 - 1)$$

When the expression is completely simplified, the coefficient of the x^2-term is ☐. CCSS A.SSE.1a

6. In the graph, each ordered pair gives the low temperature and the high temperature for each day of a recent winter week in Pinewood. Draw a vertical line on the graph that can be used to show why this relation is not a function. CCSS F.IF.1

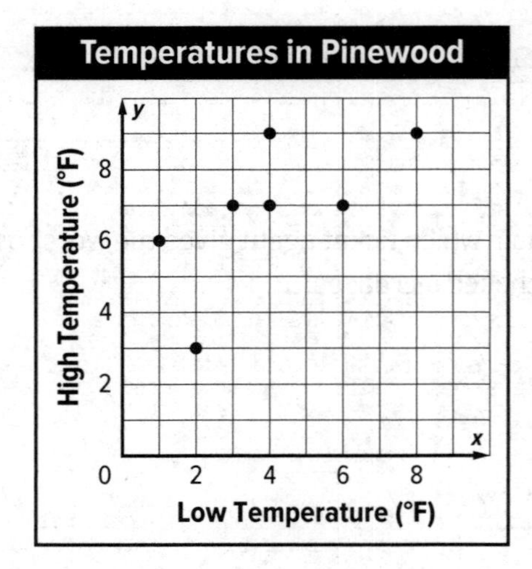

7. For the function $f(x) = \frac{x(x + 1)}{2}$ with domain $\{1, 2, 3, 4, 5\}$, what is the range? CCSS F.IF.1

Range: ☐

8. Consider each product or sum. Select *Rational* or *Irrational* for each row. Then explain why you selected *Rational* or *Irrational* for each product or sum. CCSS N.RN.3

Product or Sum	Rational	Irrational	Explanation
The product of $\sqrt{2}$ and $\sqrt{7}$.			
The product of $\sqrt{2}$ and $\sqrt{18}$.			
The sum of π and 19.			
The sum of $\sqrt{17}$ and $\frac{3}{5}$.			
The sum of $\sqrt{16}$ and 0.9.			

9. Consider each relation. Is each relation a function? Select *Yes* or *No* in each row. For any row in which you select *No*, explain why the relation is not a function. CCSS F.IF.1

Is the relation a function?	Yes	No
$\{(-3, 2), (3, 2), (2, 2)\}$		
$\{(-2, -5), (-1, 0), (0, -3), (1, 4), (1, 6)\}$		
The relation that maps each real number to 2 times the number		
The relation that maps every real number to 0		
The relation that maps each positive integer to its factors		

10. Solve the equation $0.2(4t + 10) = 6.8$. Show your work and justify each step of the solution process. CCSS A.REI.1

11. Aya plans on buying a new television with a retail price of x dollars. The store is having a sale, and there is also a rebate. The function $p(x) = 0.8x - 150$ gives the price p that Aya will end up paying for the television.

a. Interpret the meaning of the terms $0.8x$ and 150 in the function, in terms of the sale and the rebate. CCSS A.SSE.1a

b. Evaluate $p(1500)$ and describe what it means. CCSS F.IF.1, F.IF.2

c. Aya plans to end up paying between \$1400 and \$1800 for the television. Find the retail prices that she can shop for when at the store. How do these values relate to the domain and range of the function? CCSS F.IF.1, F.IF.5

Writing and Solving Linear Equations

CHAPTER FOCUS Learn about some of the Common Core State Standards that you will explore in this chapter. Answer the preview questions. As you complete each lesson, return to these pages to check your work.

What You Will Learn	Preview Question
Lesson 2.1: Accuracy and Precision	
CCSS N.Q.3 Choose a level of accuracy appropriate to limitations on measurement when reporting quantities.	CCSS SMP 6 Gasoline sells for \$3.529 per gallon. How much will 15 gallons cost rounded to the correct amount? Explain.
Lesson 2.2: Writing Equations	
CCSS A.CED.1 Create equations and inequalities in one variable and use them to solve problems. CCSS A.REI.3 Solve linear equations and inequalities in one variable, including equations with coefficients represented by letters. **Also addresses:** A.C.ED.3	CCSS SMP 2 The circumference of a circle is equal to the product of twice the radius and π. What is the equation for circumference? If you are given that the radius of a circle is 6 inches, what can you determine?
Lesson 2.3: Solving One-Step Equations	
CCSS A.REI.1 Explain each step in solving a simple equation as following from the equality of numbers asserted at the previous step, starting from the assumption that the original equation has a solution. Construct a viable argument to justify a solution method. **Also addresses:** A.REI.3, A.CED.1	CCSS SMP 3 What operation should you use to solve $23 + c = 59$? Justify your answer.
Lesson 2.4: Solving One-Step Equations	
CCSS A.REI.1 Explain each step in solving a simple equation as following from the equality of numbers asserted at the previous step, starting from the assumption that the original equation has a solution. Construct a viable argument to justify a solution method. CCSS A.CED.1 Create equations and inequalities in one variable and use them to solve problems. CCSS A.REI.3 Solve linear equations and inequalities in one variable, including equations with coefficients represented by letters.	CCSS SMP 7 Suppose you want to solve the equation $\frac{-7x-(-3)}{8} = 3$. Determine the first step and the solution of the equation.

What You Will Learn	Preview Question
Lesson 2.5: Solving Equations with the Variable on Each Side	
CCSS A.REI.3 Solve linear equations and inequalities in one variable, including equations with coefficients represented by letters. CCSS A.CED.1 Create equations and inequalities in one variable and use them to solve problems. CCSS A.REI.1 Explain each step in solving a simple equation as following from the equality of numbers asserted at the previous step, starting from the assumption that the original equation has a solution. Construct a viable argument to justify a solution method.	CCSS SMP 3 Your friend is trying to decide which gym she should join. Gym 1 charges \$165 per year and \$3 per visit. Gym 2 charges \$12 per month and \$3.50 per visit. For how many visits is the cost of membership the same? Explain. CCSS SMP 7 A square with side length $2x$ has the same perimeter as a rectangle with dimensions $x + 5$ and $x - 1$. What are the dimensions of each of the figures?
Lesson 2.6: Literal Equations	
CCSS A.CED.4 Rearrange formulas to highlight a quantity of interest, using the same reasoning as in solving equations. CCSS A.REI.3 Solve linear equations and inequalities in one variable, including equations with coefficients represented by letters. CCSS N.Q.1 Use units as a way to understand problems and to guide the solution of multi-step problems; choose and interpret units consistently in formulas; choose and interpret the scale and origin in graphs and data displays.	CCSS SMP 4 Write an equation that expresses the width of the rectangle in terms of its area and length. Then find the width of the rectangle if the area is 213.75 cm^2 and the length is 9.5 cm. 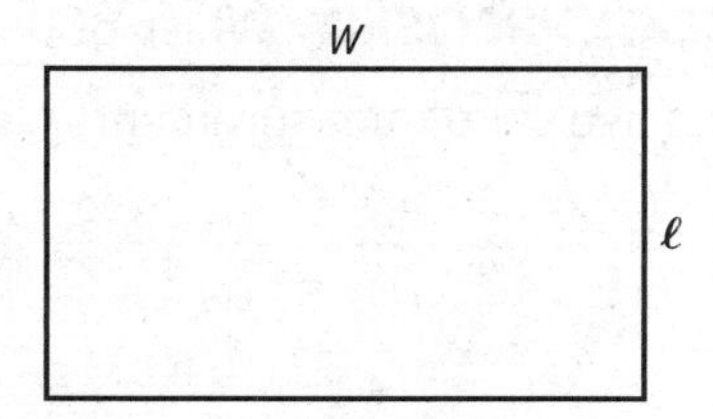 CCSS SMP 7 The equation $C = 0.1k + 9.50$ represents a customer's monthly charge for electricity, where k is the number of kilowatt-hours that the customer uses. Solve the equation for k. Determine the number of kilowatt-hours the customer can use if she budgets \$100 per month for electricity.
Lesson 2.7: Weighted Averages	
CCSS A.REI.3 Solve linear equations and inequalities in one variable, including equations with coefficients represented by letters. CCSS A.CED.1 Create equations and inequalities in one variable and use them to solve problems. CCSS A.CED.4 Rearrange formulas to highlight a quantity of interest, using the same reasoning as in solving equations. **Also addresses:** A.REI.1, N.Q.1	CCSS SMP 4 An airplane travels 1500 miles due south in 2.5 hours and 1800 miles due east in 3 hours. Write and solve an equation to find the airplane's average speed for the trip.

2.1 Accuracy and Precision

Objectives

- Choose a level of accuracy appropriate to limitations on measurement when reporting quantities.
- Evaluate the precision of a group of measurements.
- Understand the difference between accuracy and precision in measurement.

Content: N.Q.3
Practices: 2, 3, 6, 8
Use with Extend 1–3

Accuracy evaluates how close a measured value is to the actual value being measured. **Precision** describes how close measured values are to each other, without regard to the actual value.

EXAMPLE 1 Accuracy vs. Precision

EXPLORE GoYo, a new frozen yogurt store, sells yogurt and toppings by weight. The owner of the store needs to purchase a scale and tests three models by repeatedly measuring a 1-pound weight.

Scale A	Scale B	Scale C
16.12 oz	15.80 oz	15.64 oz
14.89 oz	16.15 oz	15.71 oz
16.57 oz	15.89 oz	15.68 oz
15.32 oz	16.07 oz	15.70 oz
17.08 oz	16.22 oz	15.72 oz

a. **COMMUNICATE PRECISELY** Which of the three scales provides the most precise set of measurements? Explain. CCSS SMP 6

b. **CALCULATE ACCURATELY** Which of the three scales provides the most accurate set of measurements? Explain your reasoning. CCSS SMP 6

c. **COMMUNICATE PRECISELY** Why is it important that the scale be both accurate and precise? CCSS SMP 6

d. **REASON QUANTITATIVELY** Which of the three scales do you think the owner should choose? Why? CCSS SMP 2

All measures are approximate. When evaluating measures like area or perimeter, remember that calculations should not be carried out to greater accuracy than that of the original data.

EXAMPLE 2 Evaluate Accuracy of a Derived Measure CCSS N.Q.3

a. COMMUNICATE PRECISELY Measure the length and width of the rectangle as precisely as possible and determine the range of possible values for the measures. CCSS SMP 6

in.
0 1 2 3

in.
0 1 2

b. CALCULATE ACCURATELY Determine the range of possible values for the perimeter of the rectangle. CCSS SMP 6

c. CALCULATE ACCURATELY Using the ruler above, Kiko calculated the area of the rectangle to be $3\frac{15}{64}$ square inches. Evaluate the accuracy of Kiko's calculation. CCSS SMP 6

EXAMPLE 3 Determine Appropriate Level of Accuracy

According to a calculator, $\frac{50}{17}$ is approximately 2.9411764706. Describe a real world situation that might require the indicated level of accuracy. CCSS N.Q.3

a. COMMUNICATE PRECISELY Round the value to the nearest whole number. CCSS SMP 6

b. COMMUNICATE PRECISELY Round the value to the nearest hundredth. CCSS SMP 6

c. COMMUNICATE PRECISELY Round the value to the nearest ten thousandth. CCSS SMP 6

d. REASON QUANTITATIVELY Katy plans to use the most precise value her calculator can provide: 2.9411764706. Explain why she cannot always record her answer to this level of precision. CCSS SMP 2

PRACTICE

1. **EVALUATE REASONABLENESS** Mr. Moreno's students are weighing materials for a chemistry experiment. Four students weigh the same sample using different scales:

   100 g   104 g   105 g   103.5 g

   Mr. Moreno tells the students that they each weighed the amount correctly. Explain how this is possible. CCSS N.Q.3, SMP 8

2. A measurement was taken four times. The actual measurement is 52.4 cm. Determine whether the set of measurements is *accurate*, *precise*, *both*, or *neither*. Explain your reasoning. CCSS N.Q.3

   a. COMMUNICATE PRECISELY 56.1 cm, 48.9 cm, 24.2 cm, 5 cm CCSS SMP 6

   b. COMMUNICATE PRECISELY 73.1 cm, 74.0 cm, 73.5 cm, 73.7 cm CCSS SMP 6

   c. COMMUNICATE PRECISELY 52.6 cm, 52.5 cm, 52.2 cm, 52.3 cm CCSS SMP 6

   d. CRITIQUE REASONING Nichole says that if a set of measurements is accurate, they are also precise. If you agree, explain your reasoning. If you disagree, provide a counterexample. CCSS SMP 3

3. CONSTRUCT ARGUMENTS Is the average value of a set a good indicator of the accuracy of the set? Use the data sets in the table to support your reasoning. Each set contains measures of an object with actual weight 17.65 grams. CCSS SMP 3

Set A	Set B
17.3 g	17.1 g
18.0 g	17.8 g
17.9 g	18.6 g
17.6 g	17.3 g

4. CALCULATE ACCURATELY Jayden measures and labels the dimensions of a box. CCSS SMP 6

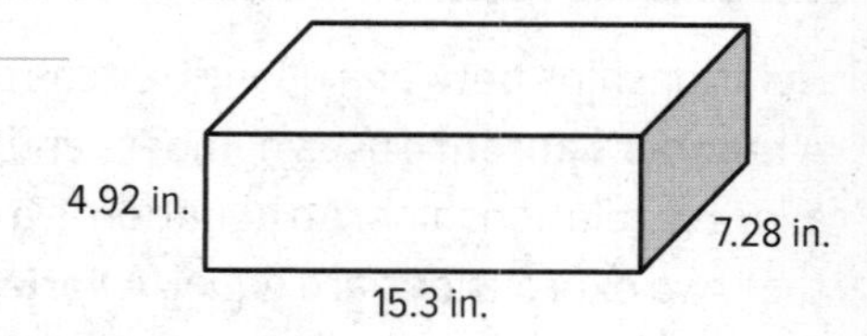

a. Calculate the areas of the faces of the box. Use appropriate levels of accuracy.

b. Determine the surface area of the box.

c. Determine a range of values that should contain the actual (true) measure of the surface area of the box. Explain your reasoning.

d. Suppose that Jayden had incorrectly measured the first dimension as 15.1 inches. Find the surface area of the box using this measure.

e. Compare the two calculations of surface area. Explain how a "minor" error in measurement can lead to the much greater error in the calculation of surface area.

5. CALCULATE ACCURATELY Manuel stops at a gas station that sells gasoline at \$3.29$\frac{9}{10}$ per gallon. He pumps 8.618 gallons of gasoline into his gas tank. How much will Manuel pay for gas? How much accuracy is possible? How much accuracy is necessary? Explain. CCSS SMP 6

2.2 Writing Equations

Objectives

- Translate between sentences and equations.
- Create and solve equations.

Content: A.CED.1, A.CED.3, A.REI.3
Practices: 1, 2, 3, 4, 6, 7
Use with Lesson 2-1

Relationships between quantities can be expressed using words or number sentences. A **number sentence** uses numbers and mathematical symbols instead of words to express relationships. An equation is a number sentence that uses an equal sign to show that two expressions are equal. A **variable** is a symbol used in algebra to represent unknown numbers or values. Any letter may be used as a variable. A **formula** is rule for the relationship between certain quantities.

EXAMPLE 1 Express Relationships Between Quantities CCSS A.CED.1

EXPLORE The table shows the number of apps six students have on their phones.

Name	Number of Apps	Name	Number of Apps
Alexander	10	Kaitlyn	5
Ichiro	2	Eduardo	3
Adela	8	Kenya	15

a. INTERPRET PROBLEMS What names could make each of these sentences true? CCSS SMP 1

________ has six more apps than ________.

________ has eight fewer apps than ______________.

________ has half as many apps as ______________.

b. INTERPRET PROBLEMS Write two true sentences that show a relationship between the number of apps the two students have. CCSS SMP 1

Kenya and Kaitlyn ______________________________

Eduardo and Ichiro ______________________________

Kenya and Alexander ______________________________

c. REASON QUANTITATIVELY Suppose Anna has half the number of apps that Ichiro and Alexander have combined. How can you determine how many apps Anna has? CCSS SMP 2

d. USE A MODEL Write an equation that can be solved to determine the number of apps Anna has, in terms of the number of apps Ichiro and Alexander have. Define the variables used in the equation. CCSS SMP 4

EXAMPLE 2 Translate Sentences into Equations CCSS A.CED.1

INTERPRET PROBLEMS Write an equation with one variable to represent each sentence. CCSS SMP 1

a. Five times the square of a number is equal to 45.

b. Three times the difference of a number and 5 is one less than the number.

c. The square of the sum of a number and 5 is equal to one more than the product of the number and 20.

EXAMPLE 3 Compare and Analyze Translations of Sentences into Equations CCSS A.CED.1

a. USE STRUCTURE What is the difference between these two sentences? 

Three times the sum of a number and 6 equals 84.

The sum of 3 times a number and 6 equals 84.

b. CRITIQUE REASONING Explain whether or not the equation is an accurate translation of the sentence. CCSS SMP 3

Five times a number is 8 less than four times the same number.

$5n = 8 - 4n$

c. CRITIQUE REASONING Explain whether or not the equation is an accurate translation of the sentence. CCSS SMP 3

Three times the difference of 4 and a number is equal to the sum of 8 and twice the number.

$$3 \cdot 4 - n = [(8 + n)2]$$

d. REASON QUANTITATIVELY Predict whether or not the equations that represent the two sentences in **part a** have different solutions. Explain your reasoning. CCSS SMP 2

EXAMPLE 4 Writing Equations CCSS A.CED.3, A.REI.3

PLAN A SOLUTION Write an equation for each of the situations described. List any constraints. Then solve each equation. CCSS SMP 1

a. Deshawn has several dimes and quarters that total $2.05. He has 3 fewer quarters than dimes. How many of each does he have?

b. It took Mrs. Nez 6 hours to drive her family 380 miles to visit their cousins. She drove part of the way on the interstate highway at 70 miles per hour and part of the way on county roads at 50 miles per hour. They spent twice as long traveling on interstate as they did on county roads. How long did they spend on each type of road?

c. Matt has three times as much money as Zita. Evan has half as much money as Zita. Altogether they have $135. How much money does each person have?

EXAMPLE 5 Writing Formulas CCSS A.CED.3

INTERPRET PROBLEMS Translate each statement into a formula. CCSS SMP 1

a. The sum s of the measures of all the angles of a polygon is the product of 180 and 2 less than the number of sides n.

b. The square of the hypotenuse of a right triangle is the sum of the squares of each of the two legs.

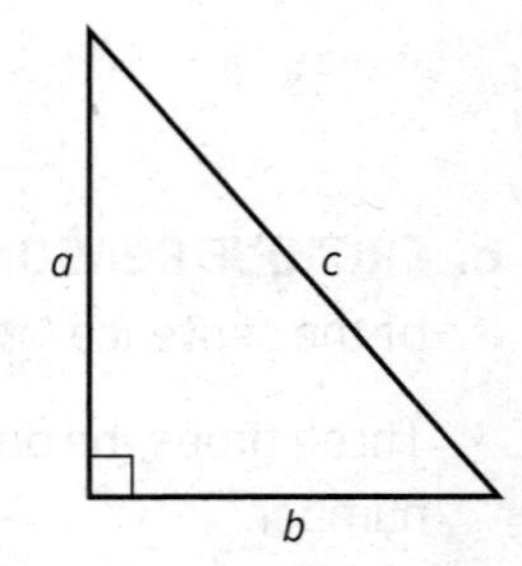

c. The average a of four numbers w, x, y, and z is equal to the quotient of the sum of the numbers and 4.

d. The temperature C in °C is $\frac{5}{9}$ the difference of the temperature F and 32 in °F.

e. The temperature F in °F is 32 more than $\frac{9}{5}$ of the temperature C in °C.

f. List any constraints on the values of the variables in the above formulas that you wrote.

EXAMPLE 6 Write Equations from Sentences CCSS A.CED.1

COMMUNICATE PRECISELY Translate each sentence into an equation. Explain the difference between the two equations. CCSS SMP 6

a. Sentence 1: Three times a number increased by 7 equals 8.
Sentence 2: Three times the sum of a number and 7 equals 8.

b. Sentence 1: Fifteen is equal to the difference of 7 and 2 times a number.
Sentence 2: Fifteen is equal to 7 less than 2 times a number.

c. Sentence 1: The square of -3 equals the quotient of 54 and a number.
Sentence 2: The opposite of the square of 3 equals the quotient of 54 and a number.

EXAMPLE 7 Write an Equation CCSS A.CED.1, A.CED.3

INTERPRET PROBLEMS Write and solve an equation to represent the given descriptions. Define the variables and list any constraints. CCSS SMP 1

a. The sale price of a messenger bag at a 25%-off sale is $15.

b. The shaded area is the difference of the square of the side length of the larger square and the square of the side length of the smaller square.

c. The product of two consecutive positive even integers is 24.

PRACTICE

1. **INTERPRET PROBLEMS Write an equation to represent the sentence.** CCSS A.CED.1, SMP 1

   a. A number increased by 7 is equal to 13 less than the square of the number.

   b. The sum of 3 times a number and 9 equals 3 times the sum of the number and 3.

   c. The quotient of 7 and twice a number equals 5.

2. **USE STRUCTURE** Describe the difference between the two statements. CCSS A.CED.1, SMP 7

   Three times the sum of a number and 6 is equal to 24.
   The sum of 3 times a number and 6 is equal to 24.

3. **CRITIQUE REASONING Explain whether or not the equation is an accurate translation of the sentence.** CCSS A.CED.1, SMP 3

   a. The square of the product of 4 and a number is equal to 8 times the sum of the number and 6.

   $(4n)^2 = 8(n + 6)$

   b. Three more than one-half a number is equal to 2 less than the number.

   $\frac{n}{\frac{1}{2}} + 3 = n - 2$

4. **PLAN A SOLUTION Write and solve an equation for each situation.** CCSS A.REI.3, A.CED.1, SMP 1

   a. Barbara is planning a party. She invited some school friends and 2 times as many friends from her swimming club as school friends. If a total of 21 friends came to her party, how many school friends did she invite?

   b. During the season, 13% of the players who signed up of for a soccer league dropped out. A total of 174 players finished the season. How many players signed up?

   c. Mr. Rhoades ordered 188 math books. The algebra books were packed in boxes of 12. The geometry books were packed in boxes of 10. He ordered one more box of algebra books than geometry books. How many boxes of each type book did he order?

5. INTERPRET PROBLEMS Translate the formula $A = \frac{b_1 + b_2}{2} \cdot h$ into words. Let A represent the area. List any constraints on the variables. CCSS A.CED.3, SMP 1

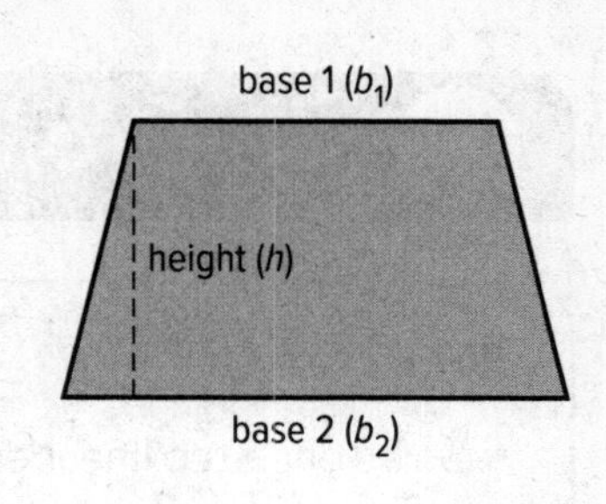

6. Consider both of the following sentences:
Forty-nine equals the square of the sum of a number and 13.
Forty-nine equals the square of a number plus 13.

a. COMMUNICATE PRECISELY Translate each sentence into an equation. Then explain the difference between the two equations. CCSS A.CED.1, SMP 6

b. PLAN A SOLUTION Use the fact that 7^2 and $(-7)^2$ equal 49, and 6^2 and $(-6)^2$ equal 36 to find the solutions to each equation. Explain how you found the solutions. CCSS A.CED.1, SMP 1

7. INTERPRET PROBLEMS Lily's profit P from making and selling n bracelets is given by the equation $4.25n - 1.65n = P$. CCSS SMP 1

a. Write a problem based on this information. List any constraints on n. CCSS A.CED.1, A.CED.3

b. Simplify the given equation by adding like terms. What does the coefficient of n mean? CCSS A.REI.3

c. If Lily makes a profit of $122.20, how many bracelets did she sell? CCSS A.REI.3

d. If Lily wants to make a profit of at least $100, will that affect the constraints on n? If so, how? If not, why? CCSS A.REI.3

2.3 Solving One-Step Equations

Objectives

- Solve one-step linear equations in one variable.
- Justify each step in solving simple linear equations in one variable.
- Create linear equations in one variable to model and solve real-world problems.

STANDARDS

Content: A.REI.1, A.REI.3, A.CED.1
Practices: 1, 2, 3, 4, 7, 8
Use with Lesson 2-2

Solving an equation means that you determine the value of the variable that makes the equation true.
Two equations are said to be **equivalent equations** if they have the same solution.

EXAMPLE 1 Analyzing Solutions CCSS A.REI.1, A.REI.3

EXPLORE Identify the operation needed to solve each equation. Then complete the template to solve the equation. Check your work. CCSS SMP 7

a. $x + 15 = 31$

Solve by ________________.

$x + 15 = 31$
$\square = \square$
$x = \square$

b. $k - 27 = 43$

Solve by ________________.

$k - 27 = 43$
$\square = \square$
$k = \square$

c. $32y = 256$

Solve by ________________.

$32y = 256$
$32y \square \square = 256 \square \square$
$y = \square$

d. $\frac{n}{10} = 7.4$

Solve by ________________.

$\frac{n}{10} = 7.4$
$\frac{n}{10} \square \square = 7.4 \square \square$
$n = \square$

e. $\frac{1}{4} + g = \frac{5}{8}$

Solve by ________________.

$\frac{1}{4} + g = \frac{5}{8}$
$\square = \square$
$g = \square$

f. $-c = \frac{6}{5}$

Solve by ________________.

$-c = \frac{6}{5}$
$-c \square \square = \frac{6}{5} \square \square$
$n = \square$

g. PLAN A SOLUTION Explain how to determine the operation needed to solve a one-step linear equation. CCSS SMP 1

KEY CONCEPT

Use the Properties of Equality to complete each sentence. Let a, b, and c be real numbers.

- By the ______________ Property of Equality, if $a = b$, then $a - c =$ __________.
- By the ______________ Property of Equality, if $a = b$ and $c \neq 0$, then $ac =$ __________.
- By the ______________ Property of Equality, if $a = b$, then $a + c =$ __________.
- By the ______________ Property of Equality, if $a = b$ and $c \neq 0$, then $a \div c =$ __________.

EXAMPLE 2 Using Linear Equations to Solve Problems

Tanaya, Chris, and Bill enjoy collecting trading cards. Tanaya has 30 cards. She has 7 fewer cards than Chris, and $\frac{3}{2}$ as many cards as Bill.

a. USE A MODEL Write an equation to find the number of cards that belong to Chris. CCSS A.CED.1, SMP 4

b. DESCRIBE A METHOD Explain how to solve the equation, justifying each step. How many cards does Chris have? CCSS A.REI.1, SMP 8

c. REASON QUANTITATIVELY Why is it important to perform exactly the same operation on each side of the equation? CCSS A.REI.1, SMP 2

d. CRITIQUE REASONING Tanaya writes the equation $30 = \frac{3}{2}b$ to show how many cards Bill has. She multiplies both sides of the equation by $\frac{3}{2}$ and decides that Bill has 45 cards. Is Tanaya correct? Justify the steps in her solution, or explain what she should have done differently. CCSS A.REI.1, SMP 3

PRACTICE

USE STRUCTURE **Solve each equation. Justify each step in your solution. Check your work.** CCSS A.REI.1, A.REI.3, SMP 7

1. $\frac{x}{9} = 24$

2. $m - 183 = -79$

3. $972 + y = 748$

4. $-\frac{4}{5}p = 32$

5. $135 = 9b$

6. $45 = \frac{3}{2}z$

7. **CRITIQUE REASONING** Jen wants to solve the equation $-8t = 0$. She reasons that because division by zero is undefined, the equation has no solution. Do you agree with Jen? Explain your reasoning and, if necessary, correct her mistake. CCSS A.REI.1, SMP 3

8. **The Lopez family is on vacation in Tennessee. They drive 210 miles from Memphis to Nashville and continued driving. By the time they reach Knoxville, they have travelled a total of 390 miles.**

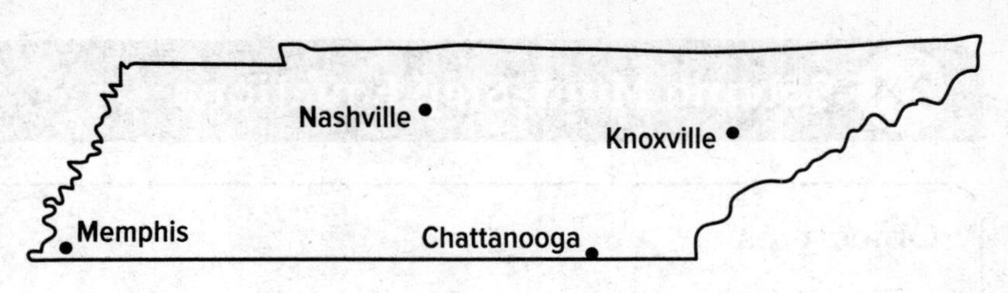

a. **USE A MODEL** Write an equation that models the distance from Nashville to Knoxville. CCSS A.CED.1, SMP 4

---

b. **PLAN A SOLUTION** Which property of equality could you use to isolate the variable in your equation? Explain your reasoning. CCSS A.REI.1, SMP 1

---

c. **PLAN A SOLUTION** How far is Knoxville from Nashville? How can you verify that your solution is accurate? CCSS A.REI.3, SMP 1

---

d. **USE A MODEL** If the Lopez family drives from Nashville to Chattanooga instead of Nashville to Knoxville, they will drive 47 fewer miles. What equation models the distance from Memphis to Chattanooga through Nashville? How far is Chattanooga from Nashville? CCSS A.CED.1, SMP 4

---

9. **REASON ABSTRACTLY** Joey uses the Addition Property of Equality to solve the equation $27 + b = 42$. Explain what Joey did and why it worked. CCSS A.REI.1, SMP 2

---

10. **CRITIQUE REASONING** Jasper argues that he can solve the equation $3x = 9$ by subtracting $2x$ from each side. Hypothesize why Jasper did this and explain why this will not work. CCSS A.REI.1, SMP 2

2.4 Solving Multi-Step Equations

Objectives

- Solve multi-step linear equations in one variable.
- Justify each step in solving simple linear equations in one variable.
- Create linear equations in one variable to model and solve real world problems.

 **STANDARDS**

Content: A.REI.1, A.REI.3, A.CED.1

Practices: 1, 2, 3, 4, 5, 6, 7

Use with Lesson 2-3 and Extend 10-2

You can solve a **multi-step equation** by undoing each operation and working backward.

EXAMPLE 1 Solution Strategies CCSS A.REI.1, A.REI.3

EXPLORE Solve each equation, writing the steps and naming the property that justifies it. CCSS SMP 7

a. $4x + 18 = 46$

$4x + 18 - 18 = 46 - 18$ __________

$\frac{4x}{4} = \frac{28}{4}$ __________

$x = 7$ __________

b. $7 - 2m = 31$

$7 - 2m - 7 = 31 - 7$ __________

__________ Division Property of Equality

__________ Simplify

c. $\frac{a+5}{3} = 13$

__________ __________

$a + 5 - 5 = 39 - 5$ __________

__________ Simplify

d. $8(q-3) = 72$

__________ __________

$q - 3 + 3 = 9 + 3$ __________

__________ Simplify

e. USE STRUCTURE Suppose the first step of **part d** was *Distribute*. What would be the next two steps? Is the solution the same? CCSS SMP 7

f. USE STRUCTURE Abid says that if the order of the steps in the solution in **part a** were reversed, the solution of the equation would remain the same. Do you agree? Justify your answer by showing the steps. CCSS SMP 7

EXAMPLE 2 Using Multi-Step Equations to Solve Problems

Kelly subscribes to a service called MySongs for downloading music. She pays a $7 fee each month to download an unlimited number of songs at $0.20 each. Last month, Kelly's bill was $23.80.

a. USE A MODEL Write an equation to find the number of songs that Kelly downloaded last month. CCSS A.CED.1, SMP 4

b. USE STRUCTURE Explain how to solve the equation, justifying each step. How many songs did Kelly download last month? CCSS A.REI.1, SMP 7

c. USE A MODEL Kelly is considering using a new service as described in the ad shown. Based on Kelly's usage last month, would Kelly save money under this new plan? CCSS A.REI.3, SMP 4

Music-Mania
Pay only $4 per month and download songs for $0.25 each!

d. REASON QUANTITATIVELY Solve the equations $7 + 0.2n = 19$ and $4 + 0.25n = 19$. Then describe what the solutions means in the context of the situation. CCSS A.CED.1, SMP 2

e. USE STRUCTURE Without calculating the value of $7 + 0.2n$ and $4 + 0.25n$ for multiple values of n, how can you determine when each service is a better deal? CCSS A.CED.1, A.REI.3, SMP 7

f. REASON QUANTITATIVELY What advice would you give Kelly? CCSS A.CED.1, SMP 2

EXAMPLE 3 Number Theory

The sum of three consecutive even integers is equal to eighty-four.

a. USE A MODEL Write an equation to model the verbal description. CCSS A.CED.1, SMP 4

b. COMMUNICATE PRECISELY Explain how to apply properties to help solve this equation. CCSS A.REI.1, SMP 6

c. USE STRUCTURE Solve the equation to find the numbers. Check your answer. 

d. USE A MODEL What if the question asked for three consecutive integers, instead of three consecutive *even* integers? Does this change the solution? CCSS A.CED.1, SMP 4

PRACTICE

USE STRUCTURE Solve each equation. Justify each step in your solution. Check your work. CCSS A.REI.1, A.REI.3, SMP 7

1. $\frac{3}{8}y + 7 = 19$

2. $12 - \frac{p}{5} = 3$

3. The sum of 4 consecutive odd integers is equal to zero.

a. USE A MODEL Write an equation to model the verbal description. CCSS A.CED.1, SMP 4

b. USE STRUCTURE Solve the equation to find the numbers. Check your answer. CCSS A.REI.3, SMP 7

4. **INTERPRET PROBLEMS** The average number of points a basketball team scored for three games was 63 points. In two games, they scored the same number of points, which was 6 points more than they scored in the third game. Write and solve an equation to find the number of points they scored in each game. CCSS A.CED.1, A.REI.3, SMP 1

5. **CRITIQUE REASONING** Emma and Jorge are each solving the equation $\frac{1}{2}n + 5 = \frac{17}{2}$. Jorge uses the Subtraction Property of Equality followed by the Multiplication Property of Equality. Emma also uses the Subtraction Property of Equality, but because n is multiplied by $\frac{1}{2}$, Emma claims that the Division Property of Equality can be used to isolate the variable. Which student is correct? Explain your reasoning. CCSS A.REI.1, SMP 3

6. **USE A MODEL** Mrs. Henry purchased a large package of big pink erasers for her algebra class. She sets aside 10 erasers to keep at her desk and then shares the remaining erasers equally among her 15 algebra students. If each student receives 4 erasers, write and solve an equation to find the number of erasers that were in the original package. CCSS A.CED.1, SMP 4

7. **USE TOOLS** Tom and Anita are dividing a garden as shown. Tom wants a triangular section that has one-third the area of Anita's section. Write and solve an equation to find the length of the base of Tom's garden. What is the area of Tom's garden? CCSS A.CED.1, A.REI.3, SMP 5

8. **COMMUNICATE PRECISELY** Write a problem that can be represented by the equation $11.9p + 23.1 = 273$. What does the variable represent? Then solve the equation. CCSS A.CED.1, SMP 6

2.5 Solving Equations with the Variable on Each Side

Objectives

- Create and solve linear equations in one variable and use them to solve problems.
- Explain each step in solving a simple equation.

CCSS STANDARDS

Content: A.REI.1, A.REI.3, A.CED.1
Practices: 1, 2, 3, 4, 7
Use with Lesson 2-4

When there are variables on both sides of an equation, you can use the Addition or Subtraction Property of Equality to solve for the equation.

EXAMPLE 1 Solve an Equation with Variables on Each Side CCSS A.REI.1, A.CED.1

EXPLORE Jamilah solves an equation by modeling with algebra tiles and drawing the steps she took. Her drawings are shown on the right.

a. USE A MODEL Create an equation for her model. CCSS SMP 4

b. USE A MODEL Explain each step in the drawing shown at right. Find the value of r. CCSS SMP 4

c. INTERPRET PROBLEMS How do you know that your solution is reasonable? Show your work to verify that the solution is correct. CCSS SMP 1

d. CONSTRUCT ARGUMENTS Jamilah's friend Raquel says she can solve the equation by moving the 4 tiles representing r from the right side of the equation to the left, and moving the 3 tiles representing 1 from the left side of the equation to the right. Help Jamilah explain to Raquel why this will not solve the equation. CCSS SMP 3

Some equations have grouping symbols, such as parentheses or brackets. You can use the Distributive Property to evaluate the grouping symbols first.

EXAMPLE 2 Solve an Equation with Grouping Symbols CCSS A.CED.1, A.REI.1

Mr. Nehru's age is 8 times the difference between 4 times his granddaughter's age and 5. At the same time, Mr. Nehru's age is $\frac{1}{2}$ the sum of 30 times his granddaughter's age and 22. How old is Mr. Nehru?

a. **REASON ABSTRACTLY** Write two expressions for Mr. Nehru's age in terms of his granddaughter's age, *g*. Describe the relationship between the two expressions, and write an equation. CCSS SMP 2

b. **USE STRUCTURE** Simplify each side of the equation. How does this equation compare to the equation you wrote in **part a**? Justify your answer. CCSS SMP 7

c. **USE STRUCTURE** Solve the equation in **part b** for *g*. Justify each step. CCSS SMP 7

d. **USE STRUCTURE** What does *g* represent in the context of this situation? Explain how to find Mr. Nehru's age. CCSS SMP 7

KEY CONCEPT

Step 1:	Simplify the expressions on each side. Use the ________ Property as needed.
Step 2:	Use the Addition and/or Subtraction Properties of Equality to get the ________ on one side of the equation and a number on the other.
Step 3:	Use the ________ or ________ Property of Equality to solve.

Some equations will have more than one letter used to represent unknown quantities. To solve for one of the variables in terms of the others, follow the same steps that you solve other multi-step equations with variables on both sides of the equation. You may need to state restrictions on values of some of the variables in order for the solution to make sense.

EXAMPLE 3 Solving Special Equations

USE STRUCTURE **Solve the equation $2ay - 6 = 5(ay + 3)$ for y. Justify each step in your solution. What, if any, restrictions are there on the value of a?** CCSS A.REI.1, A.REI.3, SMP 7

$2ay - 6 = 5(ay + 3)$ Original equation

$2ay - 6 = 5ay + 15$ ______

$-3ay - 6 = 15$ ______

$-3ay = 21$ ______

$y = \frac{21}{-3a}$ ______

$y = -\frac{7}{a}$ ______

PRACTICE

1. **USE STRUCTURE** Solve the equation. Use a property to justify each step. CCSS A.REI.1, A.REI.3, SMP 7

   $4(2a + 1) - 15 = 3(4a - 5)$ Original equation

2. **USE STRUCTURE** Solve the given equation for x. What, if any, restrictions are there on the value of a? CCSS A.REI.1, A.REI.3, SMP 7

   $5a(x + 2) - 10a = 3(ax + 6)$ Original equation

3. **USE STRUCTURE** Solve the given equation for z. What, if any, restrictions are there on the value of b? CCSS A.REI.1, A.REI.3, SMP 7

   $4z + 3 = bz + 9$ Original equation

4. USE STRUCTURE Solve the given equation for y. What, if any, restrictions are there on the value of m? CCSS A.REI.1, A.REI.3, SMP 7

$y - m = ym + 1$ Original equation

5. Mrs. Fernandez built two rectangular gardens. The length of Garden A is 2 feet longer than twice its width. The width of Garden B is 1 foot less than the width of Garden A. The length of Garden B is 2.5 times its width. The two gardens have the same perimeter. What are the dimensions of the gardens? CCSS A.CED.1, A.REI.1, A.REI.3

a. USE A MODEL Label the side lengths of the two rectangles representing the gardens. Include the units of measure. CCSS SMP 4

Garden A **Garden B**

b. USE A MODEL Write an expression for the perimeter of each garden. Simplify each expression. Name at least one property you used to simplify each expression. CCSS SMP 4

c. USE STRUCTURE Write an equation that represents the relationship between the two perimeters. Show and justify each step of the solution. CCSS SMP 7

Equation from **part b**

d. REASON QUANTITATIVELY What are the dimensions of Garden A and Garden B? CCSS SMP 2

e. INTERPRET PROBLEMS Ping found that the perimeter of Garden B is $4x - 4$. Where could Ping have made her error? If Ping solved the equation $6x - 4 = 5x - 6$, how would she know that her solution is not reasonable? CCSS SMP 1

2.6 Literal Equations

Content: A.CED.4, A.REI.3, N.Q.1
Practices: 1, 2, 3, 4, 7
Use with Lesson 2-8

Objectives

- Solve formulas for quantities of interest.
- Solve linear equations in one variable in which the coefficients are represented by letters.
- Use dimensional analysis to convert units of measure.

A **literal equation** has several variables.

$C = \pi d$ $\qquad V = \ell \cdot w \cdot h$ $\qquad P = 2\ell + 2w$

When solving problems, it is often necessary to solve the equation for one of these variables.

EXAMPLE 1 Working with Literal Equations CCSS A.CED.4, A.REI.3, N.Q.1

When traveling, it is useful to know the formula: distance = rate × time.

$$d = r \cdot t$$

In this equation, *d* represents the distance traveled, *r* represents the average rate of speed, and *t* represents the amount of time spent traveling.

a. USE A MODEL Describe a real world situation where you would want to use the equation exactly as it has been presented above. CCSS SMP 4

b. USE A MODEL Describe a real world situation in which it would be helpful to solve the equation for the variable *r*. CCSS SMP 4

c. USE STRUCTURE Solve the equation for *t*. Then use the equation to complete the table and determine how long it will take to travel 800 kilometers at different average speeds. Round answers to the nearest tenth. CCSS SMP 7

Average Speed (kilometers per hour)	Time (hours)
40	
50	
60	
70	
80	
100	
120	

EXAMPLE 2 Solve a Formula For a Specific Variable 

The formula for the surface area of a pyramid with slant height ℓ, and a square base with side length b, is $S = b^2 + 2\ell b$.

a. USE STRUCTURE Solve the formula for ℓ. Fill in the missing steps and reasons. CCSS SMP 7

$S = b^2 + 2\ell b$	Given
______________________	Subtraction Property of Equality
______________________	______________________

b. MAKE A CONJECTURE The expression $b^2 + 2\ell b$ is equivalent to $b(b + 2\ell)$. Do you think that this fact will allow you to solve the surface area equation for b? Explain your thinking. CCSS SMP 3

____

____

____

____

EXAMPLE 3 Use a Literal Equation CCSS A.CED.4, A.REI.3, N.Q.1

Your company is developing a new cat food that will be packaged in cans. You have determined the volume of food to be sold in each can, but need to find the height.

a. USE A MODEL Write the formula for the volume of a can and describe what each variable in the formula represents. Rewrite the literal equation in a form that will allow you to find the height of the can for a given radius and volume. CCSS SMP 4

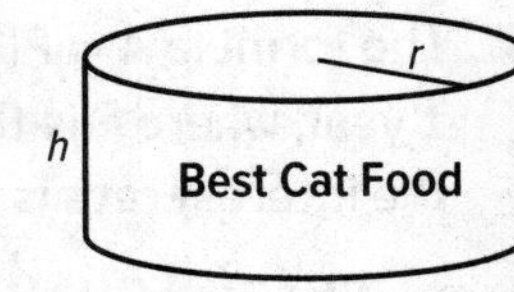

____

____

____

____

b. REASON QUANTITATIVELY You determine that a radius of 2 inches will provide the best display on retail shelves. Each can of cat food will hold 12 ounces. One ounce is approximately 1.8 cubic inches. Complete the dimensional analysis to find the height of the can. CCSS SMP 2

$h =$ ____________	State the formula.
$h = \dfrac{\square\text{ oz.}}{\pi \cdot (\square\text{ in.})^2}$	Substitute values.
$h = \dfrac{\square\text{ oz.}}{\pi \cdot \square\text{ in}^2} \times \dfrac{\square\text{ in}^3}{1\text{ oz.}}$	Apply dimensional analysis.
$h \approx$ ____________ inches	Simplify.

PRACTICE

1. **CRITIQUE REASONING** The formula $F = \frac{9}{5}C + 32$ represents the relationship between temperatures in degrees Fahrenheit F and degrees Celsius C. Sasha solves the formula for C. Her solution is shown at the right. Is Sasha's solution correct? Explain your reasoning and correct any errors you find. CCSS A.REI.3, A.CED.4, SMP 3

Sasha

$F = \frac{9}{5}C + 32$

$\frac{5}{9}F = C + 32$

$C = \frac{5}{9}F - 32$

2. **USE STRUCTURE** Jethro used dimensional analysis to convert from one rate of speed to another. Two of the conversion factors he used are $\frac{5280 \text{ ft}}{1 \text{ mi}}$ and $\frac{1 \text{ hr}}{60 \text{ min}}$. What were the units of the initial rate of speed? What were the units of the final rate of speed? Justify your answer. CCSS N.Q.1, SMP 7

3. **The formula $A = P(1 + r)$ represents the amount of money A in an account after 1 year, where P is the amount initially deposited and r is the interest rate. Note that the interest rate is written as a decimal.** CCSS A.CED.4, A.REI.3

   a. **REASON QUANTITATIVELY** Denys deposits \$2150 into a savings account. After 1 year, he has \$2182.25 in his account. Solve the equation for r, and determine the interest rate. CCSS SMP 2

   b. **USE A MODEL** The formula $A = P(1 + r)^t$ represents the amount of money A in an account after t years, where P is the amount initially deposited and r is the interest rate. Sunyi currently has \$1839.79 in an account that is compounded annually, paying an interest rate of 2.5%. She opened the account 8 years ago and has made no additional deposits since then. Solve the formula for P and find the amount of Sunyi's initial deposit. CCSS SMP 4

   c. **USE STRUCTURE** Nia says that for the formula in **part b**, A is always greater than P when r is positive and t is a positive integer. Do you agree? Explain why or why not. 

4. Julia is going on a trip to China. She gets the following information on currency exchange from an international financial institution. CCSS A.REI.3, A.CED.4

	Dollars (d) to Euros (e)	Dollars (d) to Yuan (y)	Euros (e) to Yuan (y)
United Financial	$e = 0.75d - 7.5$	$y = 6.2d - 62$	$y = 8.52e - 62.2$

a. USE A MODEL Julia is thinking about converting her dollars to euros first, and then converting the euros to yuan. Assuming that she will use this institution for both currency conversions, write an equation that represents the amount of yuan in terms of dollars. Justify each step in your solution. (Hint: Start by solving the euros to yuan equation for e.) CCSS SMP 4

b. PLAN A SOLUTION Compare this equation with the equation in the table that represents the conversion of dollars to yuan. Find the solution to this set of two equations. What would you advise Julia to do: convert her dollars directly to yuan or convert her dollars to euros and then convert the euros to yuan? CCSS SMP 1

5. The formula for the surface area of a cylinder is $S = 2\pi rh + 2\pi r^2$, where r is the radius and h is the height of the cylinder. CCSS A.REI.3, A.CED.4

a. USE STRUCTURE The surface area of a cylindrical can is 72π cm^3 and the radius is 4 cm. Solve the formula for h and find the height of the can. CCSS SMP 7

b. USE A MODEL The formula for the lateral surface area of a cylinder is $L = 2\pi rh$. A label covering the lateral surface area of another can has a height of 7.5 cm and a lateral area of 52.5π cm^2. Solve for r and find the radius of the can. CCSS SMP 4

2.7 Weighted Averages

Objectives

- Create and solve linear equations in one variable, explaining each step.
- Use units as a guide in the solution of multi-step problems.

CCSS STANDARDS

Content: A.REI.1, A.REI.3, N.Q.1, A.CED.1, A.CED.4
Practices: 1, 2, 3, 4, 5, 6, 7
Use with Lesson 2-9

The **weighted average** M of a set of data is found by multiplying each data value by its weight and then finding the mean of the new data set.
Mixture problems are problems in which two or more parts are combined into a whole.
Rate problems are problems in which an object moves at a certain speed or rate.

EXAMPLE 1 Using Weighted Averages to Solve Problems CCSS A.REI.1, A.REI.3, A.CED.1

EXPLORE Jesse is going to the movies with his family and friends. Tickets for seniors are \$7 each, tickets for adults are \$9 each, and tickets for children are \$6 each. Eighteen people, including three of his grandparents who are older than 65, go to the movie and spend \$141 on tickets. How many seniors, adults, and children went to see the movie?

a. **USE TOOLS** Complete the table to organize the information given. CCSS SMP 5

	Number of People	Cost of a Ticket	Total Cost
Seniors	3		
Adults under 65	x		
Children under 12			
Total	18	-----	

b. **USE A MODEL** Write and solve an equation that represents the numbers of seniors, adults, and children that went to the movies. Show your steps and justify your reasoning for each step. CCSS SMP 4

c. **USE A MODEL** Another person joined this group at the movies. If possible, write an equation that represents the amount all nineteen people paid. Otherwise, explain why it can't be done. CCSS SMP 4

EXAMPLE 2 Solve a Mixture Problem CCSS A.REI.1, A.REI.3, A.CED.1

Solution A is 70% acid and Solution B is 15% acid. How much of each solution is needed to make 110 gallons of a 50% solution?

a. CRITIQUE REASONING Kendall read the problem statement and wrote the following equation to model the mixture problem. Is her equation a good model? Explain why or why not. CCSS SMP 3

$$0.7x + 0.15y = 110$$

b. USE TOOLS Complete the table using the information given. CCSS SMP 5

	Number of Gallons	Amount of Acid (gallons)
Solution A	x	
Solution B		
50% Solution	110	55

c. USE STRUCTURE How many gallons of 70% solution and gallons of 15% solution are needed? Justify your reasoning. How do you know your answer is reasonable? CCSS SMP 7

You need ______ gallons of 70% solution and ______ gallons of 15% solution.

d. REASON QUANTITATIVELY Ten gallons of a new solution are added to the 110 gallons. If the 120 gallons are a 50% solution, what do you know about the added solution? If the 120 gallons are less than a 50% solution, what is true about the added solution? Explain your answer. CCSS SMP 2

EXAMPLE 3 Analyze a Rate Problem CCSS A.REI.3, N.Q.1, A.CED.1, A.CED.4

A train leaves the station traveling north at 40 miles per hour. A second train leaves the station at the same time traveling south at 50 miles per hour. If neither train stops, how far from the station will each train be when the trains are 300 miles apart?

a. USE STRUCTURE How will analysis of the units in this problem help you create a good model? CCSS SMP 7

b. PLAN A SOLUTION If the distance in miles traveled by one train is x, what is the other distance? Use this information to write an equation of the form $d = rt$ for each train. CCSS SMP 1

c. PLAN A SOLUTION Write a single equation that can be solved to find the distance traveled by the first train when the trains are 300 miles apart. Explain each step in your answer. CCSS SMP 1

d. COMMUNICATE PRECISELY How far is the distance of the second train from the station when the trains are 300 miles apart? CCSS SMP 6

PRACTICE

1. REASON ABSTRACTLY Alex created the table shown to solve a mixture problem. State the problem that Alex wants to solve. CCSS N.Q.1, SMP 2

	Pounds of Candy	Price per Pound	Total Cost
Hard Candy	x	3	$3x$
Gummy Candy	$10 - x$	8	$8(10 - x)$
Candy Mixture	10	5	50

2. A plane traveled the 3908-mile distance from Honolulu to Tokyo at a speed of 217 miles per hour. The average speed for the round trip was 265 miles per hour. CCSS A.REI.1, A.REI.3, N.Q.1, A.CED.1

a. **USE A MODEL** If the entire trip took 29.5 hours, what was the plane's speed flying from Tokyo back to Honolulu? Is your solution reasonable? CCSS SMP 4

b. **INTERPRET PROBLEMS** Rodrigo makes the following calculation to approximate the speed of the plane flying from Tokyo to Honolulu. Explain his thinking. CCSS SMP 1

$$\frac{18(217)}{11.5} \approx 339.7 \text{ miles per hour}$$

3. Marta is riding her bicycle on a bike path next to a 6-mile long canal. She would like to bicycle from one end to the other and back again. In one direction, she will ride with the wind; in the other direction, she will ride against the wind. After bicycling for 16 minutes, she has traveled a distance of 4 miles. CCSS A.REI.3, N.Q.1, A.CED.1, A.CED.4

a. **PLAN A SOLUTION** If Marta has 1 hour and 5 minutes to bicycle, how quickly must Marta ride on her return trip? Assume that she will continue for 2 miles to one end of the bike path before returning. Explain. CCSS SMP 1

b. **USE A MODEL** If she does this, what will be Marta's average speed? Explain why finding the weighted average is necessary. CCSS SMP 4

4. REASON ABSTRACTLY Carol says she can make a 50% solution from adding a 40% solution and a 10% solution. Do you agree? Justify your answer. CCSS A.REI.3, N.Q.1, A.CED.1, SMP 2

Performance Task

Analyzing Weather Data

Provide a clear solution to the problem. Be sure to show all of your work, include all relevant drawings, and justify your answers.

This table shows data gathered via a weather balloon. The data in the first column shows the balloon's altitude in feet, and the second column shows the air temperature in degrees Fahrenheit. Investigate the relationship between the data sets.

Altitude (ft)	Temperature (°F)
0	59
5,000	41.17
10,000	23.36
15,000	5.55
20,000	−12.26
25,000	−30.05
30,000	−47.83
35,000	−65.61
40,000	−69.7

Part A

Suppose you wanted to create a comparable metric table of data. If the altitude is measured in kilometers and the temperature is measured in degrees Celsius, what would you need to do to create such a table?

Part B

Use the given formulas to the convert 7,000 feet to kilometers and 34 degrees Fahrenheit to degrees Celsius. Show all work and round to the nearest tenth. Which of the formulas below could be rewritten to make these conversions easier? Show the revised formula.

$$k = 0.000305f$$

$$F = \frac{9}{5}C + 32$$

Part C

Complete the metric data table. Show each value rounded to the nearest tenth.

Altitude (km)	Temperature (°C)

Part D

Describe the relationship between altitude and temperature based on the data. Is the relationship the same in both data sets? Explain.

Performance Task

Planning for a Parade

Provide a clear solution to the problem. Be sure to show all of your work, include all relevant drawings, and justify your answers.

A club is decorating cars for a parade. Each of the club's eight members will be responsible for decorating three cars. Among other items, each car will be decorated with 20 helium-filled balloons. Each balloon is in the shape of a sphere with a radius of 4 in. You need to determine the amount of helium needed for all the balloons.

Part A

This is the formula for the volume of a sphere. What is the volume of helium needed for 1 balloon in terms of π? What is the unit of measure?

$$V = \frac{4}{3}\pi r^3$$

Part B

How many balloons are needed for the parade? Write the numerical expression and simplify it.

Part C

What is the total volume of helium needed for all the balloons in terms of π?

Part D

Helium tanks come in these sizes: 50 $ft.^3$, 80 $ft.^3$, and 120 $ft.^3$. The volume measurements indicate the maximum amount of helium that the tank can hold. How many and what size tanks should be used?

Standardized Test Practice

1. Write an equation to represent the following: CCSS A.CED.1, A.REI.1

Five less than p is 13.

[ ]

Solve the equation for p.

$p =$ [ ]

2. Consider the following rectangle: CCSS A.CED.1, A.REI.1

If the perimeter of the rectangle is 40 units, write an equation to determine the width, w.

[ ]

What is the width of the rectangle?

3 units | 26 units
13 units | 33 units

3. Solve for t: CCSS A.REI.3

$5t - 3 = 27$

$t =$ [ ]

4. David measured the width of a book with a ruler. Which of the following could have been his measurement? CCSS N.Q.3

104.5 mm | 11.4 cm
5.5 in. | 0.18 yd

5. Solve for f: CCSS A.CED.4

$d = gf - 2$

$f =$ [ ]

6. Arturo is making 5 pounds of snack mix from raisins and almonds. The raisins cost \$3 per pound, and the almonds cost \$8 per pound. He wants the cost of the snack mix to be \$6.50 per pound. How many pounds of each item should he use? CCSS A.REI.1, N.QN.1

[ ] pounds of raisins

[ ] pounds of almonds

7. Anne weighed her kitten using three different scales. The results are shown in the table. CCSS N.Q.3

Scale A	Scale B	Scale C
1.8 kg	1.763 kg	1.77 kg

The most precise of the three scales is [ ]

8. The area of a triangle is given by the formula $A = \frac{1}{2}bh$, where h is the height of the triangle and b is the length of its base. If you know the area and height of a triangle, what formula gives the length of its base? CCSS A.CED.4

[ ]

9. Solve for x: CCSS A.REI.3

$-2x + 11 = 3x - 4$

$x = -3$ | $x = 3$
$x = 7$ | $x = 15$

10. The product of 6 and a number when added to 5 is equal to 1 less than 9 times the number. CCSS A.CED.1

Write an equation to represent this situation.

[ ]

Solve the equation to determine the number.

The number is [ ]

11. On a homework assignment, Jack wrote the following steps to solve $2x + 6 = 20$.

$2x + 6 = 20$
$x + 6 = 10$
$x = 4$

a. Explain Jack's error. CCSS A.REI.1

b. Show the correct steps to solve $2x + 6 = 20$. Explain each step. CCSS A.REI.1, A.REI.3

12. Ms. Jones gives math grades using a weighted average. Homework is worth 30%, tests are worth 50%, and quizzes are worth 20%. The table shows homework, test and quiz scores for three students. Complete the table to show their final grades. CCSS A.REI.1, N.QN.1

Student	Homework Grade (%)	Test Grade (%)	Quiz Grade (%)	Final Grade (%)
Bianca	94	88	100	
Noelle	82	78	91	
Julius	90	97	95	

13. Solve each equation. Show your work and explain each step. CCSS A.REI.1, A.REI.3

a. $5x - 7 = 9x + 1$

b. $3(x - 3) = 5(x - 2)$

14. On Saturday, Britani babysat for 6 hours. She was paid an hourly wage, and also given a $5 tip. Her total pay was $53.

a. Let *h* represent her hourly wage. What equation represents the situation described above? 

b. Solve your equation for *h*. How can you check your answer? 

3 Writing and Solving Linear Inequalities

CHAPTER FOCUS Learn about some of the Common Core State Standards that you will explore in this chapter. Answer the preview questions. As you complete each lesson, return to these pages to check your work.

What You Will Learn	Preview Question
Lesson 3.1: Writing Inequalities	
CCSS A.CED.1 Create equations and inequalities in one variable and use them to solve problems. CCSS A.CED.3 Represent constraints by equations or inequalities, and by systems of equations and/or inequalities, and interpret solutions as viable or nonviable options in a modeling context. CCSS A.REI.3 Solve linear equations and inequalities in one variable, including equations with coefficients represented by letters.	CCSS SMP 4 The sum of twice a number and 5 is less than or equal to −2. Write an inequality representing this situation.
Lesson 3.2: Solving Inequalities by Addition and Subtraction	
CCSS A.CED.1 Create equations and inequalities in one variable and use them to solve problems. CCSS A.REI.3 Solve linear equations and inequalities in one variable, including equations with coefficients represented by letters. CCSS A.REI.1 Explain each step in solving a simple equation as following from the equality of numbers asserted at the previous step, starting from the assumption that the original equation has a solution. Construct a viable argument to justify a solution method.	CCSS SMP 2 A bank account is free as long as a balance of at least $250 is maintained. You currently have a balance of $337.58. You are planning to buy gifts for some friends. Write and solve an inequality to determine the most you can spend on gifts without dropping below the $250 minimum balance.
Lesson 3.3: Solving Inequalities by Multiplication or Division	
CCSS A.REI.1 Explain each step in solving a simple equation as following from the equality of numbers asserted at the previous step, starting from the assumption that the original equation has a solution. Construct a viable argument to justify a solution method. CCSS A.REI.3 Solve linear equations and inequalities in one variable, including equations with coefficients represented by letters. CCSS A.CED.1 Create equations and inequalities in one variable and use them to solve problems.	CCSS SMP 4 The temperature falls 3° each hour. If the temperature is currently 20°, write and solve an inequality to determine how many hours it will take for the temperature to drop below 14°. Explain your process.

What You Will Learn	Preview Question
Lesson 3.4: Solving Multi-Step Inequalities	
CCSS A.REI.3 Solve linear equations and inequalities in one variable, including equations with coefficients represented by letters. **Also addresses:** A.REI.1, A.CED.1	CCSS SMP 7 Solve the inequality. $3(y + 3) > 5 + 4y$
Lesson 3.5: Solving Compound Inequalities	
CCSS A.REI.1 Explain each step in solving a simple equation as following from the equality of numbers asserted at the previous step, starting from the assumption that the original equation has a solution. Construct a viable argument to justify a solution method. CCSS A.REI.3 Solve linear equations and inequalities in one variable, including equations with coefficients represented by letters. CCSS A.CED.1 Create equations and inequalities in one variable and use them to solve problems.	CCSS SMP 1 Amelia normally sells her paintings for \$55. She sometimes discounts the price, reducing the price by as much as \$11. Some paintings are larger, and she charges as much as \$9 more than the regular price. Write a compound inequality representing the prices at which Amelia has sold her paintings. CCSS SMP 7 Renata is trying to solve an online puzzle to win prizes. One part of the puzzle requires entering a combination of digits to unlock a vault. The only thing that Renata knows is that the digits satisfy the compound inequality $2x - 7 > 5 - x > 2x - 19$. What are the digits that can be used in the combination?
Lesson 3.6: Graphing Inequalities in two Variables	
CCSS A.CED.3 Represent constraints by equations or inequalities, and by systems of equations and/or inequalities, and interpret solutions as viable or nonviable options in a modeling context. CCSS A.REI.12 Graph the solutions to a linear inequality in two variables as a half plane (excluding the boundary in the case of a strict inequality), and graph the solution set to a system of linear inequalities in two variables as the intersection of the corresponding half planes.	CCSS SMP 1 Graph the inequality $y \geq 2x - 3$ and determine if $(3, -1)$ is a solution. Explain your reasoning. y, 4, 2, −2, O, 2, 4, x, −2

3.1 Writing Inequalities

Objectives

- Represent constraints by inequalities.
- Write inequalities in one variable.

STANDARDS

Content: A.CED.1, A.CED.3, A.REI.3
Practices: 2, 3, 4, 8
Use with Lesson 5-2

A mathematical statement that contains $<$, $>$, $\le$, or $\ge$ is an **inequality**. An inequality can be solved by using methods similar to those used for solving equations. Inequalities can be written for real-life problems by using the constraints defined in the problem statement. The following key words can be used to determine the correct inequality symbol to use when writing an inequality.

Symbol	$<$	$\le$	$>$	$\ge$
Keywords	Less than	Less than or equal to	Greater than	Greater than or equal to
	Fewer than	At most	More than	At least
		No more than		No less than

EXAMPLE 1 Create Inequalities Using Addition and Subtraction CCSS A.CED.1, A.CED.3

a. USE A MODEL Kevin has 3.5 gallons of gas in his car. He puts more gas in the car, but the tank is still not full. Kevin's car has an 18 gallon tank. Write an inequality to model the number of gallons of gas Kevin added to his car. What is the constraint in this situation? CCSS SMP 4

b. USE A MODEL Kevin bought a new car with a gas tank that holds 2 gallons more than the gas tank in his old car. Write an inequality to model the number of gallons of gas that can be added if Kevin has 3.5 gallons of gas in his new car. CCSS SMP 4

c. USE A MODEL Jennifer collects stamps and currently has 37 stamps in her collection. If she collects 10 more stamps, she will have at most 8 more stamps than Radha. Write an inequality to model the number of stamps in Radha's collection. CCSS SMP 4

EXAMPLE 2 Create Inequalities Using Multiplication and Division CCSS A.CED.1, A.CED.3, SMP 4

a. USE A MODEL Rose wants to buy a rug for her living room. She does not have the exact dimensions of her living room, but she knows that the width of her living room is at least half of the length. Write an inequality to model the width of Rose's living room.

b. USE A MODEL The area of Rose's garden is more than twice the area of the living room. Write an inequality to model the area of Rose's garden.

c. USE A MODEL Jake drops a ball on a concrete floor from a height *h*. The ball bounces to a height no more than 80 percent of *h*. Write an inequality to model the height of the ball on the first bounce.

d. USE A MODEL Melissa can read x pages per hour. After reading for 4 hours, she is unable to finish a 160-page book. Write an inequality to model the number of pages she can read every hour. What is the constraint in this situation?

EXAMPLE 3 Create Inequalities with Multiple Operations CCSS A.CED.1, A.CED.3

a. USE A MODEL On a science test, Daniel, scores at most five more than half of what John scores. Write an inequality to model Daniel's test score. CCSS SMP 4

b. USE A MODEL David plans to construct a swimming pool in his back yard. David will surround the pool with a fence. The fence is to be no closer than 10 feet to any side of the pool. He has budgeted for 200 feet of fencing. Write inequalities that model the constraints on the fenced in area. CCSS SMP 4

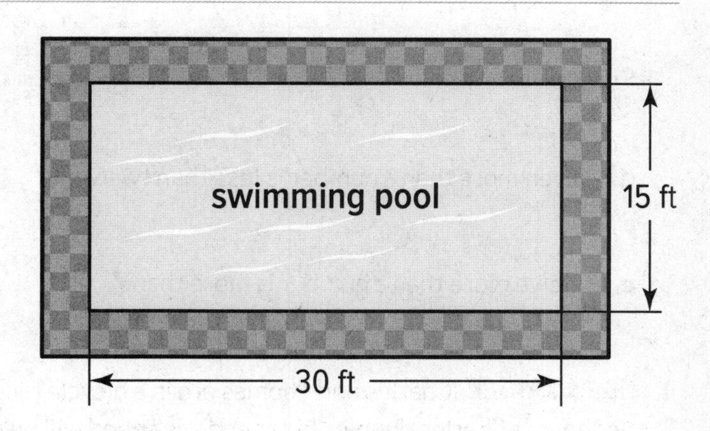

c. REASON QUANTITATIVELY Carmella is planning a party for some of her friends. She has beverages for at most 20 people and sandwiches for 5 more people than she had originally planned to invite. Write an inequality that expresses the constraints of her situation. What is the greatest number of people that she can invite? CCSS SMP 2

PRACTICE

1. **USE A MODEL** Tim added 500 gumballs to a giant gumball machine, the total number of gumballs did not exceed 2000. Write an inequality to model the number of gumballs initially present in the machine. CCSS A.CED.1, SMP 4

2. **CRITIQUE REASONING** Alex added 4 quarts of water to a 10 quart bucket. The bucket already contained some water. He says that an inequality that models the amount of water initially in the bucket is $x - 4 \leq 10$. Is he correct? CCSS A.CED.1, SMP 3

3. **EVALUATE REASONABLENESS** A table shows some inequalities. Match each inequality with the situation it models. CCSS A.CED.1, SMP 8

   a. Twelve is more than 7 fewer than a number.

   b. Seven less than a number is at least twelve.

   c. Twelve less than a number is at most 7.

   d. Seven more than a number is less than twelve.

   e. Twelve more than a number is more than 7.

Inequalities
$x + 7 < 12$
$12 > x - 7$
$x + 12 > 7$
$7 \geq x - 12$
$x - 7 \geq 12$

4. **USE A MODEL** Charles and Thomas are in a bicycle race. Thomas is cycling at 26 miles per hour. If Charles doubles his speed, his speed will be more than Thomas' speed by at least 5 miles per hour. Write an inequality to model Charles' original speed. CCSS A.CED.1, SMP 4

5. **USE A MODEL** The number of apartments in Mandy's building is at least one more than twice as many as the number in Jack's apartment building. Write an inequality to model the number of apartments in Mandy's building. CCSS A.CED.3, SMP 4

6. **USE A MODEL** Naresh has $1000 saved up for a vacation. He budgets $550 for transportation and $45 per day for food. Write and solve an inequality to model how many days he can spend on vacation. CCSS A.CED.3, A.REI.3, SMP 4

7. **USE A MODEL** Jake has \$190. He buys a video game and a T-shirt. He wants to use the rest of his money to buy shoes. Each pair of shoes costs \$40. Write and solve an inequality to determine how many pairs of shoes he can purchase. What is the constraint in this situation? CCSS A.CED.3, A.REI.3, SMP 4

video game	\$55
T-shirt	\$20
Shoes	\$40

8. **USE A MODEL** Ben has \$50. He buys a pizza for \$15, and two cans of soda for \$3. He wants to use the rest of the money to take burgers to a party. Each burger costs \$4. Write an inequality to model the number of burgers he can purchase. CCSS A.CED.3, SMP 4

9. **USE A MODEL** Sally has \$250. She wants to purchase head phones for \$60 and 3 birthday gifts. CCSS A.CED.3, SMP 4

   a. Write an inequality to model how much she can spend on each birthday gift, if each gift costs the same amount.

   b. What is the greatest amount that Sally can spend on each gift?

   c. Sally realizes that she has \$50 more than she originally thought. What is the inequality that represents how much she can spend on each birthday gift?

   d. What is the greatest amount that Sally can spend on each gift now?

10. **USE A MODEL** Ryan bought a 4-pack of shaving gel. Twelve weeks later Ryan was finishing his last can of shaving gel. Write an inequality to model the number of weeks one can of shaving gel lasts Ryan. CCSS A.CED.1, SMP 4

11. **USE A MODEL** Last week, Marco rode his bike 5 miles more than Bob. Altogether they rode at least 20 miles. CCSS A.CED.1, SMP 4

   a. Write an inequality that expresses the constraints of the situation.

   b. Solve the inequality to express the number of miles that each person rode.

3.2 Solving Inequalities by Addition and Subtraction

Objectives

- Explain each step in solving a simple inequality by using addition and subtraction.
- Solve linear inequalities in one variable by using addition and subtraction.
- Create inequalities in one variable and use them to solve problems.

Content: A.REI.1, A.REI.3, A.CED.1
Practices: 1, 2, 4, 7
Use with Lesson 5-1

Solving linear inequalities is similar to solving linear equations. To solve an **inequality** that contains addition or subtraction, undo the operation by subtracting or adding the same quantity on both sides. Add or subtract either constants or variables to isolate the variable on one side of the inequality and the constant on the other side.

Addition and Subtraction Properties of Inequality

For all numbers a, b, and c, the following are true.

1. If $a > b$, then $a + c > b + c$ and $a - c > b - c$.
2. If $a < b$, then $a + c < b + c$ and $a - c < b - c$.

EXAMPLE 1 Write and Solve an Inequality CCSS A.CED.1

USE A MODEL James has $30. He buys a shirt and is left with more than $16. At most, how much did he pay for the shirt? CCSS SMP 4

EXAMPLE 2 Explain Steps in Solving a Linear Inequality CCSS A.REI.1, A.REI.3

PLAN A SOLUTION Solve each inequality. CCSS SMP 1

a. $x + 8 > 17$

b. $25 \leq z - 10$

EXAMPLE 3 Solve a Linear Inequality by Adding and Subtracting CCSS A.REI.1, A.REI.3

USE STRUCTURE Solve each inequality. CCSS SMP 7

a. $4x \leq 3x - 5$

b. $2y < y + 10$

c. $8 - 5z \leq -6z$

d. $3x - 10 > 2x - 3$

e. $6 < -y - 5$

PRACTICE

USE STRUCTURE Solve each inequality. CCSS A.REI.3, SMP 7

1. $a + 7 > 32$

2. $14n \leq 13n - 20$

3. $-a + 7 > 32$

4. $-14n \leq -13n - 20$

REASON QUANTITATIVELY Solve each inequality. Then graph the solution on a number line. CCSS A.REI.3, SMP 2

5. $p - 17 \geq 10$

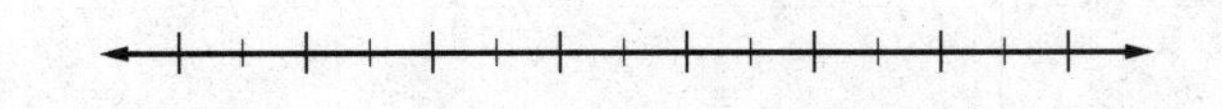

6. $8s + 12 < 7s$

PLAN A SOLUTION Define a variable, write an inequality, and solve. CCSS A.REI.3, A.CED.1, SMP 1

7. A number increased by five is at most twelve.

8. Seventeen less than a number is at least three.

9. Eight is more than a number increased by twelve.

10. Five is less than a number minus two.

11. Eighteen minus a number is no more than 10.

12. **USE A MODEL** Four times the number of baseball cards in Ted's collection is more than five times that number minus 15. Define a variable and write an inequality to represent Ted's marbles. Solve the inequality and interpret the results. CCSS A.REI.3, A.CED.1, SMP 4

13. **USE A MODEL** In a mathematics exam with a maximum score of 100 Mary loses less than 27 points. The table shows the grade that matches the exam score. Compares points to grades and identify which grades Mary can get. CCSS A.CED.1, A.REI.3, SMP 4

Grade	Points
A	92–100
B	83–91
C	74–82
D	65–73
F	64 and below

   a. Define a variable. Then write and solve an inequality.

   b. Interpret the solution to your inequality. What do you know about Mary's grade on the exam?

14. **USE A MODEL** Darnell deposits $20 into his savings account, and his balance is at least $150. Find the balance in Darnell's savings account before he deposited the money. Define a variable, write an inequality, and solve. CCSS A.CED.1, A.REI.3, SMP 4

15. **USE A MODEL** The product of a number and 6 is more than the sum of 10 and 5 times the number. Define a variable, write an inequality, and solve. CCSS A.CED.1, A.REI.3, SMP 4

16. USE A MODEL Janet adds 8 kg of newspapers to a box, and it weighs less than 20 kg. Find the weight of the box before Janet added the newspapers. Define a variable, write an inequality, and solve. CCSS A.CED.1, A.REI.3, SMP 4

USE STRUCTURE Match each inequality to the graph of its solution. CCSS A.REI.3, SMP 7

17. $x + 3 > 8$ **18.** $2x + 5 \leq 3x$ **19.** $7x - 5 < 6x$ **20.** $x - 3 \leq 2$

a.

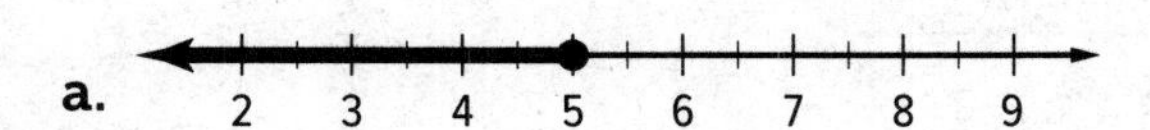

b.

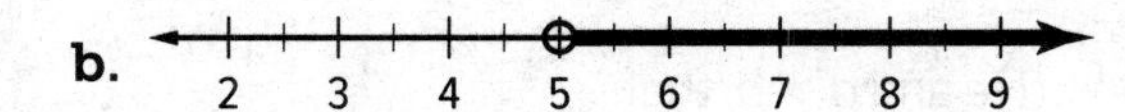

c.

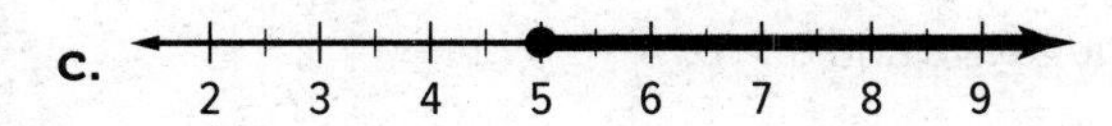

d.

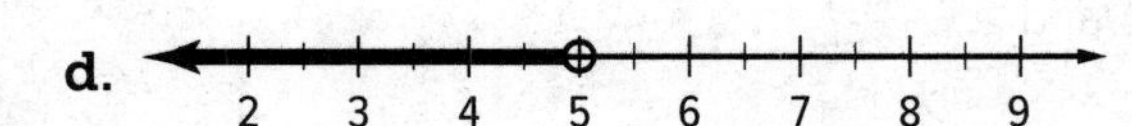

21. USE A MODEL Airport officials told Mandy that her suitcase exceeded the maximum weight allowance. She removed a 5 kg item from her suitcase, but the suitcase still weighed more than 22 kg. What was the original weight of Mandy's suitcase? Define a variable, write an inequality, and solve. CCSS A.CED.1, A.REI.3, SMP 4

22. USE A MODEL Linda has 83 stamps in an album. She gave at least 25 stamps to her friend Sara. How many stamps does Linda have in her album? CCSS A.CED.1, A.REI.3, SMP 4

23. USE A MODEL Mariah told her friend that the number of minutes she has left on her cell phone this month can be modeled by the inequality $500 - x \geq 200$. Write a problem about Mariah's cell phone minutes that can be modeled by this inequality. Solve the inequality and state the solution to the inequality in terms of the number of minutes left on Mariah's cell phone. CCSS A.CED.1, A.REI.3, SMP 4

3.3 Solving Inequalities by Multiplication and Division

STANDARDS

Content: A.REI.1, A.REI.3, A.CED.1
Practices: 1, 2, 3, 4, 7
Use with Lesson 5-2

Objectives

- Explain each step in solving a linear inequality by using multiplication and division.
- Solve linear inequalities in one variable by using multiplication and division.
- Write inequalities in one variable and use them to solve problems.

Multiplication and Division Properties of Inequality

For all numbers a, b, and c, the following are true.

1. If c is positive and $a < b$, then $ac < bc$ and $\frac{a}{c} < \frac{b}{c}$ and if c is positive and $a > b$, then $ac > bc$ and $\frac{a}{c} > \frac{b}{c}$.
2. If c is negative and $a < b$, then $ac > bc$ and $\frac{a}{c} > \frac{b}{c}$, and if c is negative and $a > b$, then $ac < bc$ and $\frac{a}{c} < \frac{b}{c}$.
3. These properties also hold true for inequalities involving $\leq$ and $\geq$.

EXAMPLE 1 Properties of Inequalities CCSS A.REI.1

EXPLORE The number line shows the graphs of two numbers a and b with $a < b$.

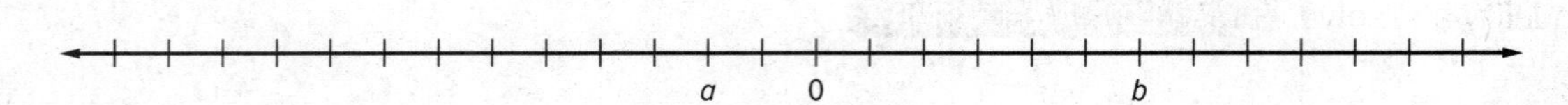

a. REASON ABSTRACTLY Graph $2a$ and $2b$ on the number line and write an inequality that shows the relationship between $2a$ and $2b$. CCSS SMP 2

____

b. MAKE A CONJECTURE Graph the points $-2a$ and $-2b$ on the number line and write an inequality that shows the relationship between $-2a$ and $-2b$. How does this inequality compare to your answer to the original inequality $a < b$? CCSS SMP 3

____

c. MAKE A CONJECTURE Graph the points $\frac{a}{2}$ and $\frac{b}{2}$ on the number line. Also graph the points $-\frac{a}{2}$ and $-\frac{b}{2}$ on the number line. Write inequalities showing the relationship between $\frac{a}{2}$ and $\frac{b}{2}$ and between $-\frac{a}{2}$ and $-\frac{b}{2}$. What do you notice? CCSS SMP 3

____

d. CRITIQUE REASONING Would the inequalities you wrote in **parts a, b,** and **c** be different if a and b were both positive numbers? What if a and b were both negative numbers? Hint: Use a number line to help answer the question. CCSS SMP 3

____

____

EXAMPLE 2 Solve a Linear Inequality by Multiplying or Dividing

REASON QUANTITATIVELY Solve each inequality. Justify each step. CCSS A.REI.1, A.REI.3, SMP 2

a. $\frac{y}{-2} \leq 3$

b. $-54 < -6a$

c. Solve the inequality for x: $-px > 5$

REAL-WORLD EXAMPLE 3 Solve Problems Using Inequalities CCSS A.CED.1, SMP 4

a. USE A MODEL Nick has an $18 gift card for a health food store where a smoothie costs $2.50. How many smoothies can Nick buy with his gift card? Explain.

b. USE A MODEL Sofia bought 3 T-shirts for the same price. Her credit card statement, showed that she owed no more than $45. How much did Sofia spend for each T-shirt? Explain.

PRACTICE

USE STRUCTURE **Solve each inequality.** CCSS A.REI.3, SMP 7

1. $-9a > 81$

2. $\frac{u}{8} \leq -2$

3. $16 \leq -\frac{4}{3}p$

4. $-2.2b < -7.7$

USE STRUCTURE **Solve each inequality for x.** CCSS A.REI.3, SMP 7

5. $4 > -ax$

6. $\frac{3}{b}x < -8$

REASON QUANTITATIVELY **Define a variable, write an inequality, and solve the problem.** CCSS A.REI.3, SMP 2

7. The product of a number and 7 is not less than 21.

8. The quotient of a number and -6 is at least 5.

9. The product of $-\frac{4}{5}$ and a number is at most -16.

10. Ten is no more than the quotient of a number and 4.

11. **USE A MODEL** Stacey has a budget of \$2000 for an event at a golf club. The golf club charges \$82.00 per guest. Write and solve an inequality to determine the maximum number of guests Stacey can invite. CCSS A.CED.1, SMP 4

12. **REASON QUANTITATIVELY** A sales clerk earns a 7% commission of sales. How much must she sell in order to earn at least \$1000 in commissions? Explain. CCSS A.CED.1, SMP 2

13. INTERPRET PROBLEMS A small boat carries 150 feet of anchor rope. The length of the anchor rope must be at least seven times the depth of the water. To the nearest whole foot, what is the maximum depth for this boat? Define a variable, write and solve an inequality and interpret the results. CCSS A.CED.1, SMP 1

14. USE A MODEL The table shows the average amount of food consumed at an animal shelter in a week. If mixed seeds are sold in 5 pound bags, what is the least number of mixed seed bags the shelter will purchase in a year? CCSS A.CED.1, SMP 4

Food Consumed at the Shelter in a Week	
Type of Food	**Amount of Food (lb)**
Grapes	4
Mixed seed	10
Cat Chow	10
Peanuts	5

USE STRUCTURE Match each inequality to the graph of its solution. CCSS A.REI.3, SMP 7

15. $-0.5t \geq 1.5$

a.

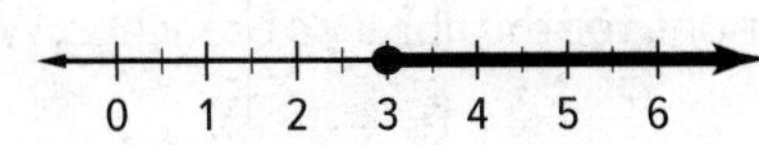

16. $\frac{1}{9}t \leq -3$

b.

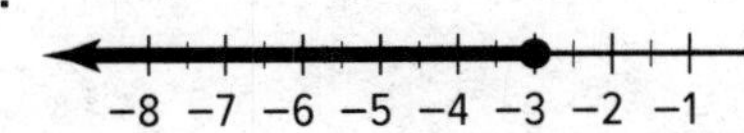

17. $-13.5 \leq -4.5t$

c.

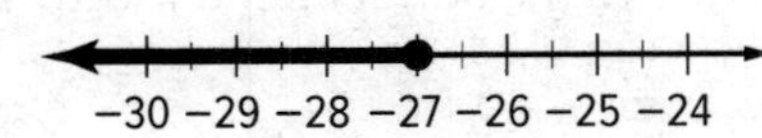

18. $-\frac{t}{6} \leq -\frac{1}{2}$

d.

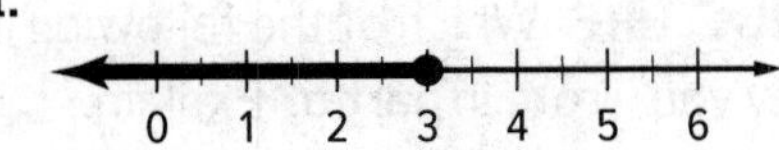

19. CRITIQUE REASONING Two students solved a homework problem differently. Is either of them correct? Explain your reasoning. CCSS A.REI.1, SMP 3

Student A	*Student B*
$3m \geq -21$	$3m \geq -21$
$\frac{3m}{3} \geq \frac{-21}{3}$	$\frac{3m}{3} \geq \frac{-21}{3}$
$m \leq -7$	$m \geq -7$

3.4 Solving Multi-Step Inequalities

STANDARDS

Content: A.REI.1, A.REI.3, A.CED.1
Practices: 1, 2, 3, 4, 6, 7, 8
Use with Lesson 5-3

Objectives

- Create linear inequalities in one variable.
- Solve linear inequalities in one variable.
- Explain each step in solving a linear inequality

You can solve multi-step inequalities by using properties of inequalities and undoing each operation. If the inequality simplifies to a statement that is never true, the solution is the empty set, ∅. If the solution results in a statement that is always true, the solution is the set of all real numbers.

EXAMPLE 1 Create and Solve a Multi-Step Inequality

EXPLORE A city library has 5 branches. A librarian is buying an online newspaper subscription for each of the branches. The pricing is shown. The librarian has a budget of \$270 for the subscriptions.

Metro News Subscriptions
\$7.25 per month \$4 off your first month for new subscriptions!

a. USE A MODEL Write an expression for the total cost of one subscription for m months. Then write an expression for the total cost of all the subscriptions for the library's branches. What are the constraints on the variable? CCSS A.CED.1, SMP 4

b. USE A MODEL Write an inequality to determine the number of months all 5 branches can subscribe to the newspaper and stay within budget. CCSS A.CED.1, SMP 4

c. USE STRUCTURE Which of the following inequalities, if any, is equivalent to the inequality you wrote in **part b**? Explain. CCSS A.REI.3, SMP 7

i. $36.25m - 4 \leq 270$ **ii.** $36.25m - 20 \leq 270$
iii. $7.25m - 4 \leq 54$ **iv.** $5(7.25m) \leq 274$

d. USE STRUCTURE Solve the inequality you chose in **part c** and explain the steps for solving it. Graph the solution on the number line, assuming you can only subscribe to the newspaper for a whole numbers of months. CCSS A.REI.3, SMP 7

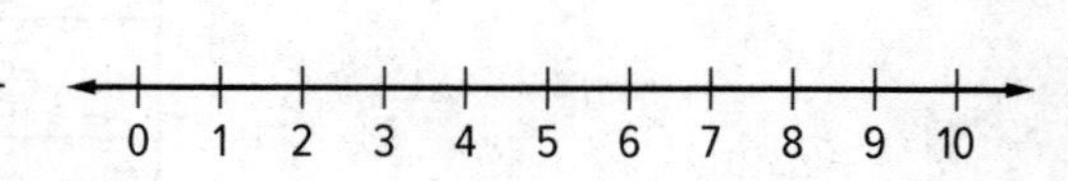

e. INTERPRET PROBLEMS Explain how you know your solution is reasonable. CCSS A.REI.3, SMP 1

When you solve an inequality, you should be able to construct an argument to justify your solution. To do so, you must explain each step of the solution process using a property of numbers or a property of inequalities.

EXAMPLE 2 Explain the Steps of a Solution

Follow these steps to solve $-2(3x - 5) > -x - 2x + 31$ and justify the solution.

a. REASON QUANTITATIVELY Complete the steps of the solution below. Provide a property to justify each step. CCSS A.REI.1, A.REI.3, SMP 2

$-2(3x - 5) > -x - 2x + 31$	Original inequality
$-2(3x - 5) > -3x + 31$	Combine like terms

b. COMMUNICATE PRECISELY Write the solution set, then graph it on the number line. CCSS A.REI.3, SMP 6

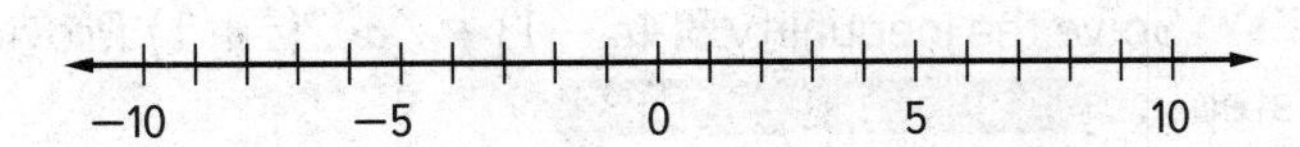

c. CRITIQUE REASONING A student said it is possible to solve the inequality by dividing both sides by −2 first. Do you agree? Explain. CCSS A.REI.3, SMP 3

d. DESCRIBE A METHOD In **part a**, you solved a multi-step inequality and justified the steps of the solution. How does this compare to the process for solving a multi-step equation? CCSS A.REI.1, A.REI.3, SMP 8

EXAMPLE 3 Solve a Problem

Jerome wants to set up a rectangular garden and put a fence around it. He has 32 feet of fencing available, so the perimeter of the garden cannot exceed 32 feet.

a. USE A MODEL Write an inequality that relates the length, width, and perimeter of the garden. CCSS A.CED.1, SMP 4

b. USE A MODEL Show how you can write an inequality in one variable, w, if Jerome decides that the length must be 4 feet longer than the width. What property do you use to write the new inequality? Explain. CCSS A.CED.1, SMP 4

c. REASON QUANTITATIVELY Solve the inequality you wrote in **part b** and then interpret the solution. CCSS A.REI.3, SMP 2

d. REASON QUANTITATIVELY What are the possible lengths of the garden? CCSS A.REI.3, SMP 2

PRACTICE

1. REASON QUANTITATIVELY Solve the inequality $3(4c - 1) + 7 > 2(c + 1)$. Provide a property to justify each step. CCSS A.REI.1, A.REI.3, SMP 2

$3(4c - 1) + 7 > 2(c + 1)$ Original inequality

2. CONSTRUCT ARGUMENTS What is the solution set of the inequality $2(2x + 4) < 4(x + 1)$? Why? How is the solution set related to the solution set of $2(2x + 4) \geq 4(x + 1)$? Explain. CCSS A.REI.3, SMP 3

3. CONSTRUCT ARGUMENTS Eric says that 15 more than 6 times the number of pencils he has is less than 20. What can you conclude about the number of pencils Eric has? Justify your answer. CCSS A.CED.1, REI.3, SMP 3

4. USE A MODEL Mei got scores of 76, 80, and 78 on her last three history exams. Write and solve an inequality to determine the score she needs on the next exam so that her average is at least 82. CCSS A.CED.1, REI.3, SMP 4

5. INTERPRET PROBLEMS A triangular carpet has sides of length a feet, b feet, and c feet. The maximum perimeter is 20 feet. CCSS A.CED.1, REI.3, SMP 1

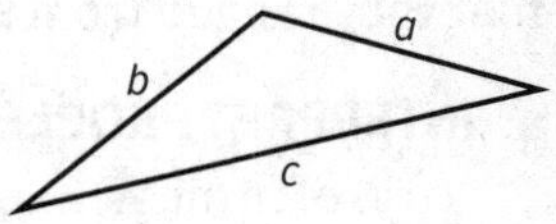

a

b

c

a. Side b is 2 feet longer than a and c is 2 feet longer than b. Which side is the shortest?

b. What are the possible lengths of the shortest side of the carpet? Explain.

6. The figure shows the solution set of an inequality.

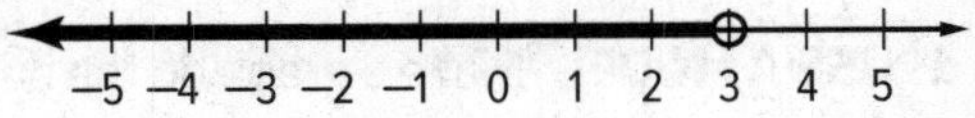

−5 −4 −3 −2 −1 0 1 2 3 4 5

a. REASON ABSTRACTLY Write an inequality that has the given solution set. Solving the inequality should require the Distributive Property, Addition Property of Inequalities, and the Division Property of Inequalities. CCSS A.CED.1, SMP 2

b. REASON QUANTITATIVELY Solve the inequality you wrote in **part a** and provide a property to justify each step. CCSS A.REI.1, A.REI.3, SMP 2

Original inequality

3.5 Solving Compound Inequalities

Objectives

- Create compound inequalities in one variable.
- Solve compound inequalities in one variable and graph their solution set.

CCSS STANDARDS

Content: A.REI.1, A.REI.3, A.CED.1
Practices: 1, 2, 3, 4, 6, 7
Use with Lesson 5-4

When two inequalities are considered together, they form a **compound inequality**.

EXAMPLE 1 Investigate a Compound Inequality

EXPLORE Jason is a driver for A-Plus Taxi. The fares for the taxi company are shown. Jason reviewed all of the fares he collected today and found that they ranged from \$11 to \$39.

A-Plus Taxi • Fares
Flat rate: \$3 upon entering the taxi Plus \$2 per mile traveled

a. INTERPRET PROBLEMS Write an expression that represents the total fare for a trip of *m* miles. Explain how you wrote the expression. CCSS A.CED.1, SMP 1

b. USE A MODEL Write and solve an inequality that Jason can use to determine the minimum distance he traveled today during any of his trips. Interpret the solution. CCSS A.CED.1, A.REI.3, SMP 4

c. USE A MODEL Write and solve an inequality that Jason can use to determine the maximum distance he traveled today during any of his trips. Interpret the solution. CCSS A.CED.1, A.REI.3, SMP 4

d. USE A MODEL Make a graph on the number line that shows the possible range of distances for all of the trips Jason made today. CCSS A.CED.1, A.REI.3, SMP 4

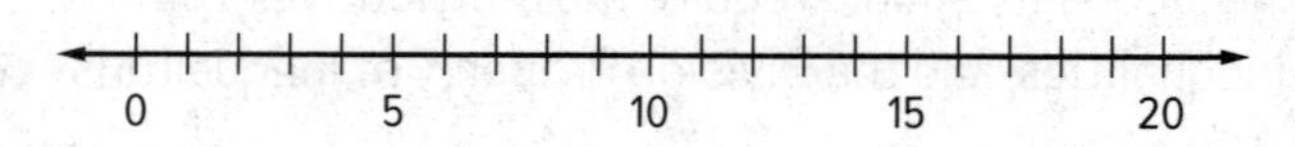

e. INTERPRET PROBLEMS Explain how your graph in **part d** is related to the graphs of the solutions of the inequalities in **parts b** and **c.** CCSS A.CED.1, A.REI.3, SMP 1

f. COMMUNICATE PRECISELY Write the solution set that corresponds to the graph in part d. CCSS A.CED.1, A.REI.3, SMP 6

A compound inequality containing the word *and*, such as $m \geq 4$ and $m \leq 18$, is only true if both inequalities are true. The graph of the compound inequality is the overlap, or **intersection**, of the graphs of the two inequalities.

EXAMPLE 2 Solve and Graph an Intersection

Follow these steps to solve $-16 < 3x - 1 < 20$ and graph the solution set.

a. **USE STRUCTURE** Show how to write the given compound inequality as two separate inequalities that are connected by the word *and*. CCSS A.CED.1, SMP 7

b. **REASON QUANTITATIVELY** Solve each of the inequalities you wrote in **part a.** Use the spaces below. Provide a property to justify each step. CCSS A.REI.1, A.REI.3, SMP 2

Step	Property	Step	Property
	Original inequality		Original inequality

c. **INTERPRET PROBLEMS** Graph each of the solutions from **part b** on the first two number lines below. Then use the third number line to graph the solution of the compound inequality. CCSS A.REI.3, SMP 1

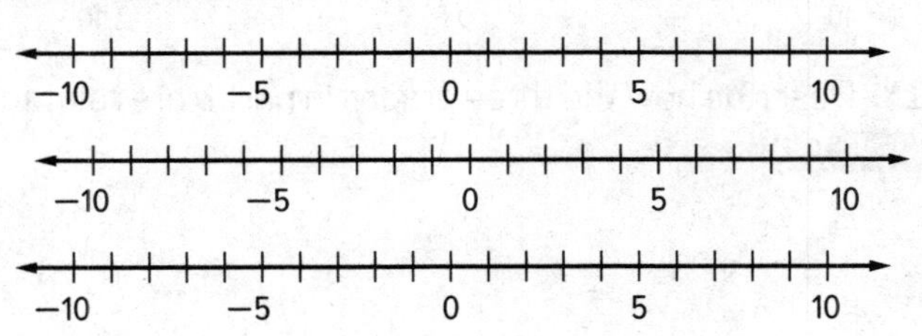

d. **COMMUNICATE PRECISELY** Describe how the three graphs in **part c** are related to each other. CCSS A.REI.3, SMP 6

e. **INTERPRET PROBLEMS** Explain how you can check that your graph of the solution of the compound inequality is reasonable. CCSS A.REI.3, SMP 1

A compound inequality containing the word *or*, such as $y < 1$ or $y > 5$, is true if at least one of the inequalities is true. The graph of the compound inequality is the combination, or **union**, of the graphs of the two inequalities.

EXAMPLE 3 Solve and Graph a Union

Follow these steps to solve $6p + 3 \leq -9$ or $6p + 3 \geq 39$ and graph the solution set.

a. USE STRUCTURE Solve each of the given inequalities. Use the spaces below. Provide a property to justify each step. CCSS A.REI.1, A.REI.3, SMP 2

Step	Property	Step	Property
________	Original inequality	________	Original inequality
________	________	________	________
________	________	________	________
________	________	________	________

b. INTERPRET PROBLEMS Graph each of the solutions from **part a** on the first two number lines below. Then use the third number line to graph the solution of the compound inequality. CCSS A.REI.3, SMP 1

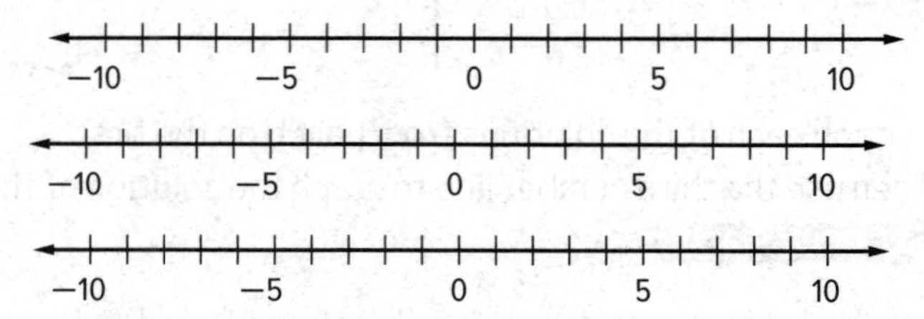

c. COMMUNICATE PRECISELY Describe how the three graphs in **part b** are related to each other. CCSS A.REI.3, SMP 6

____

____

d. REASON ABSTRACTLY Explain how you can use your graph to determine which of the following values, if any, are solutions of the compound inequality: $\{-5, -2, 0, 2, 5\}$. CCSS A.REI.3, SMP 2

____

____

e. REASON ABSTRACTLY What is the solution set of $6p + 3 \leq -9$ and $6p + 3 \geq 39$? Explain. CCSS A.REI.3, SMP 2

____

____

f. REASON ABSTRACTLY Explain how you could determine the solution set to the inequalities in **part e** without performing any arithmetic operations. CCSS A.REI.3, SMP 2

____

____

EXAMPLE 4 Solve a Problem

As part of her fitness routine, Shauntay rides her bike and then takes a walk every Saturday morning. She always bikes at a constant rate of 12 miles per hour. She always walks 2 miles. Looking back at her fitness journal for the year, Shauntay finds that the total distance she biked and walked on Saturday mornings ranged from 5 miles to 17 miles.

a. USE A MODEL Shauntay wants to know the amount of time t in hours that she biked on Saturday mornings this year. Write a compound inequality that she can use to answer this question. CCSS A.CED.1, SMP 4

b. CALCULATE ACCURATELY Solve the compound inequality and then graph the solution on the number line below. CCSS A.REI.3, SMP 6

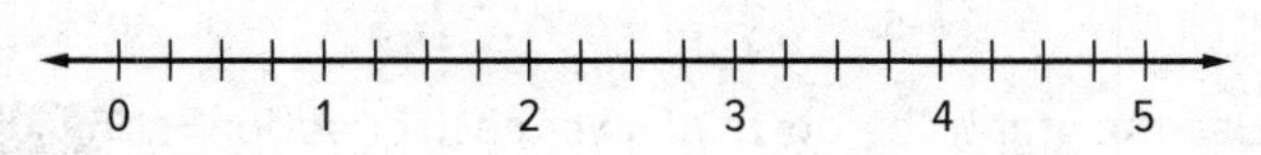

c. REASON QUANTITATIVELY Interpret your solution. What does the solution tell you about the time that Shauntay rides her bike on Saturday mornings? CCSS A.REI.3, SMP 2

PRACTICE

1. REASON QUANTITATIVELY Solve the compound inequality $-20 \leq 2(3y - 7) < 34$. Use the spaces below. Provide a property to justify each step. Then write the solution set and graph the solution set on the number line. CCSS A.REI.1, A.REI.3, SMP 2

Step	Property	Step	Property
	Original inequality		Original inequality

Solution set: ______

−10 −5 0 5 10

2. **REASON QUANTITATIVELY** Solve the compound inequality $-3x + 7x - 1 < -5$ or $-3x + 7x - 1 > 11$. Use the spaces below. Provide a property to justify each step. Then write the solution set and graph the solution set on the number line. CCSS A.REI.1, A.REI.3, SMP 2

Step	Property	Step	Property
________	Original inequality	________	Original inequality
________	________	________	________
________	________	________	________
________	________	________	________
________	________	________	________

Solution set: ____________________

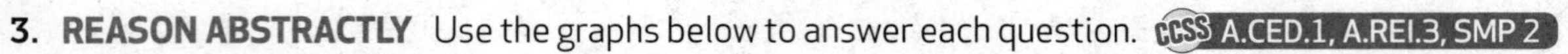

3. **REASON ABSTRACTLY** Use the graphs below to answer each question. CCSS A.CED.1, A.REI.3, SMP 2

a. Write a compound inequality whose solution is the union of the two graphs. Then explain how the compound inequality can be expressed as a single inequality.

____

b. Write a compound inequality whose solution is the intersection of the two graphs. Then explain how the compound inequality can be expressed as a single inequality.

____

c. How are the graphs of the solution sets for the inequalities in **part a** and **part b** related to the given graphs?

____

____

____

4. **USE A MODEL** Jocelyn is planning to place a fence around the triangular flower bed shown. The fence costs \$1.50 per foot. Assuming that Jocelyn spends between \$60 and \$75 for the fence, what is the shortest possible length for a side of the flower bed? Use a compound inequality to explain your answer. CCSS A.CED.1, A.REI.3, SMP 4

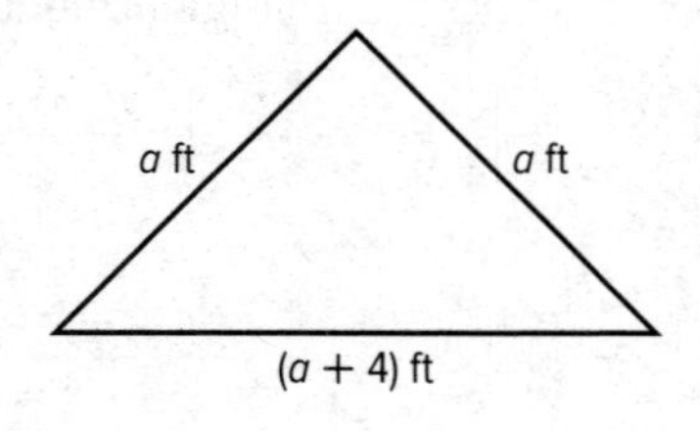

____

____

____

____

____

5. USE A MODEL It costs \$1000 to rent a bus that holds 100 students. A school is planning to rent one of these buses for a field trip to an aquarium. The trip will also have a cost of \$15 per student for the tickets to the aquarium. Given that the total expense for the trip must be between \$2000 and \$3000, find the minimum and maximum number of students who can go on the trip. Explain. CCSS A.CED.1, A.REI.3, SMP 4

6. CRITIQUE REASONING Arturo said that if k is a real number, then the solution set of the compound inequality $x < k$ or $x > k$ is all real numbers. Do you agree? Why or why not? CCSS A.REI.3, SMP 3

7. REASON ABSTRACTLY For each solution set, write a compound inequality whose solution set is the given set. CCSS A.CED.1, SMP 2

a. $\{x \mid 4 \leq x\}$

b. $\{4\}$

8. CALCULATE ACCURATELY Write the solution set of the following compound inequality. Then graph the solution set on the number line below: $-x + 1 < 8$ and $-x + 1 < 3$ and $-x + 1 > -4$. CCSS A.REI.3, SMP 6

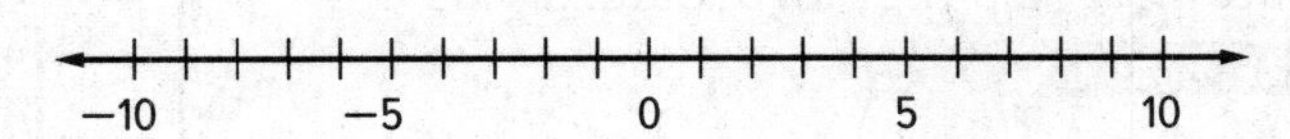

9. The Triangle Inequality states that in any triangle, the sum of the lengths of any two sides is greater than the length of the third side. In the figure, this means $a + b > c$, $a + c > b$, and $b + c > a$. CCSS A.CED.1, SMP 2

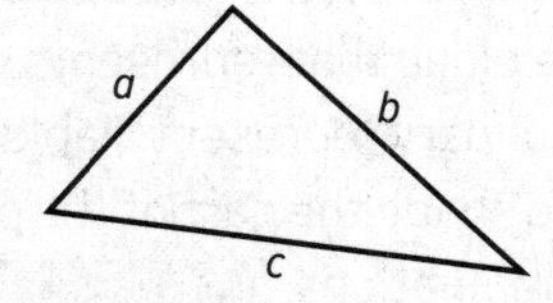

a. REASON ABSTRACTLY Suppose a triangle has a side that is 5 meters long and a perimeter of 14 meters. Let one of the unknown side lengths be x. Write a compound inequality that you can use to determine the value of x. Explain. CCSS A.REI.3, SMP 6

b. COMMUNICATE PRECISELY What can you conclude about the two unknown side lengths of the triangle? Explain. CCSS A.REI.3, SMP 6

3.6 Graphing Inequalities in Two Variables

Objectives

- Represent constraints by inequalities in two variables.
- Graph and interpret the solutions to a linear inequality in two variables.

CCSS STANDARDS

Content: A.CED.3, A.REI.12
Practices: 1, 2, 3, 4, 5, 6, 7, 8
Use with Lesson 5-6

EXAMPLE 1 Represent and Graph a Real-World Constraint

EXPLORE Isaiah is making a large batch of granola to sell at a fund-raising event. He plans to buy raisins and walnuts from the bulk bins at the supermarket, and he can spend at most $18. The prices are shown.

ShopCo Bulk Prices	
Raisins	$2 per pound
Walnuts	$3 per pound

a. USE A MODEL Write an expression that represents the total cost of *r* pounds of raisins and *w* pounds of walnuts. Then use this expression to write an inequality that compares the total cost of the raisins and walnuts to the amount Isaiah can spend. CCSS A.CED.3, SMP 4

b. REASON QUANTITATIVELY Determine whether each combination of raisins and walnuts is a viable combination for Isaiah to purchase. Place a check mark in the last row of the table to show the viable combinations. CCSS A.CED.3, SMP 8

Raisins (lb)	0	1	1.5	2	2	4	6	7	8	9
Walnuts (lb)	6	5	1.5	5	3	4	2	0	1	0
Viable?										

c. FIND A PATTERN Plot the viable combinations on the coordinate plane shown. Look for a pattern to help you plot additional viable combinations. CCSS A.REI.12, SMP 7

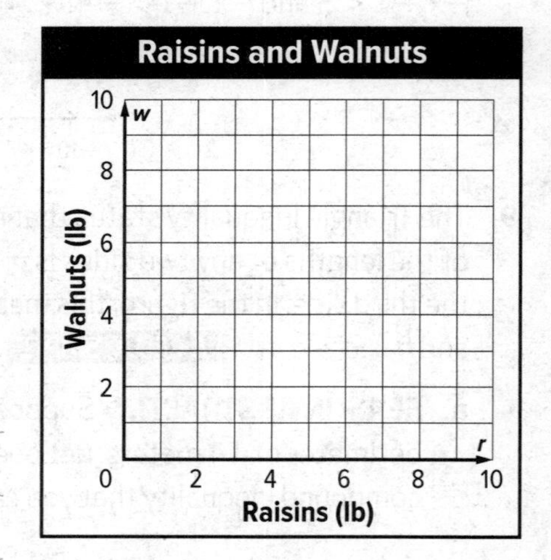

d. COMMUNICATE PRECISELY What is the boundary of this region of the plane that represents viable combinations? Do points on the boundary represent viable combinations? Graph the boundary line and shade the part of the plane that represents all of the viable combinations. CCSS A.CED.3, SMP 6

e. COMMUNICATE PRECISELY Describe the part of the plane that represents all of the viable combinations that Isaiah can purchase. CCSS A.CED.3, SMP 6

f. REASON ABSTRACTLY Explain how you can use your graph to determine whether 5 pounds of raisins and 3 pounds of walnuts is a viable combination. CCSS A.CED.3, SMP 2

The graph of a linear inequality in two variables is the set of points in the plane that represent all of the possible solutions of the inequality. If an inequality contains $\leq$ or $\geq$, the points on the boundary represent solutions, and the graph of the boundary is a solid line. If an inequality contains $<$ or $>$, the points on the boundary do not represent solutions and the graph of the boundary is a dashed line.

EXAMPLE 2 Graph an Inequality CCSS A.REI.12

Follow these steps to graph $2x + y \geq 3$.

a. PLAN A SOLUTION Describe the main steps you can use to graph the solution of the inequality. CCSS SMP 1

b. USE STRUCTURE Graph the solution of the inequality on the coordinate plane at the right. CCSS SMP 7

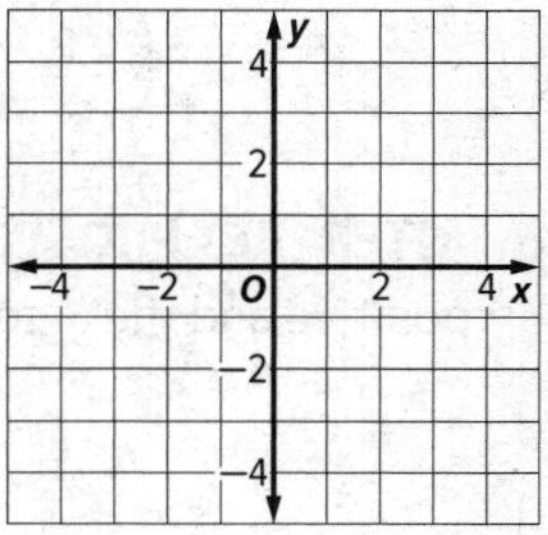

c. INTERPRET PROBLEMS Explain how you can check that you graphed the solution correctly. CCSS SMP 1

d. REASON ABSTRACTLY How would the graph be different if the inequality were $2x + y > 3$? CCSS SMP 2

e. REASON ABSTRACTLY How would the graph be different if the inequality were $2x + y \leq 3$? CCSS SMP 2

f. CRITIQUE REASONING A student claimed that if the point (a, b) is a solution of the inequality, then a and b cannot both be negative numbers. Do you agree? Why or why not? CCSS SMP 3

EXAMPLE 3 Graph an Inequality CCSS A.REI.12

Follow these steps to graph $-4x + 12y < 0$.

a. PLAN A SOLUTION Describe the main steps you can use to graph the solution of the inequality. CCSS SMP 1

b. USE TOOLS Graph the solution of the inequality on the coordinate plane shown. CCSS SMP 5

c. USE STRUCTURE Is it possible to use the origin, (0, 0), as a test point to determine which half plane to shade? Explain why or why not. CCSS SMP 7

d. INTERPRET PROBLEMS Explain why any ordered pair in which the x-coordinate is positive and the y-coordinate is negative is a solution of the inequality. CCSS SMP 1

e. REASON ABSTRACTLY Write an inequality that has a solution set that is represented by all the points that are *not* shaded in your graph. Justify your answer. CCSS SMP 2

f. FIND A PATTERN On a separate sheet of graph paper, graph the solutions of the inequalities $3x + y < 0$ and $2x - y < 0$. In general, what can you conclude about the solution of an inequality of the form $ax + by < 0$, where $a \neq 0$ and $b \neq 0$? CCSS SMP 8

EXAMPLE 4 Solve a Real-World Problem

Ticket Prices	
Student ticket	\$7.50
Guest ticket	\$12.00

Audra is selling tickets to a school jazz concert. Ticket prices are shown. Audra's goal is to sell at least \$120 in tickets.

a. **USE A MODEL** Write an inequality that represents the number of student tickets s and guest tickets g that Audra must sell to reach her goal. CCSS A.CED.3, SMP 4

b. **USE STRUCTURE** State any conditions that must be placed on the variables for the solutions to make sense in the context of the problem and explain. CCSS A.CED.3, SMP 7

c. **USE STRUCTURE** Graph the solution of the inequality on the coordinate plane at the right. CCSS A.REI.12, SMP 7

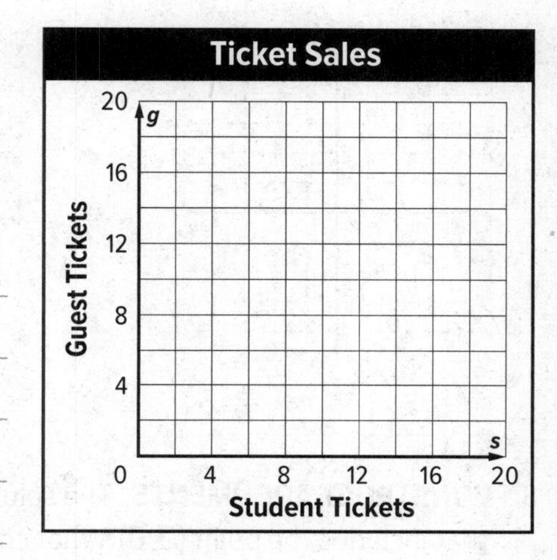

d. **USE A MODEL** What are three different combinations of tickets that Audra can sell to reach her goal. Explain how you used the graph to find these combinations. CCSS A.CED.3, SMP 4

e. **USE A MODEL** Describe two different ways to determine whether selling 6 student tickets and 6 guest tickets will allow Audra to reach her goal. CCSS A.CED.3, SMP 7

f. **USE STRUCTURE** Is every point of the shaded region of your graph a solution of the problem? Explain. CCSS A.CED.3, SMP 7

g. **REASON ABSTRACTLY** Audra makes her goal by selling twice as many guest tickets as student tickets. Explain how you can use the graph to find three possible combinations of tickets that Audra may have sold. CCSS A.CED.3, SMP 2

PRACTICE

USE STRUCTURE **Graph each inequality. Then name three ordered pairs in the solution set.** CCSS A.REI.12, SMP 7

1. $-x + 2y \leq 2$

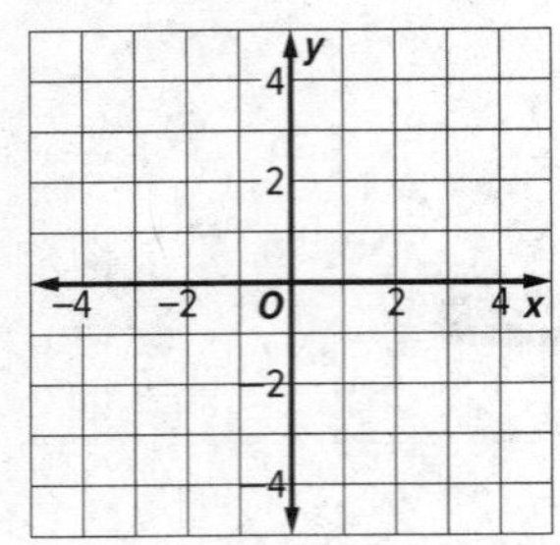

2. $6x + 2y \geq -4$

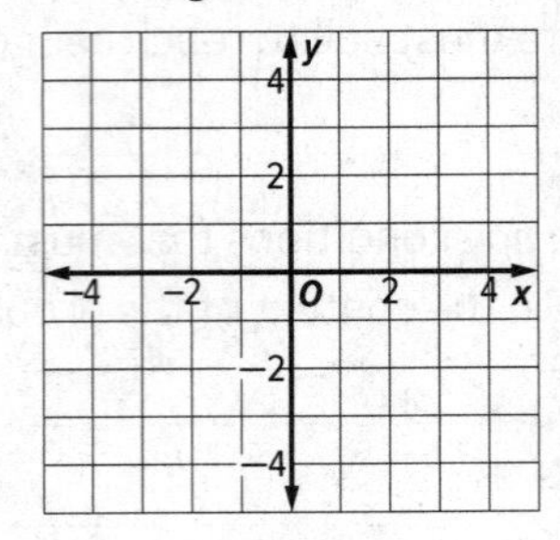

3. $-2x + 3y > 6$

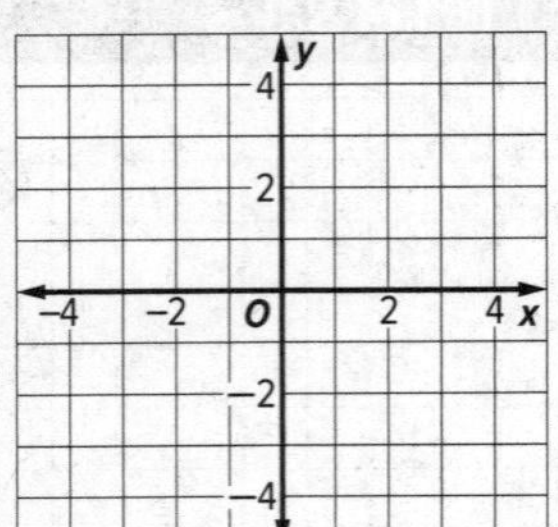

4. $2y < -5x$

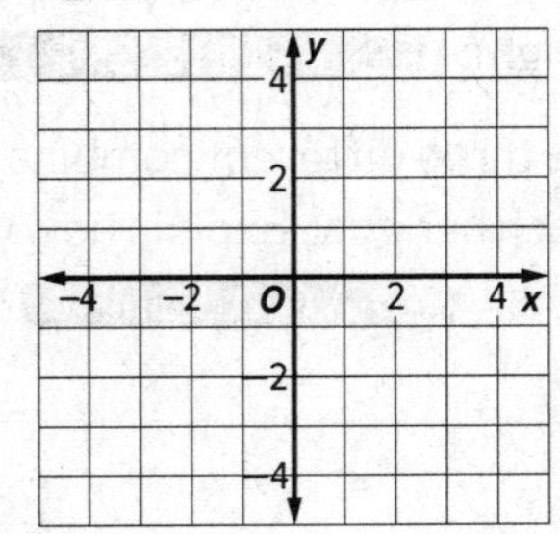

5. **CONSTRUCT ARGUMENTS** The solution of the inequality $ax + by < c$ is a half plane that includes the point (0, 0). What conclusion can you make about the value of c? Justify your answer. CCSS A.REI.12, SMP 3

6. Kumiko has a $50 gift card for a Web site that sells song downloads and video downloads. Videos cost $2.50 each, and songs cost $1.25 each.

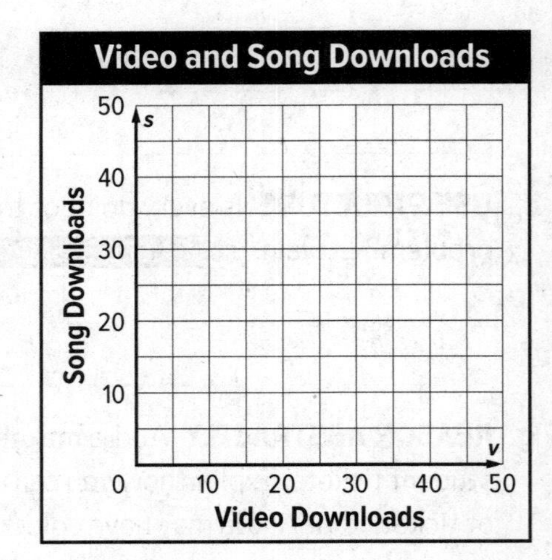

a. **REASON QUANTITATIVELY** Write an inequality that represents the number of songs s and the number of videos v that Kumiko can download and describe any constraints. CCSS A.CED.3, SMP 2

b. **USE STRUCTURE** Graph the solution of the inequality on the coordinate plane shown. CCSS A.REI.12, SMP 7

c. **USE A MODEL** Use your graph to find three different combinations of songs and videos that Kumiko can download. CCSS A.CED.3, SMP 4

d. REASON QUANTITATIVELY Kumiko decides to download the same number of songs as videos, and she decides to spend as much of the $50 as possible. How many songs and videos does she download? How much money does she have left on her card? CCSS A.CED.3, SMP 2

7. A café sells peach smoothies and berry smoothies. The café makes a profit of $2.25 for each peach smoothie that is sold and a profit of $2 for each berry smoothie that is sold. The owner of the café wants to make a total profit of more than $90 per day from sales of smoothies.

a. USE A MODEL Write an inequality that represents the number of peach smoothies p and berry smoothies b that the café needs to sell. Describe the constraints on the variables. CCSS A.CED.3, SMP 4

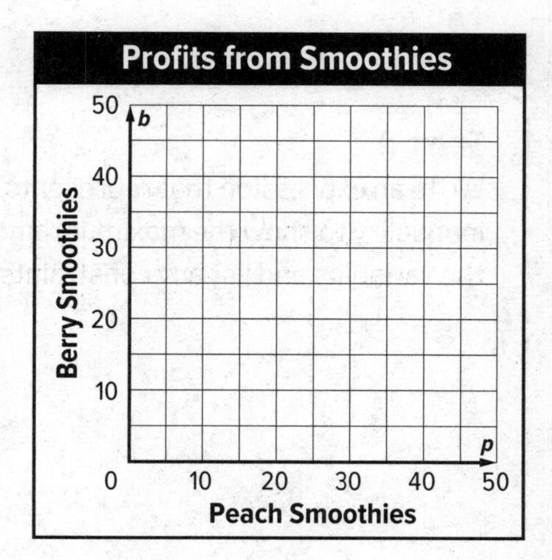

b. USE STRUCTURE Graph the solution of the inequality on the coordinate plane at the right. CCSS A.REI.12, SMP 7

c. USE A MODEL On Monday, the café sold 20 peach smoothies and made the daily profit goal. What can you say about the number of berry smoothies that were sold on Monday? CCSS A.CED.3, SMP 4

d. REASON QUANTITATIVELY On Tuesday, the Café made the daily profit goal by selling the minimum number of smoothies. How many smoothies did they sell? Explain. CCSS A.CED.3, SMP 2

8. Oleg is training for a triathalon. One day, he jogged for 2 hours at x miles per hour. Then he bicycled for 2 hours at y miles per hour. Finally he swam a distance of 2 miles. The total number of miles did not exceed 30 miles.

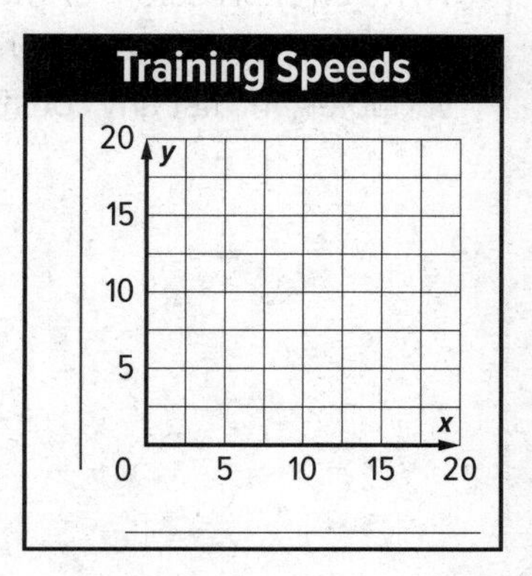

a. USE STRUCTURE Write an inequality to represent the distance that he traveled that day. Describe the constraints on the variables. CCSS A.CED.3, SMP 7

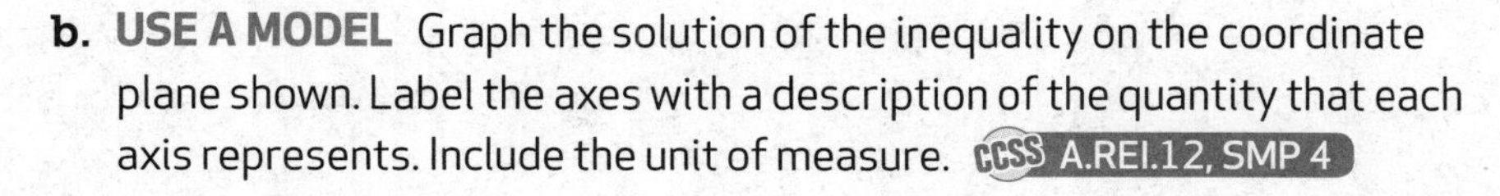

b. USE A MODEL Graph the solution of the inequality on the coordinate plane shown. Label the axes with a description of the quantity that each axis represents. Include the unit of measure. CCSS A.REI.12, SMP 4

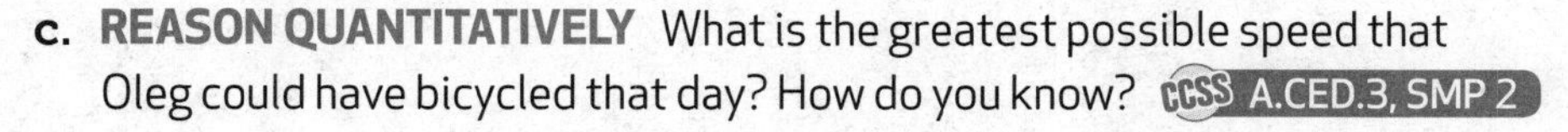

c. REASON QUANTITATIVELY What is the greatest possible speed that Oleg could have bicycled that day? How do you know? CCSS A.CED.3, SMP 2

Performance Task

Analyzing Car Mileage

Provide a clear solution to the problem. Be sure to show all of your work, include all relevant drawings, and justify your answers.

A family has a hybrid car that gets 40 miles per gallon (mpg) and a pickup truck that gets 17 mpg. They drive at least 500 miles per week using both vehicles but don't want to use more than 20 gallons of gas. Make a recommendation for using each vehicle.

Part A

Write an expression that represents the total amount of gasoline used. Then write an inequality to show the maximum amount of gasoline that the family wants to use. Define the variables and list any constraints on the variables.

Part B

Write an expression for the total number of miles for both vehicles per week. Then write an inequality that represents the distance the family travels each week. Define the variables and list any constraints on the variables.

Part C

Graph both inequalities and identify the region where they overlap. What does this overlapping region represent?

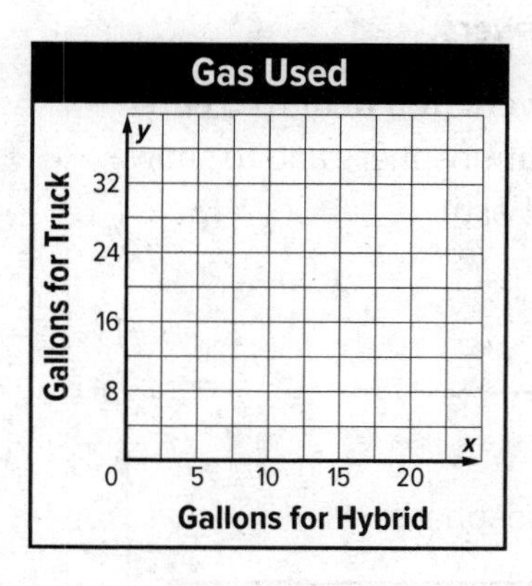

Part D

Based on your analysis of the graphs, make a recommendation for how much to use each vehicle. Provide several options for them to consider. Include scenarios where the hybrid is used more, the truck is used more, and each vehicle is used equally.

Performance Task

Earth's Atmosphere and Satellite Orbits

Provide a clear solution to the problem. Be sure to show all of your work, include all relevant drawings, and justify your answers.

The Earth's atmosphere is made up of different layers. You want to create a vertical model to show the various layers of the atmosphere and to show which layer various satellites are in when orbiting Earth.

Part A

Write inequalities for each the layer of the atmosphere.

Layer	Miles Above Sea Level	Inequalities
Troposphere	0 to 7	
Stratosphere	7 to 31	
Mesosphere	31 to 50	
Thermosphere	50 to 440	
Exosphere	440 to 6200	

Part B

Model the layers of the atmosphere on the vertical number line. Include appropriate labels for each layer.

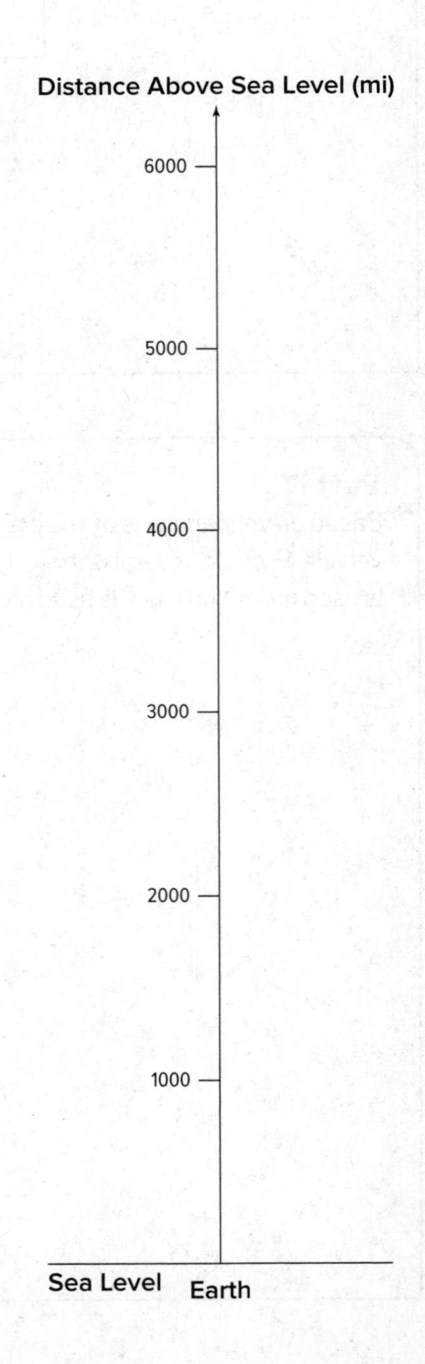

Part C

Satellites are classified by the distance above Earth that they orbit. The table shows approximate distances from Earth for three kinds of satellites. Write inequalities that show the range for each kind of satellite.

Satellite Orbits	Miles Above Sea Level	Inequalities
Low Earth (LEO)	111 up to 1243	
Medium Earth (MEO)	1243 up to 22,223	
Geosynchronous (GEO)	22,223 or more	

Part D

Compare the range of the satellite orbits with the atmosphere, creating a vertical number-line model of satellite orbits. Label the interval for the atmosphere. Then write a description of the layers of the atmosphere where each kind of satellite orbits Earth. Which satellites orbit Earth within the atmosphere?

Distance Above Sea Level (mi)

25,000

20,000

15,000

10,000

5000

Sea Level

Earth

Standardized Test Practice

1. Mindy's empty suitcase weighs 8 pounds. The airline allows luggage to weigh no more than 40 pounds. Write an inequality that shows the total weight w of the items Mindy can pack in her suitcase. CCSS A.CED.3

2. Solve the inequality. CCSS A.CED.1

$3p > -6 + p$

3. Write the compound inequality shown on the graph. CCSS A.CED.3

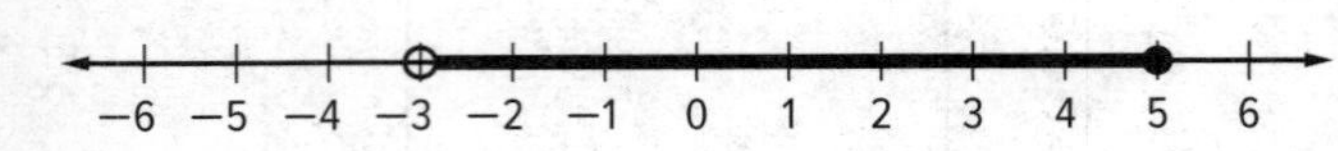

4. Solve the inequality, and graph its solution. CCSS A.CED.3

$-\frac{1}{2}a \leq 2$

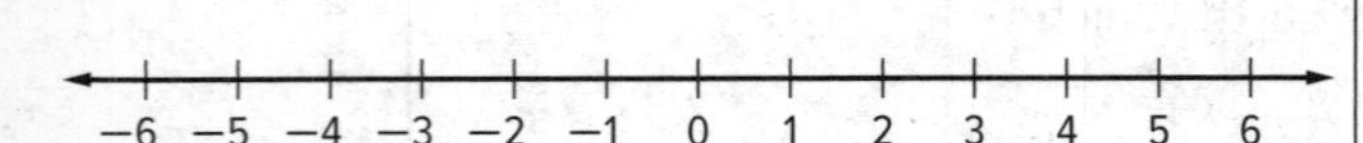

5. On Tuesday, the high temperature was 79°F, and the low temperature was 56°F. Write a compound inequality that represents the possible temperatures t on that day. CCSS A.CED.3

6. Solve the inequality, and graph its solution. CCSS A.CED.3

$3c + 1 < 7$ OR $c - 5 \geq -1$

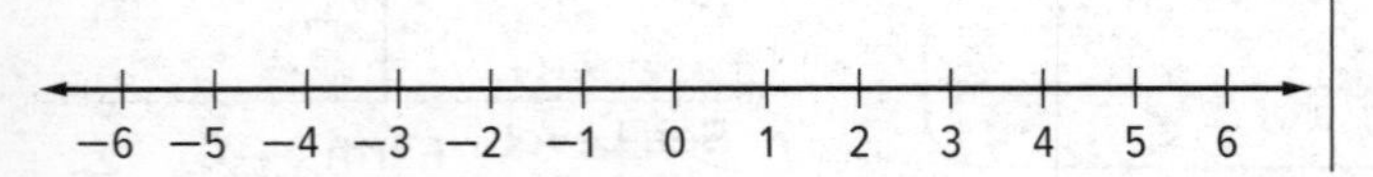

7. The area of the rectangle shown below is at least 45 square units and at most 100 square units. Write an inequality that shows the possible values for *m*. CCSS A.CED.3

m

5

8. Solve the inequality. CCSS A.CED.3

$-9 \leq x - 4 < 1$

9. Which of the following inequalities are shown on the graph? CCSS A.CED.3

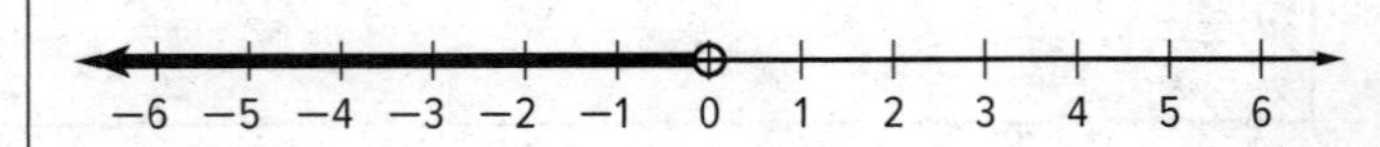

$a - 5 < 5$

$6(z - 1) \leq -6$

$-2x > 0$

$b - 3 < -3$ OR $5b \leq -10$

10. What is the solution to the inequality? CCSS A.REI.3

$4(x + 1) \geq 5x + 7$

$x \geq -3$	$x \leq 3$
$x \leq -3$	$x \geq 3$

11. Graph the solution to $y \leq 3x$. CCSS A.REI.12

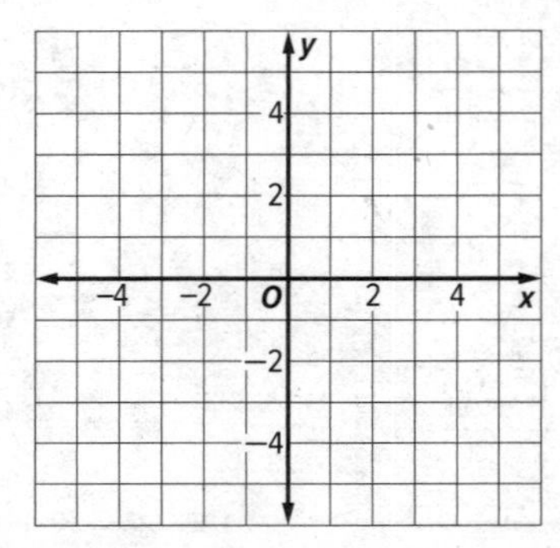

12. Javier has $10 to spend at the school store. He wants to buy one T-shirt and some pens. The table below shows the cost of items at the school store.

	Notebook	T-Shirt	Pen	Pencil
Price ($)	1.50	6.25	0.30	0.15

a. Write an inequality that shows the number of pens p that Javier can buy. CCSS A.CED.1

b. Solve the inequality. Explain each step. 

c. Explain the solution in the context of the problem. 

13. Graph the solution to $-3x + 3y + 7 > 13$. CCSS A.REI.12

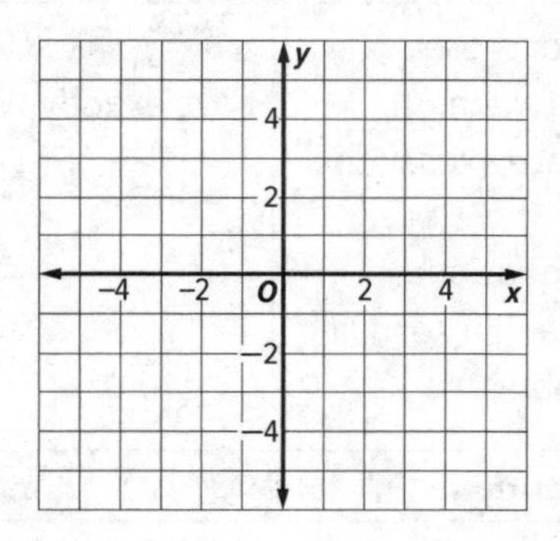

14. Solve $-3(x + 1) \leq x + 5$. Show your work and explain each step. Graph the solution. 

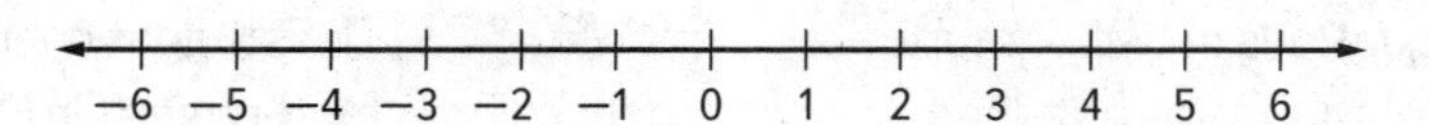

15. Jocelyn said that she was thinking of a number and that the number satisfied two inequalities: $4x - 7 \geq x + 14$ and $-x - 20 \leq -3x - 6$. What was Jocelyn's number? Explain. CCSS A.CED.1

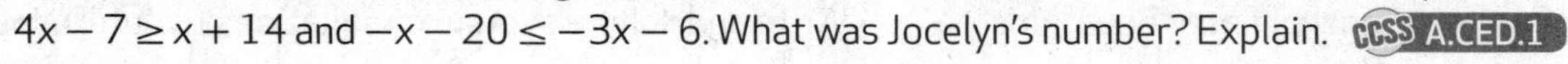

4 Systems of Linear Equations and Inequalities

CHAPTER FOCUS Learn about some of the Common Core State Standards that you will explore in this chapter. Answer the preview questions. As you complete each lesson, return to these pages to check your work.

What You Will Learn	Preview Question
Lesson 4.1: Writing Systems of Equations	
CCSS A.CED.2 Create equations in two or more variables to represent relationships between quantities; graph equations on coordinate axes with labels and scales. CCSS A.CED.3 Represent constraints by equations or inequalities, and by systems of equations and/or inequalities, and interpret solutions as viable or nonviable options in a modeling context.	CCSS SMP 4 The sum of two numbers is 24. The difference of the two numbers is 4. Write a system of equations representing this situation.
Lesson 4.2: Solving Systems by Graphing	
CCSS A.CED.2 Create equations in two or more variables to represent relationships between quantities; graph equations on coordinate axes with labels and scales. CCSS A.REI.6 Solve systems of linear equations exactly and approximately (e.g., with graphs), focusing on pairs of linear equations in two variables.	CCSS SMP 3 Can a system of two linear equations have exactly two solutions? Explain.
Lesson 4.3: Solving Systems by Substitution	
CCSS A.CED.2 Create equations in two or more variables to represent relationships between quantities; graph equations on coordinate axes with labels and scales. CCSS A.REI.6 Solve systems of linear equations exactly and approximately (e.g., with graphs), focusing on pairs of linear equations in two variables.	CCSS SMP 2 You have 44 dimes and quarters. The total value of the coins is $6.80. How many dimes and quarters do you have? Write and solve a system of equations to represent this situation.
Lesson 4.4: Solving Systems by Elimination	
CCSS A.CED.2 Create equations in two or more variables to represent relationships between quantities; graph equations on coordinate axes with labels and scales. CCSS A.REI.6 Solve systems of linear equations exactly and approximately (e.g., with graphs), focusing on pairs of linear equations in two variables.	CCSS SMP 7 The student council is selling pens and pencils for a fund-raiser. On Monday they sold 29 pens and 13 pencils for a total of $14.85. On Tuesday they sold 22 pens and 13 pencils for a total of $12.05. Write and solve a system of equations to determine how much each pen and pencil costs.

What You Will Learn	Preview Question
Lesson 4.5: Solving Systems by Elimination Using Multiplication	
CCSS A.REI.5 Prove that, given a system of two equations in two variables, replacing one equation by the sum of that equation and a multiple of the other produces a system with the same solutions. CCSS A.CED.2 Create equations in two or more variables to represent relationships between quantities; graph equations on coordinate axes with labels and scales. CCSS A.REI.6 Solve systems of linear equations exactly and approximately (e.g., with graphs), focusing on pairs of linear equations in two variables.	CCSS SMP 1 Solve the system using elimination. Explain your solution process. $2x + 3y = 9$ $x + 5y = 8$
Lesson 4.6: Solving Systems of Inequalities	
CCSS A.REI.12 Graph the solutions to a linear inequality in two variables as a half plane (excluding the boundary in the case of a strict inequality), and graph the solution set to a system of linear inequalities in two variables as the intersection of the corresponding half planes.	CCSS SMP 5 Graph the system of inequalities. $y < 2x - 1$ $y \geq 4x - 7$ Provide the coordinates of a point in the solution region. 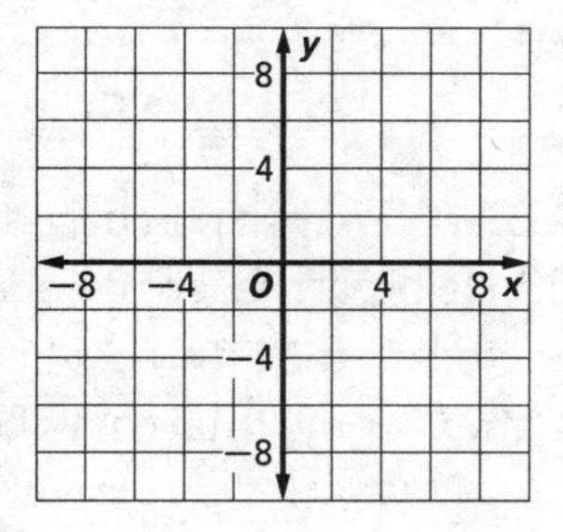 CCSS SMP 5 Graph the system of inequalities. $x + y > 3$ $y \leq -x - 2$ 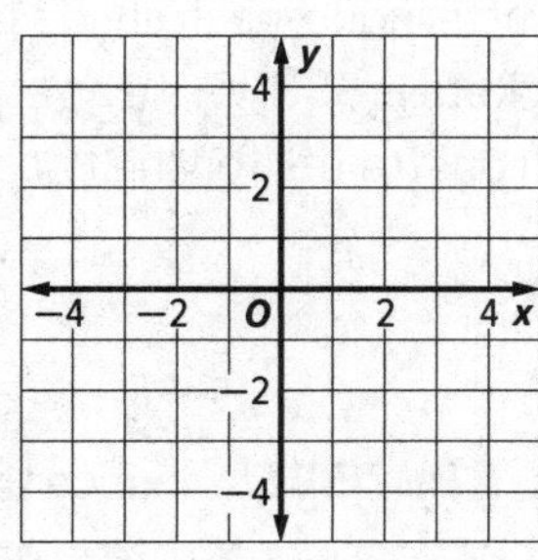 Provide the coordinates of a point in the solution region.

4.1 Writing Systems of Equations

Objectives

- Create systems of equations in two or more variables to represent relationships between quantities.
- Represent constraints by equations.

CCSS STANDARDS

Content: A.CED.2, A.CED.3
Practices: 1, 3, 4, 6
Use with Lesson 6-1

A **system of equations** is a set of equations with the same variables. Each equation expresses a relationship among the variables, which places **constraints**, or restrictions, on the values that the variables represent. Real-world situations often require that there be additional constraints on possible values of the variables.

EXAMPLE 1 Identify the Constraints of a Problem

At a movie theater, an adult ticket costs more than a child's ticket. Let x represent the cost for one adult ticket and y represent the cost for one child's ticket.

a. INTERPRET PROBLEMS List the constraints on the variables using equations. Describe the kinds of numbers that could be values for x and y in this situation. Explain the reasons for the constraints on these variables. CCSS A.CED.3, SMP 1

b. PLAN A SOLUTION Aimee purchased two adult tickets and three child's tickets and spent $34.00. Heidi purchased one adult ticket and two child's tickets and spent $20.00. Write an equation for the total cost of movie tickets for two adults and three children. Write an equation for the total cost of movie tickets for one adult and two children. CCSS A.CED.2, SMP 1

c. COMMUNICATE PRECISELY Mr. Hu and Ms. Claremont want to take their science classes to see a movie. Four teachers (adults) and 28 students (children) will attend. At regular price, the total cost of the tickets is $100. If the group goes to the movie on a Tuesday afternoon, each child's ticket is half off the regular price, and the total cost of the tickets is $58. Write an equation for each of the possible total costs. Describe the constraints for the two possible costs. CCSS A.CED.2, A.CED.3, SMP 6

d. COMMUNICATE PRECISELY Since the cost of either ticket is greater than 0, how can you use the equations you wrote in **part c** to determine a constraint for the maximum cost of an adult ticket and a child's ticket? Explain your reasoning. CCSS A.CED.3, SMP 6

EXAMPLE 2 Identify Constraints and Equations CCSS A.CED.2, A.CED.3, SMP 3

CRITIQUE REASONING Read each situation and identify whether the given system of equations correctly matches the information given in the problem. Correct any errors.

a. Arun sells 3 shirts and 2 pairs of trousers for \$140. He sells one shirt and a pair of trousers for \$55. He wrote the equations $3x - 2y = 140$ and $x - y = 55$ to represent the total amount of each sale.

b. Amanda has \$20 to buy groceries for a class dinner. Amanda left the grocery store with \$7.37 and a total of 6 bags of onions and potatoes. She wrote the following system of equations to represent her purchases. Critique Amanda's account to the class. Is there anything you would change?

Sale Prices

Item	Cost/Bag
Onions	\$1.29
Potatoes	\$2.92

x = number of bags of onions
y = number of bags of potatoes

$x + y = 6$
$1.29x + 2.92y = 20$

EXAMPLE 3 Writing System of Equations CCSS A.CED.2, A.CED.3, SMP 1, SMP 6

REASON QUANTITATIVELY For each situation, define variables and the constraints on the variables. Then express the given information as a system of equations.

a. A grocery store sells a large box of cereal for \$4 and a small box for \$2. One weekend, the store sold 78 boxes of cereal for \$228. Hugo needs to find the number of large boxes sold and the number of small boxes sold.

b. A cab company charges an initial fare and a cost per mile for each trip. The distance and cab fare for 2 passengers are shown in the table. A third passenger wants to know the initial fare and cost per mile.

	Distance (mi)	Cab Fare ($)
John	18.1	45.63
Carlos	22.3	55.29

PRACTICE

1. INTERPRET PROBLEMS A game show awards 60 points during 2 rounds of questions. In the first round, each question is worth 2 points, and in the second round, each question is worth 3 points. A total of 25 questions are asked in the show. A contestant wants to know how many questions are asked in each round. CCSS SMP 1

a. Define the variables and any constraints on the variables. CCSS A.CED.2

b. Write a system of equations to represent the relationships between the questions in each round. CCSS A.CED.3

2. CRITIQUE REASONING Martha sold 45 items worth $191.25 at her hot dog stand on Monday. Bridget wants to find the number of hot dogs and the number of burgers that Martha sold that day. She wrote the system of equations shown to represent the situation.

Martha's Hot Dog	$1.25
Martha's Burger	$8.75

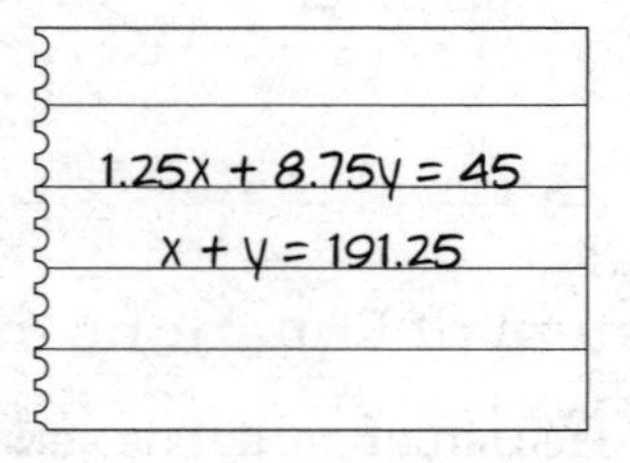

Is Bridget correct? Correct any errors or omissions. CCSS A.CED.2, A.CED.3, SMP 3

3. **INTERPRET PROBLEMS** Will and Baljit assemble 29 centerpieces for the tables for a school dance. They spend a total of 10 hours on the work. Will assembles 3.5 centerpieces per hour, and Baljit assembles 2.5 per hour. CCSS SMP 1

   a. Define the variables and identify any constraints on them. CCSS A.CED.3

   b. Write a system of equations to represent the situation. CCSS A.CED.2

   c. Will and Baljit assemble more of the centerpieces for another school function. This time they only make a total of 12 centerpieces in a total of 4 hours. Write a system of equations to represent the situation. CCSS A.CED.2

4. The table shows the prices of some metals at a specialty metal shop. Each purchase must be a whole number of ounces. CCSS A.CED.2, A.CED.3

Metal	Cost ($)
Aluminum	0.84 per pound
Copper	3.01 per pound
Silver	19.51 per ounce

   a. **INTERPRET PROBLEMS** A jeweler plans to purchase 20 pounds of metal, of which only 2.5% will be silver. Define variables and write equations to represent how much of each metal will be purchased. State any constraints on the variables. CCSS SMP 1

   b. **USE A MODEL** A necklace weighs 3 ounces, and costs $40.74 to make. Define variables and write a system of equations to represent this situation for two metals that you choose. CCSS SMP 4

5. **INTERPRET PROBLEMS** Linda invests a total of $5000 into two bank accounts. One account yields a simple interest of 1.05% and the other yields a simple interest of 1.01%. At the end of the month, she earns an interest of $3.50. Write a system of equations to find the money she deposited into each account. List all the constraints. CCSS A.CED.2, A.CED.3, SMP 1

6. **USE A MODEL** Adam and Josh competed in a bicycle race, which Josh won by 2 minutes. Adam rode his bike at a rate of 15 mph and Josh at 20 mph. Write a system of equations and any other constraints needed to find the time that Adam and Josh took to complete the race. CCSS A.CED.2, A.CED.3, SMP 4

4.2 Solving Systems by Graphing

Objectives

- Create equations in two variables.
- Solve systems of linear equations in two variables by graphing.

CCSS STANDARDS

Content: A.CED.2, A.REI.6
Practices: 1, 3, 4, 5, 6
Use with Lessons 6-1, 6-5

To solve a system of two linear equations by graphing, graph both equations on the same coordinate grid. The coordinates of the point of intersection of the two lines are the solution of the system of equations. Verify the solution by substituting the values for x and y in both of the equations.

EXAMPLE 1 Use a Graph to Solve a System of Equations CCSS A.REI.6, SMP 1

EXPLORE Consider the graphs of the equations $y = 1.5x - 3$ and $y = -2x + 7.5$.

a. INTERPRET PROBLEMS Find the solution of the system of equations. Why is it not obvious?

b. Substitute the point of intersection from **part a** into both equations to check your solution. Does your answer make both equations true?

c. Kaneo graphed these equations and estimated that the point of intersection was at (3, 1.4). Evaluate his approximation. What could Kaneo do to make his approximation more accurate?

EXAMPLE 2 Solve a System of Linear Equations by Graphing

A hardware store sells widgets for x dollars each and gadgets for y dollars each.

a. USE A MODEL Estefan bought 1 widget and 2 gadgets and paid \$5. Define the variables. Write and graph an equation to represent the cost of widgets and gadgets. CCSS A.CED.2, SMP 4

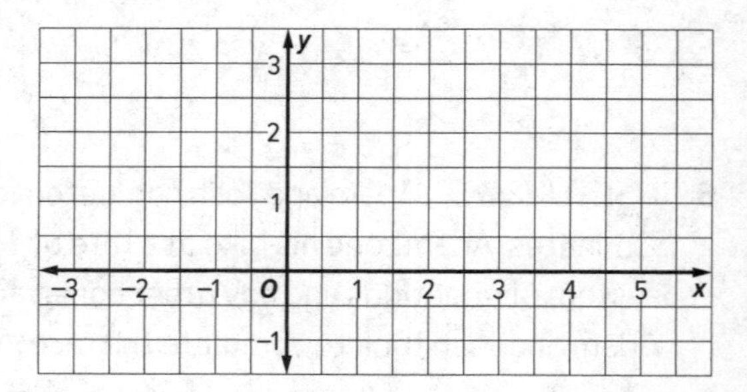

b. **USE A MODEL** Ben bought 2 widgets and returned 5 gadgets. He paid the store \$1. Use this information to write an equation to represent the cost of widgets and gadgets. Graph the equation on the same grid that you used for **part a.** CCSS A.CED.2, SMP 4

c. **USE A MODEL** How much does each widget and gadget cost? Check your solution. CCSS A.REI.6, SMP 4

d. **CONSTRUCT ARGUMENTS** Seth graphed his equations for this problem and found a point of intersection at (7, −1). What constraints on the variables alert him that he made an error? Explain. CCSS A.REI.6, SMP 3

EXAMPLE 3 Solve Systems of Linear Equations CCSS A.REI.6, SMP 3

CONSTRUCT ARGUMENTS Graph each system of equations and identify all solutions. Give a reason for the number of solutions.

a. $x - 2y = 2$
$1.5x - 3y = 3$

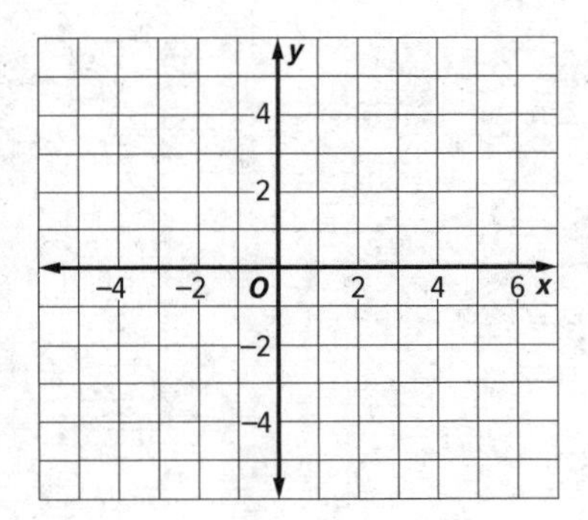

b. $-2x + y = 6$
$x + 2y = 7$

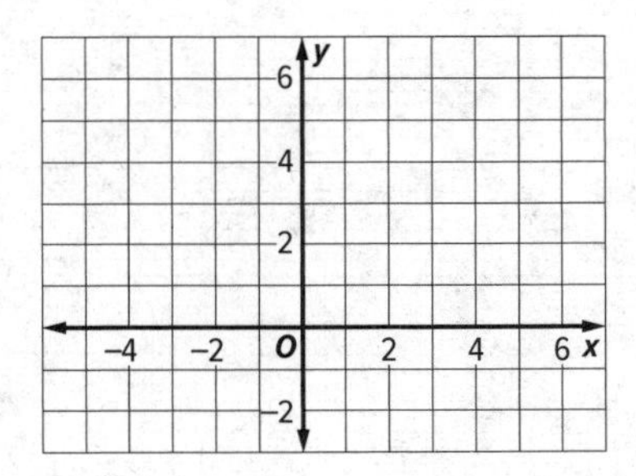

c. $3x - y = 1$
$y - 6 = 3x$

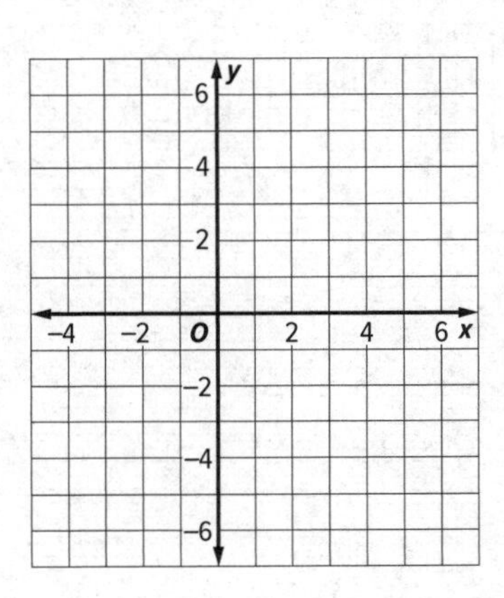

PRACTICE

INTERPRET PROBLEMS Solve each system of equations graphically and write the solution. Verify the answer is correct. CCSS A.REI.6, SMP 1

1. $2x - y = -3$
$2x + y = -1$

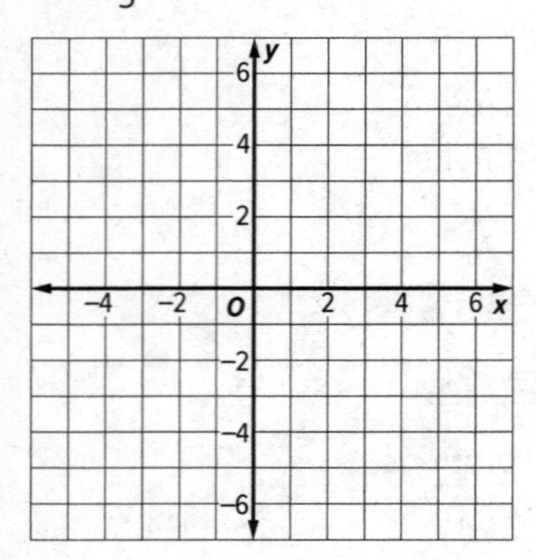

2. $2x + y = 4$
$y = -2x - 2$

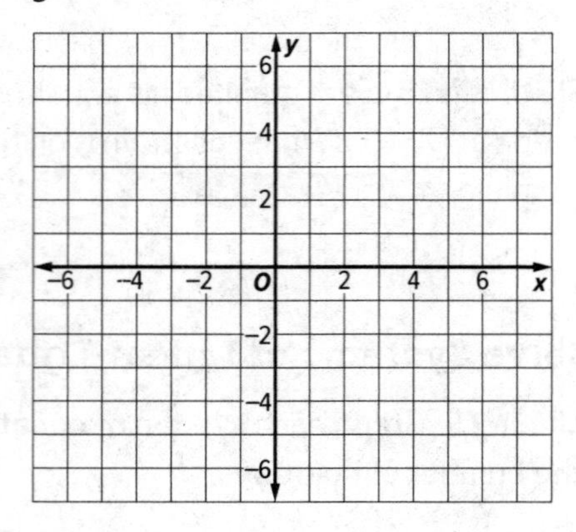

3. $2y = 5 + x$
$3x - 6y = -15$

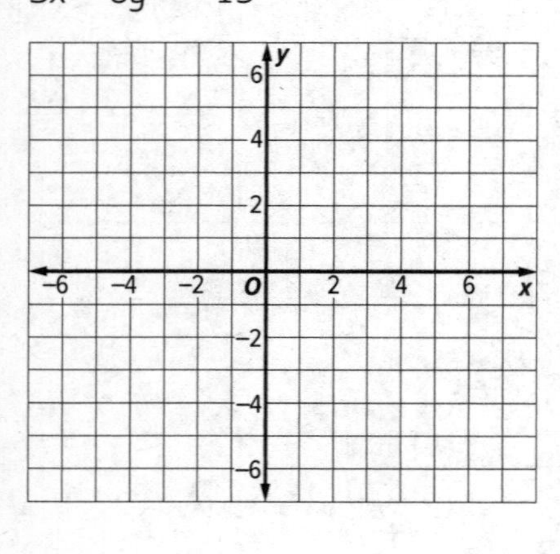

4. $2x - y = 5$
$x + y = -2$

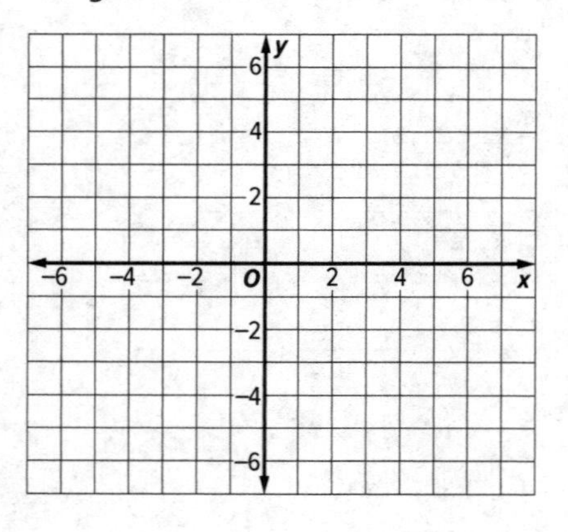

USE A MODEL Define variables and write a system of equations to model each problem. Then solve the system graphically to answer the question. CCSS A.CED.2, A.REI.6, SMP 4

5. Gustavo sets up tables for a caterer on weekends. Each round table can seat 8 people. Each square table can seat 10 people. One weekend, his boss asked him to set up tables for 124 people. He uses 2 more round tables than square tables. How many round tables did Gustavo use?

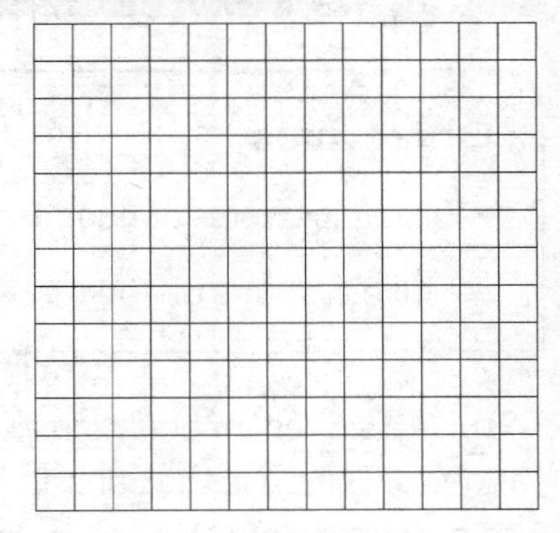

6. Cayley walks to and from school some days. On other days, she rides her bicycle. When she walks to school, she needs to leave home 15 minutes earlier than when she rides her bicycle. Cayley walked 3 days last week and rode her bike on 2 days. If Cayley spent a total of 2 hours 20 minutes going to and from school last week, how long does it take Cayley to walk to school?

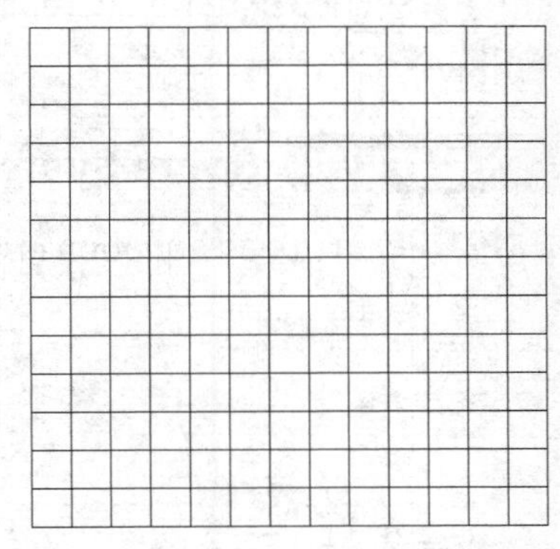

7. **CRITIQUE REASONING** Al uses a calculator to graph the system $y = 1.21x - 3$ and $5y - 5 = 6x$. He concludes that the system has no solution. Would you agree? Explain. CCSS A.REI.6, SMP 3

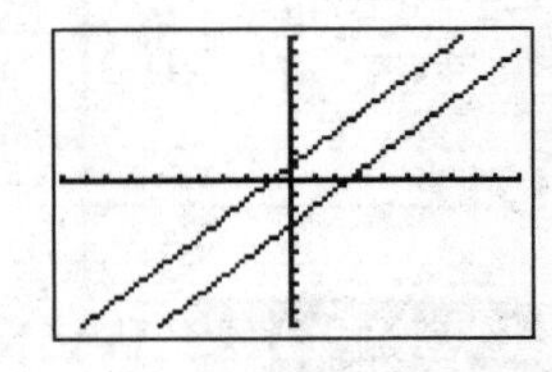

8. **CRITIQUE REASONING** Maureen says that if a system of linear equations has three equations, then there could be exactly two solutions to the system. Do you agree? Explain. Use a graph or graphs to justify your reasoning. CCSS A.REI.6, SMP 3

4.3 Solving Systems by Substitution

CCSS STANDARDS

Content: A.CED.2, A.REI.6
Practices: 1, 2, 3, 4, 6, 7
Use with Lessons 6-2, 6-5

Objectives

- Create systems of equations in two variables.
- Solve systems of equations in two variables by using substitution.

To solve a system of equations by substitution, first solve one equation for one of the variables. Then, use the Substitution Property of Equality and substitute that expression into the second equation to replace the variable and obtain a single equation with one variable.

EXAMPLE 1 Solve Systems of Equations Using Substitution

EXPLORE Write an equation in one variable that you could use to solve for x. CCSS A.REI.6, SMP 1

a. $y = -2x + 3$
$3x + 4y = -6$

b. $5x - y = 2$
$3x + 2y = 18$

c. $3x + 3y = 15$
$4x + 3y = 20$

d. $3x + 4y = 8$
$5x + 2y = -3$

EXAMPLE 2 Write and Solve Systems of Equations in Two Variables

Jamal and Marco each chose a number. Jamal says: *My number is a whole number. The sum of my number and* 3 *times Marco's number is* 6. Marco says: 4 *times the sum of Jamal's number and* 6 *times my number equals* 32 *minus twice Jamal's number.*

a. REASON QUANTITATIVELY Create a system of linear equations that you can use to find the two numbers. Remember to define the variables and to list any constraints. CCSS A.REI.6, SMP 2

Variables	**Constraints**	**Equations**

b. USE STRUCTURE Solve the system of equations that you wrote in **part a**, using substitution. What are Jamal's and Marco's numbers? Show your work.

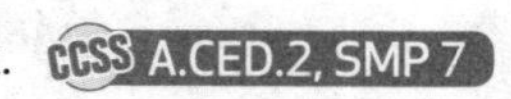

CCSS A.CED.2, SMP 7

EXAMPLE 3 Solve Real-World Problems Using a System of Equations

a. USE A MODEL The table shows Sal and Kurt's current savings and the amount that each plans to add to savings each week. Let y represent the total savings after x weeks. Write a system of equations to represent the total savings for each boy after x weeks. CCSS A.CED.2, SMP 4

Name	Current Savings	Weekly Contributions
Sal	\$60	\$4
Kurt	\$80	\$3

b. INTERPRET PROBLEMS Solve the system of equations and interpret its meaning. CCSS A.REI.6, SMP 1

c. USE A MODEL If \$250 is required to purchase a new tablet, which person will be able to buy a tablet first? At the end of which week will he be able to buy the tablet? CCSS A.REI.6, SMP 4

d. CRITIQUE REASONING Kurt decides to increase his weekly savings to \$3.50. He claims that doing so will mean that their savings will be equal sooner than if he saved only \$3 per week. Do you agree? Justify your answer. CCSS A.REI.6, SMP 3

EXAMPLE 4 Special Cases of Systems of Equations

INTERPRET PROBLEMS A city is planning some rectangular gardens. Write and solve a system of equations to find the dimensions of each garden. Describe any constraints. CCSS A.CED.2, A.REI.6, SMP 4

a. The perimeter is 20 meters, and the length is 10 meters greater than the width.

b. The perimeter is 48 yards. Five times the sum of the length and width is 120 yards.

PRACTICE

1. **CALCULATE ACCURATELY** Solve the systems of equations by using substitution. CCSS A.REI.6, SMP 6

a. $x = y - 6$
$2x - y = 6$

b. $x - 3y = -13$
$x + y = 3$

c. $2x + y = 2.4$
$x - 2y = 1.7$

d. $2x - 3y = 26$
$3x - 4.5y = 39$

e. $2x + 3y = 16$
$3x + 6y = 30$

f. $4x + 3y = 25$
$3x = x + 2y - 19$

g. $2y + 3x = 7$
$3x - 2y = -1$

h. $\frac{4}{3}x + y = 8$
$x = 3y + 6$

2. **USE A MODEL A zoo keeps track of the number of visitors to each exhibit. The table shows the number of visitors for two exhibits on one day.** CCSS A.CED.2, A.REI.6, SMP 4

	Big Cats	Petting Zoo
Adults	x	21
Children	1024	y

a. Three times the total number of adults was 17 less than the number of children who visited the petting zoo. Write an equation to model this relationship.

b. The total number of children who visited the zoo that day was 681 less than 10 times the number of adults who visited the big cats. Write an equation to model this relationship.

c. Solve the system of equations to find the total number of visitors to those two exhibits on that day.

d. Adults pay $25 and children pay $10 for tickets to the zoo. The total ticket sales one day were $47,750. The number of children who visited the zoo was 270 more than 6 times the number of adults who visited. Write and solve a system of equation to find the total number of visitors to the zoo that day.

3. USE STRUCTURE A two-digit number is reduced by 45 when the digits are interchanged. The digit in the ten's place of the original number is 1 more than 3 times the digit in the unit's place. Define variables and write a system to find the original number. CCSS A.CED.2, A.REI.6, SMP 7

4. USE A MODEL The perimeter of $\triangle ABC$ is 24.3 centimeters. The ratio of b to c is $4:5$. Write and solve a system of equations to find the values of b and c. CCSS A.CED.2, A.REI.6, SMP 4

A
c
b
B 9 cm C

5. INTERPRET PROBLEMS The table shows the number of boys and girls in the freshmen and sophomore classes. The ratio of freshmen to sophomores is $5:6$. In all, there are 4 more boys than girls. Write and solve a system to find the total number of students in the two classes. Describe your solution process.
CCSS A.CED.2, A.REI.6, SMP 1

	Boys	Girls
Freshmen	x	48
Sophomores	60	y

6. USE A MODEL Bella bought 4 notebooks and 5 pens for $15. Baxter bought 7 notebooks and 3 pens from the same store for $20.50. CCSS A.CED.2, A.REI.6, SMP 4

a. Define variables and write and solve a system of equation by substitution. Describe your solution process.

b. Susan has $17 to spend. Will she have enough money to purchase 5 notebooks and 4 pens? Explain your reasoning.

7. USE A MODEL A rectangular area has a diagonal walkway with two congruent triangular gardens on each side. The perimeter of the rectangle is 260 feet. The length of the rectangle is 10 feet longer than twice the width. CCSS A.CED.2, A.REI.6, SMP 4

a. Draw and label a diagram that you can use to write a system of linear equations to model this situation.

b. Write and solve equations to find the dimensions of the rectangle.

c. Find the area of each garden.

4.4 Solving Systems by Elimination

Objectives

- Create systems of equations in two variables to represent relationships between quantities.
- Solve systems of equations using elimination.

STANDARDS

Content: A.CED.2, A.REI.6
Practices: 1, 2, 3, 4, 7, 8
Use with Lessons 6-3, 6-5

EXAMPLE 1 Investigate a System of Equations

EXPLORE A store is having a back-to-school sale, with prices as shown at the right. Nick bought some pens and some notebooks. He spent a total of \$14. Sushila bought the same number of pens as Nick and twice as many notebooks. She spent a total of \$20.

Back-to-School Sale!	
Gel pens	\$2 each
Notebooks	\$3 each

a. USE A MODEL Define variables and write a system of equations to represent this situation. CCSS A.CED.2, SMP 4

b. USE STRUCTURE Write the two equations in the spaces provided below. Align the terms that have the same variables. Then subtract Nick's equation from Sushila's equation, working term by term from left to right. CCSS A.CED.2, SMP 7

Sushila's Equation →

Nick's Equation →

Difference of the Equations →

c. USE STRUCTURE What happens when you subtract one equation from the other in **part b**? Why is this useful? CCSS A.REI.6, SMP 7

d. REASON QUANTITATIVELY Solve the equation in the last row of **part b**. What does the solution tell you? CCSS A.REI.6, SMP 2

e. REASON QUANTITATIVELY How many pens and notebooks did each person buy? Explain. CCSS A.REI.6, SMP 2

When you solve a system by **elimination**, you add or subtract the equations in order to eliminate one of the variables. Once you have done this, you can solve the resulting equation for the variable that remains. Finally, you can substitute the value of this variable in one of the original equations to solve for the other variable.

EXAMPLE 2 Solve a System Using Elimination CCSS A.REI.6

Solve the system of equations.

$4x + y = 11$

$2x - y = 7$

a. **PLAN A SOLUTION** Should you add or subtract the equations? Why? CCSS SMP 1

b. **USE STRUCTURE** What is the result of performing the operation you identified in **part a**? Explain how to use this result to solve the system. CCSS SMP 7

c. **INTERPRET PROBLEMS** Graph the system of equations on the coordinate plane at the right. Explain how to use the graph to check your solution. CCSS SMP 1

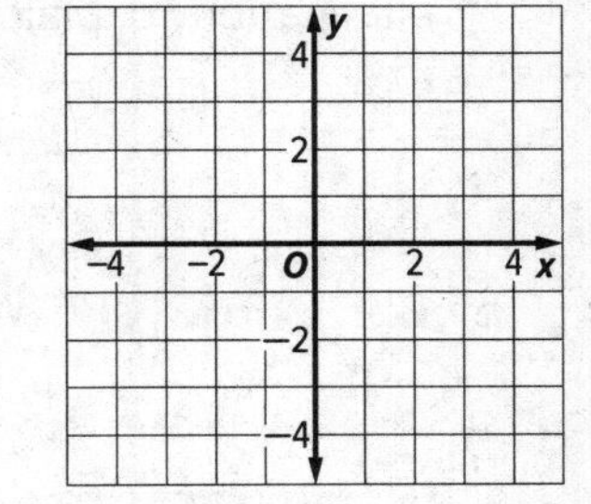

d. **INTERPRET PROBLEMS** What is a different way to check that your solution is correct? CCSS SMP 1

e. **USE STRUCTURE** Explain how you can solve the system of equations by substitution. CCSS SMP 7

f. **DESCRIBE A METHOD** Which method is best for solving this system: graphing, substitution, or elimination? Justify your choice. CCSS SMP 8

EXAMPLE 3 Use a System to Solve a Problem

The table shows the time Erin spent jogging and walking this weekend and the total distance she covered each day. Erin always jogs at the same rate and always walks at the same rate.

Day	Time Jogging	Time Walking	Total Distance
Saturday	15 min.	30 min.	3.5 mi.
Sunday	1 h	30 min.	8 mi.

a. USE A MODEL Write a system of equations to represent this situation. CCSS A.CED.2, SMP 4

b. USE STRUCTURE Explain how to solve the system by elimination. Then interpret the solution. CCSS A.REI.6, SMP 7

PRACTICE

1. Consider the system of equations $\begin{cases} 0.4x - 2y = 6 \\ 0.8x + 2y = 0 \end{cases}$. CCSS A.REI. 6

a. PLAN A SOLUTION What is the first step to solving the system of equations by elimination? Explain your reasoning. CCSS SMP 1

b. USE STRUCTURE What is the solution? CCSS SMP 7

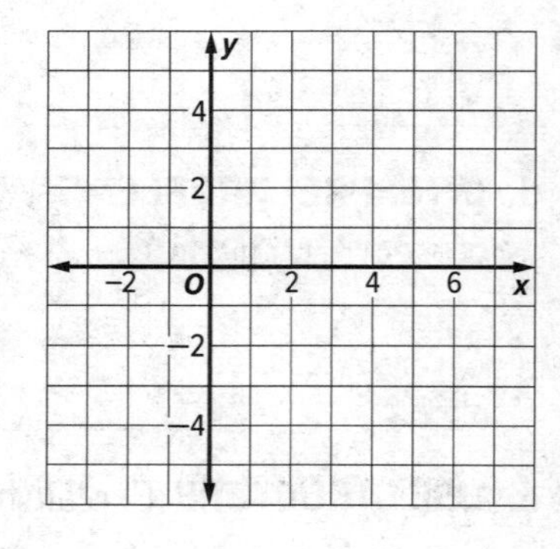

c. INTERPRET PROBLEMS Graph the equations on the coordinate plane at the right. Explain how you can use the graph to check your solution. CCSS SMP 1

d. INTERPRET PROBLEMS How would the solution change if the second equation was $x - 5y = 15$? Explain. CCSS SMP 1

e. INTERPRET PROBLEMS How would the solution change if the second equation was $0.4x - 2y = 0$? Explain. CCSS SMP 1

USE STRUCTURE **Solve each system of equations by elimination.** CCSS A.REI.6, SMP 7

2. $3x - 3y = -15$
$3x + 5y = 17$

3. $4m - n = 14$
$3m - n = 12$

4. $0.3x - 2y = -28$
$0.8x + 2y = 28$

5. $-2a + b = 1$
$2a + 9b = -51$

6. $\frac{1}{2}q + 4r = -2$
$\frac{1}{6}q - 4r = 10$

7. $\frac{1}{2}x + \frac{1}{3}y = -1$
$-\frac{1}{2}x + \frac{2}{3}y = 10$

8. Jeremy and Kendrick each bought snacks for their friends at a skating rink. The table shows the number of bags of popcorn and the number of plates of nachos each person bought, as well as the total cost of the snacks.

Name	Bags of Popcorn	Plates of Nachos	Total Cost
Jeremy	4	2	$18.50
Kendrick	7	2	$26.75

a. USE A MODEL Write a system of equations that you can use to find the prices of the popcorn and the nachos. Describe what each variable represents. CCSS A.CED.2, SMP 4

b. USE STRUCTURE Solve the system of equations and explain what your solution represents. CCSS A.REI.6, SMP 7

9. Marisol works for a florist that sells two types of bouquets, as shown at the right. On Monday, Marisol used 96 tulips to make the bouquets. On Tuesday, she used 192 tulips to make the same number of Spring Mix bouquets as Monday, but 3 times as many Garden Delight bouquets.

Seasonal Bouquets	
Spring Mix	12 tulips
Garden Delight	16 tulips

a. USE A MODEL Write a system of equations that you can use to find how many bouquets of each type Marisol made. Describe what each variable represents. CCSS A.CED.2, SMP 4

b. CONSTRUCT ARGUMENTS Find the total number of tulips Marisol used to make Garden Delight bouquets on Monday and Tuesday. Explain your answer. CCSS A.REI.6, SMP 3

4.5 Solving Systems by Elimination Using Multiplication

Objectives

- Create systems of equations in two variables to represent relationships between quantities.
- Solve systems of equations by elimination using multiplication.

CCSS STANDARDS

Content: A.CED.2, A.REI.5, A.REI.6

Practices: 1, 2, 3, 4, 7, 8

Use with Lessons 6-4, 6-5

EXAMPLE 1 Write a Proof CCSS A.REI.5

EXPLORE Given the system of two equations in two variables at the right, prove that you can replace one of the equations by a sum of that equation and a multiple of the other equation to produce a new system with the same solution as the first system.

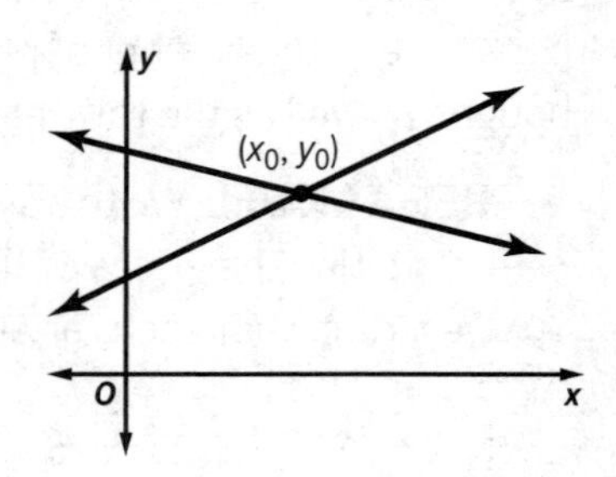

a. INTERPRET PROBLEMS Suppose the point (x_0, y_0) is a solution to the given system of equations. Write two equations by substituting the coordinates of the solution into the given system of equations. CCSS SMP 1

____

b. REASON ABSTRACTLY Create a new system of equations as follows: (1) Multiply both sides of the first equation in the system by k. (2) Add the resulting equation to the second equation in original system. (3) Use the sum to replace the second equation in the original system. What properties justify Steps 1 and 2? CCSS SMP 2

____

____

① $ax + by = c$
$dx + ey = f$

____

② $dx \quad + ey \quad = f$

____

③ $ax + by = c$

____

c. CONSTRUCT ARGUMENTS Show that (x_0, y_0) is a solution to the new system of equations. To do this, show that (x_0, y_0) is a solution to each equation in the system. The first equation has not changed, so (x_0, y_0) is still a solution. What do you need to do in order to show that (x_0, y_0) is a solution to the second equation? CCSS SMP 3

____

d. CONSTRUCT ARGUMENTS Explain why the relationship you identified in **part c** is true. CCSS SMP 3

____

____

____

e. CONSTRUCT ARGUMENTS Explain why, in terms of the graph of $ax + by = c$, multiplying the equation by k has no effect on the solution to the system of equations. CCSS SMP 3

____

____

To solve any system by elimination, you can multiply at least one of the equations by a constant to get two equations that contain opposite terms. Then you add the equations to eliminate one of the variables. Solve the resulting equation for the remaining variable. Finally, substitute the value of this variable into one of the original equations to solve for the other variable.

EXAMPLE 2 Solve a System Using Elimination CCSS A.REI.6

Follow these steps to solve this system of equations.

$3x + 4y = 5$

$x - 8y = -17$

a. **PLAN A SOLUTION** Explain how you can multiply one of the equations in the system by a constant to get two equations that contain opposite terms. Is there another possible answer? Explain. CCSS SMP 1

b. **USE STRUCTURE** Perform one of the operations you identified in **part a** to obtain a new system. Then solve the new system. CCSS SMP 7

c. **EVALUATE REASONABLENESS** Graph the system of equations on the coordinate plane at the right. Explain how to use the graph to check your solution. CCSS SMP 8

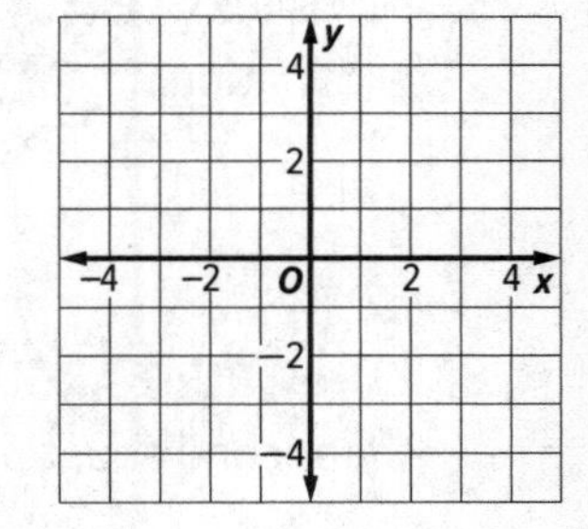

d. **EVALUATE REASONABLENESS** What is a different way to check that your solution is correct? CCSS SMP 8

e. **CRITIQUE REASONING** Kaelyn solves the system by multiplying the second equation in the system by 3. She then adds the equations and solves for y, getting $y = 2.3$. Explain the error she made and correct it. CCSS SMP 3

EXAMPLE 3 Use a System to Solve a Problem

Roberto is buying posters for his college dorm room. The prices of the posters are shown at the right. Roberto decides to buy 9 posters, and he spends a total of $58.10.

Poster Sale	
Large posters	$8.75 each
Small posters	$5.80 each

a. USE A MODEL Write a system of equations to find the number of large and small posters that Roberto buys. Explain what the variables represent. CCSS A.CED.2, SMP 4

b. USE STRUCTURE Show how to solve the system by elimination. Then interpret the solution. CCSS A.REI.6, SMP 7

c. EVALUATE REASONABLENESS Explain how you know your answer is correct. CCSS A.REI.6, SMP 8

PRACTICE

1. Consider the system of equations $\begin{cases} -2x + 3y = -5 \\ 3x - 4y = 6 \end{cases}$.

a. PLAN A SOLUTION Describe two different ways to solve the system by elimination. CCSS A.REI.6, SMP 1

b. CONSTRUCT ARGUMENTS Explain why multiplying the first equation by 6 and the second equation by 5 and then adding is not useful for solving the system. CCSS A.REI.6, SMP 3

c. USE STRUCTURE Solve the system by elimination. CCSS A.REI.6, SMP 7

d. EVALUATE REASONABLENESS Graph the equations on the coordinate plane at the right. Explain how to use the graph to check your solution. CCSS A.REI.6, SMP 8

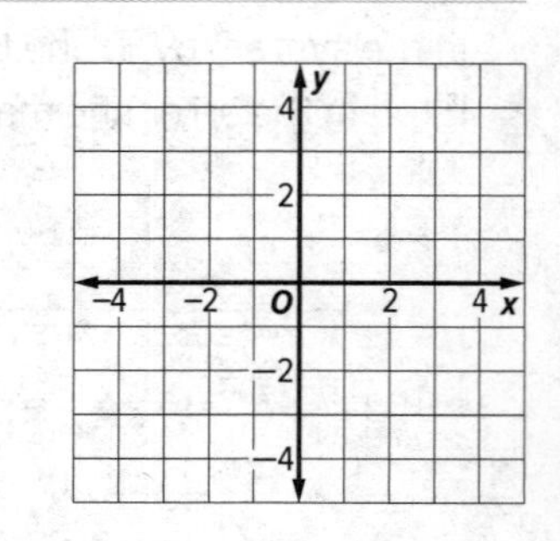

USE STRUCTURE **Solve each system by elimination.** CCSS A.REI.6, SMP 7

2. $2x - 3y = 14$
   $6x + y = 2$

3. $5c - 3d = 49$
   $-2c + d = -19$

4. $-m + 3n = 17$
   $4m - 6n = -38$

5. $3p + 5q = -20$
   $2p - 3q = -7$

6. $0.4x + 0.5y = 3.8$
   $0.8x - 0.2y = 0.4$

7. $3x + y = \frac{1}{2}$
   $-6x + \frac{1}{2}y = -\frac{9}{4}$

8. Megan works as a cashier at a grocery store. At the end of the day, she has a total of 125 five-dollar bills and ten-dollar bills. The total value of these bills is $990.

   a. **USE A MODEL** Write a system of equations that you can use to find the number of five-dollar bills and the number of ten-dollars bills. Tell what the variables represent. CCSS A.CED.2, SMP 4

   b. **USE STRUCTURE** Solve the system of equations and explain what your solution represents. CCSS A.REI.6, SMP 7

9. The owner of a juice stand wants to make a new juice drink. He would like to mix Tropical Breeze and Kona Cooler to make 10 quarts of a new drink that is 40% pineapple juice.

Juice Drinks	
Tropical Breeze	20% pineapple juice
Kona Cooler	50% pineapple juice

   a. **PLAN A SOLUTION** Complete the table. CCSS A.CED.2, SMP 1

	Tropical Breeze	Kona Cooler	Total
Amount of Juice (qt)	t	k	10
Amount of Pineapple Juice (qt)			

   b. **USE A MODEL** Write a system of equations that the owner of the juice stand can solve to determine the amount of each drink he should use to make the new drink. CCSS A.CED.2, SMP 4

   c. **USE STRUCTURE** Solve the system and explain what your solution represents. CCSS A.REI.6, SMP 7

   d. **EVALUATE REASONABLENESS** Explain how you know your answer is correct. CCSS A.REI.6, SMP 8

4.6 Solving Systems of Inequalities

Objectives

- Solve systems of linear inequalities by graphing.
- Apply systems of linear inequalities to a real-world context.

Content: A.REI.12
Practices: 1, 2, 4, 5, 7
Use with Lesson 6-6

A solution of a system of linear inequalities in two variables is any ordered pair that satisfies all of the inequalities in the system. Graphically, the solution to the system is the intersection of the shaded regions of each inequality.

EXAMPLE 1 Use Technology to Graph a System of Linear Inequalities CCSS REI.12, SMP 5

USE TOOLS You can use technology, such as a graphing calculator, to graph the system of linear inequalities.

$$y \geq 3 - x$$
$$y \leq x + 1$$

a. Enter the expression $3 - x$ into your graphing calculator. How can you adjust the setting so that the graph represents the inequality $y \geq 3 - x$?

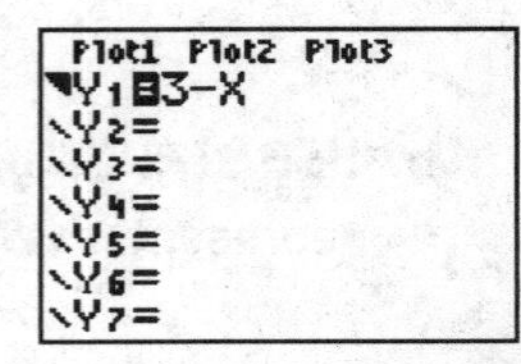

b. Graph the inequality. How does the graphing calculator represent the solution?

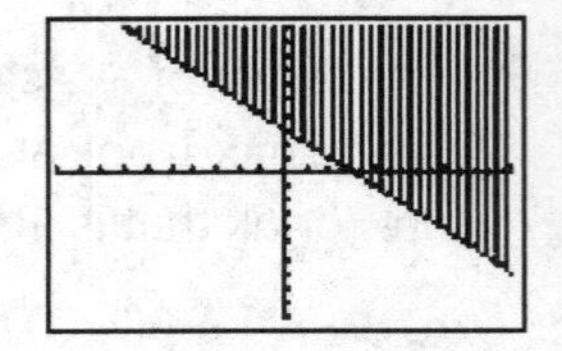

c. Repeat **parts a** and **b** to graph $y < x + 1$. In the calculator screens shown, how is each type of inequality distinguished? How is the solution shown on the plane?

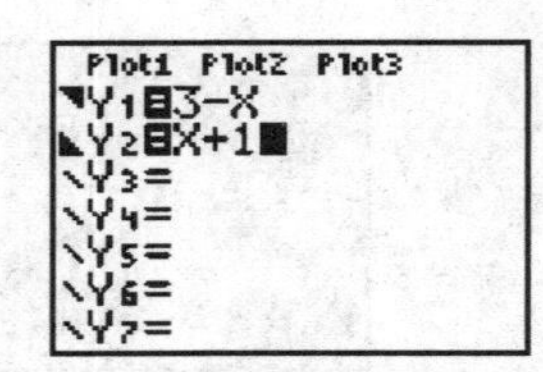

d. Could the calculator distinguish the solutions if the symbols $\leq$ and $\geq$ were replaced by $<$ and $>$? Explain.

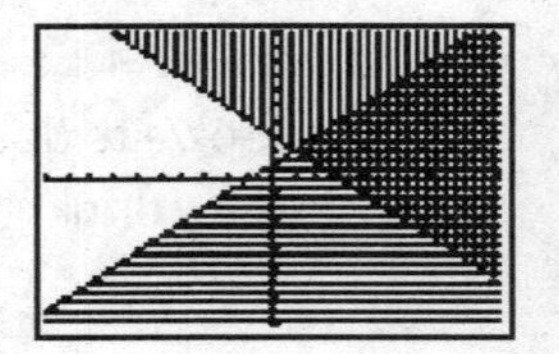

e. Using the calculator screen show, describe the region that represents solutions to the system of linear inequalities. Explain.

$y < 3 - x$

$y > x + 1$

EXAMPLE 2 Graph a System of Linear Inequalities 

USE TOOLS **Graph and describe the solution set of each system of linear inequalities.**

a. $x + 2y < 4$
$x - 2y \geq -4$

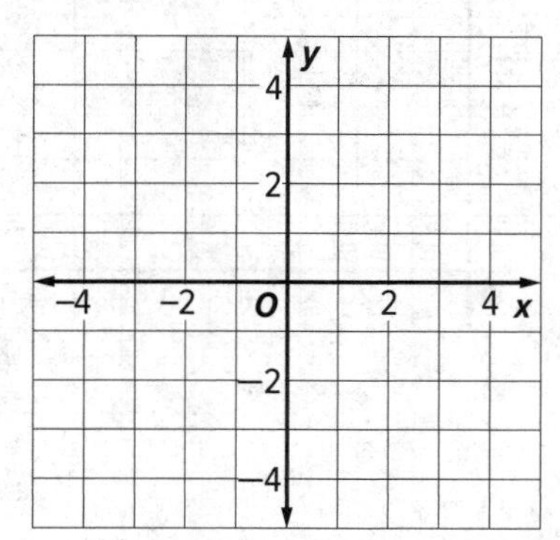

b. $y + 2 < 3x$
$3x + y > 1$

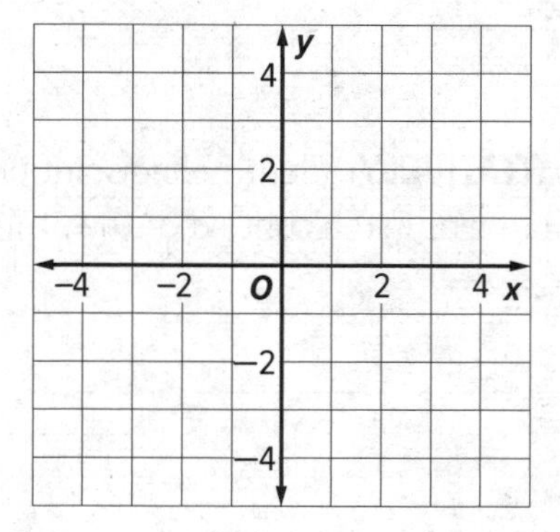

c. $x - 2y < -6$
$x - 2y \geq -4$

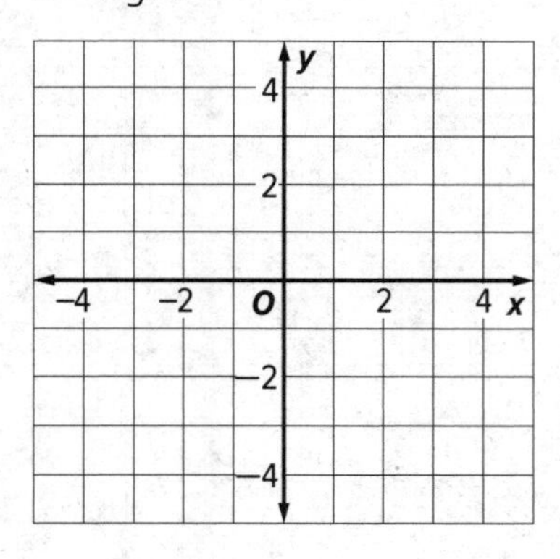

d. $x + 3y \leq 3$
$x + 3y > -3$

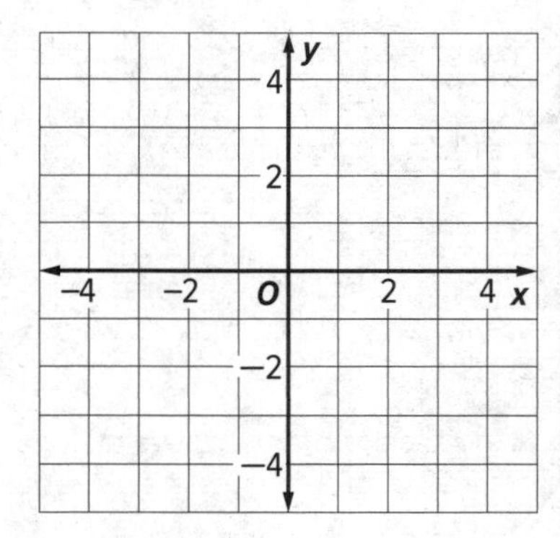

EXAMPLE 3 Write and Solve a System of Linear Inequalities CCSS REI.12

The drama club is sponsoring a trip to the theater. Adult theater tickets cost \$12, and student tickets cost \$8. The total cost for tickets must not exceed \$480. The bus can hold up to 50 passengers. There must be at least one adult chaperone for every six students.

a. USE A MODEL Write a system of inequalities to model this problem and include constraints. Let x represent the number of students and y represent the number of adults. State any other restrictions on the values of the variables. CCSS SMP 4

b. USE TOOLS Graph the solution of the system of inequalities. Describe the solution region in words. CCSS SMP 5

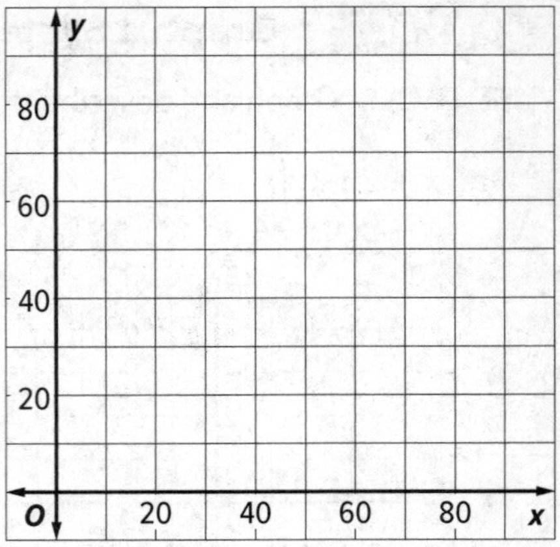

c. REASON QUANTITATIVELY Identify one solution of the system. Describe what it means in the context of the problem. CCSS SMP 2

EXAMPLE 4 Write a System of Inequalities from a Graph CCSS REI.12, SMP 2

REASON ABSTRACTLY Write a system of inequalities whose solution is represented by the shaded region of the graph. If this is not possible, explain why.

a.

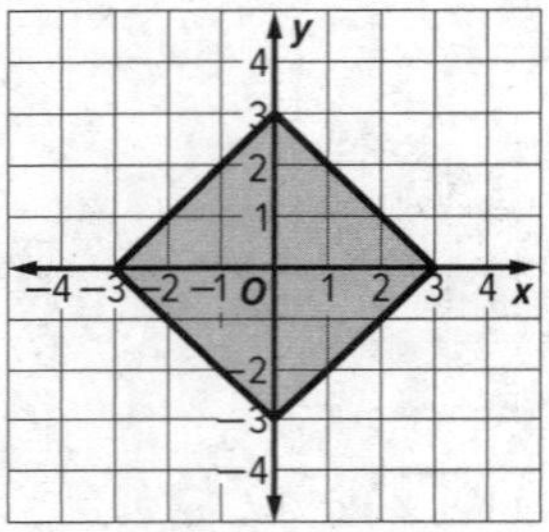

b.

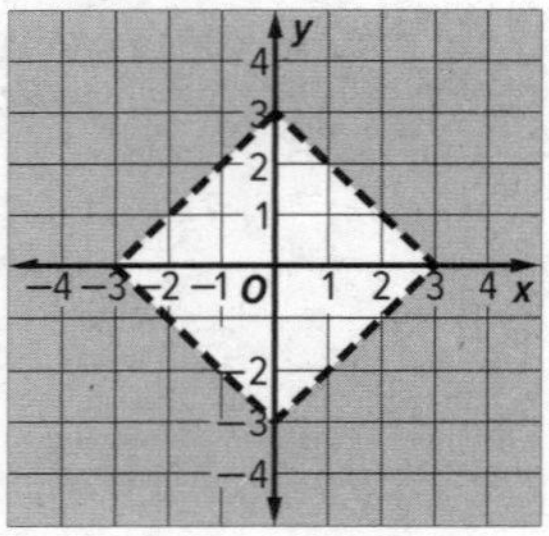

c.

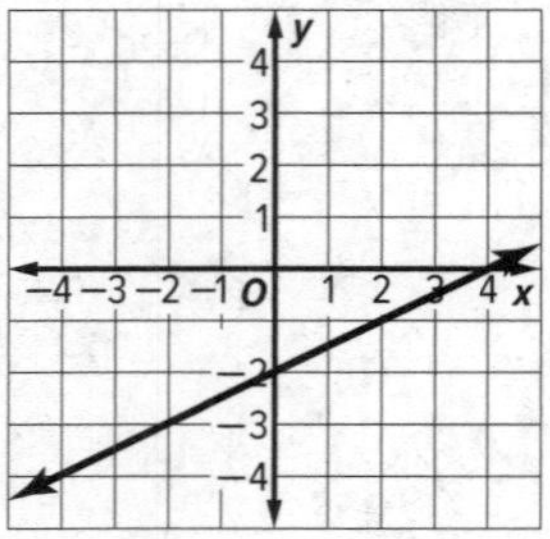

d. (0, −2)

PRACTICE

USE TOOLS **Graph and describe the solution set of each system of linear inequalities.** CCSS A.REI.12, SMP 5

1. $2x + y \geq 2$
$x - y > 1$

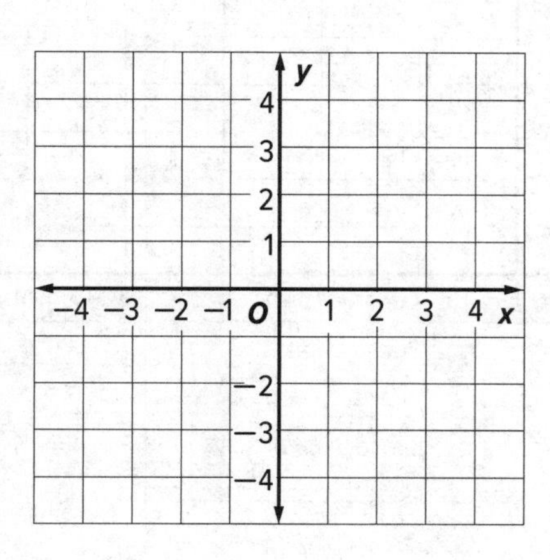

2. $x - 4y < -8$
$x - 4y \geq 4$

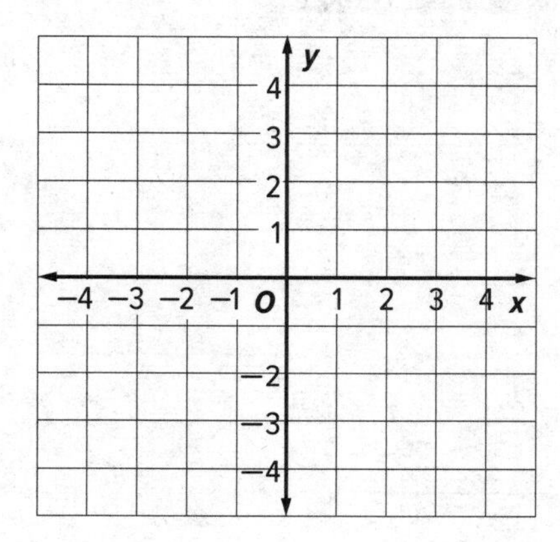

3. $x + y \geq 0$
$2x + y > -2$
$x + 3y > -3$

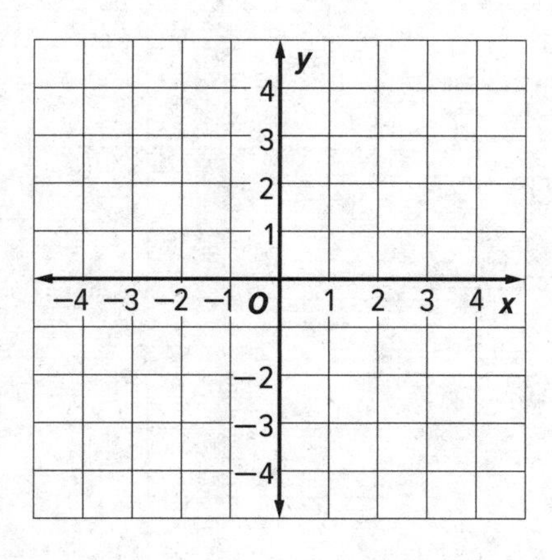

4. $y \geq 2$
$2x + y < -1$
$2x - y > 1$

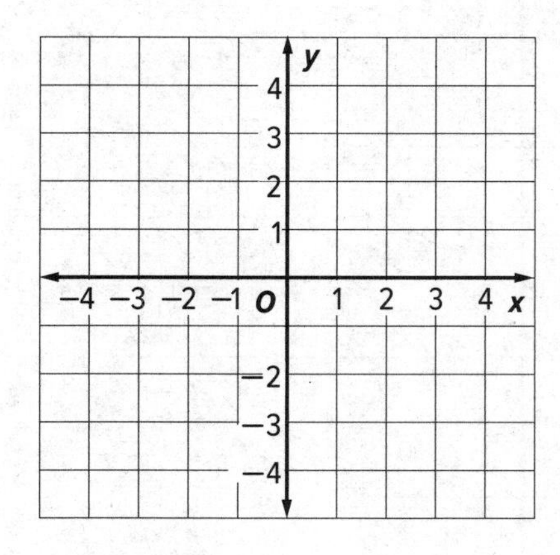

5. REASON ABSTRACTLY **Write a system of inequalities to represent the graph.** CCSS A.REI.12, SMP 2

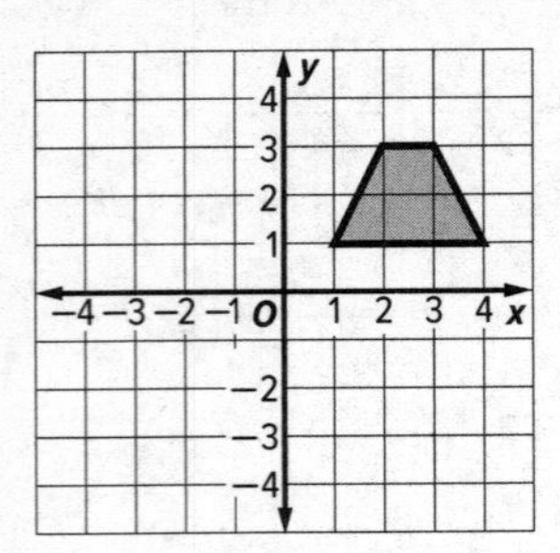

6. USE A MODEL Linda has \$30 to purchase pens and pencils for her class. A three-pack of pens costs \$2, and a dozen pencils costs \$3. Everyone needs at least one pen and two pencils. There are 15 students in her class including Linda. Write and graph a system of inequalities that models this situation and identify the solution set. CCSS A.REI.12, SMP 4

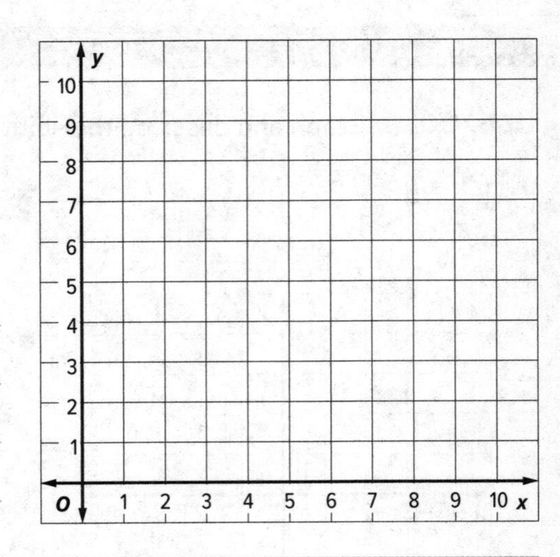

7. USE TOOLS Is (2.5, 1) a solution to the following system of inequalities?

$$2x + 3y > 8$$
$$4x - 5y \geq 2$$

Explain how to use a graph to decide. Then explain how you can tell if the point is a solution without graphing the inequalities. CCSS A.REI.12, SMP 5

8. PLAN A SOLUTION The solution of a system of inequalities is described as the points inside a right triangle, with one vertex at the origin. Write a system that could have this solution. Describe your method. CCSS A.REI.12, SMP 1

____

____

____

USE STRUCTURE Match each system of inequalities with its graph below. CCSS A.REI.12, SMP 7

9. $y \geq x$
$y \geq -x$
$y < 4$

______

10. $y \geq x$
$y \geq -x$
$2y - 4 < x$

______

11. $x > 1$
$x < 4$
$y < 3$
$y > -1$

______

12. $y < x + 4$
$y < -x + 4$
$y \geq x$
$y \geq -x$

______

a.

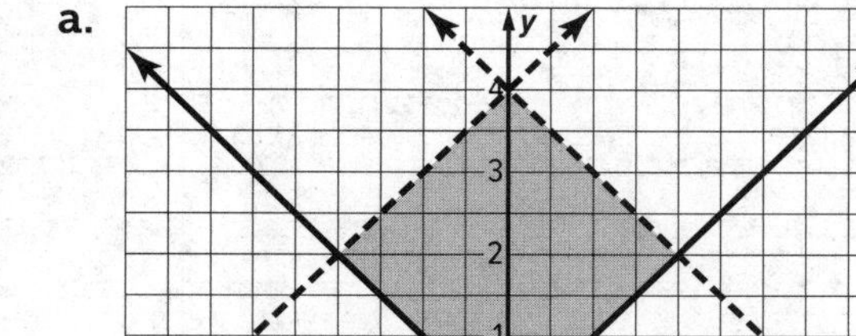

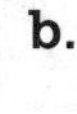

b.

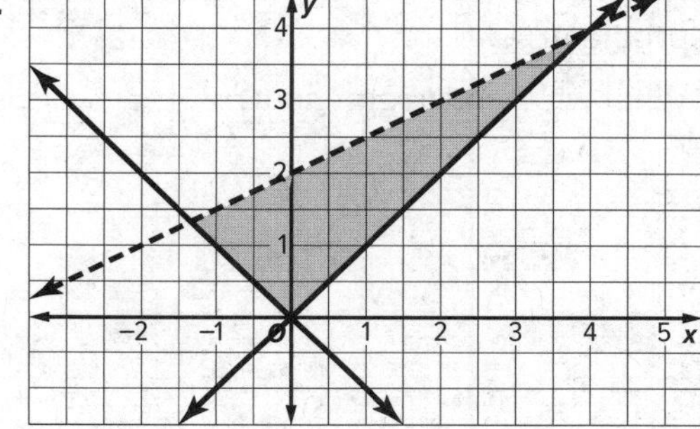

c.

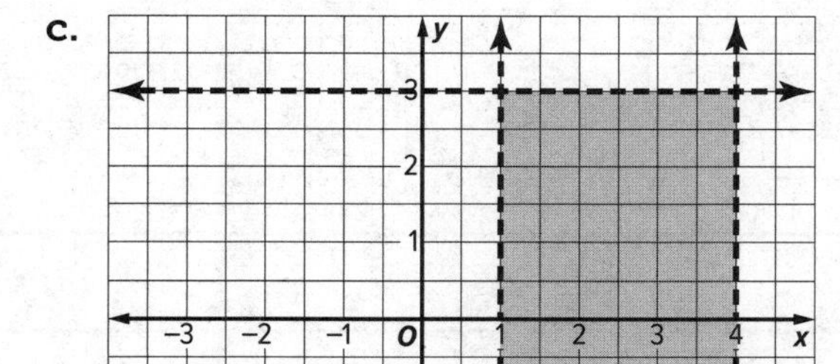

d.

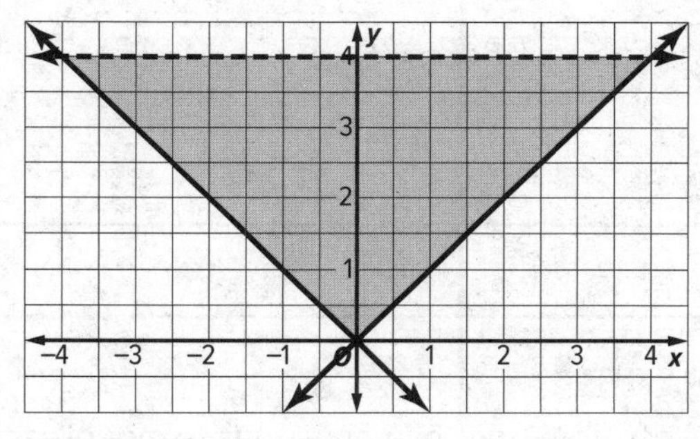

13. REASON ABSTRACTLY Write a system of inequalities that defines all of the points in the described region. CCSS A.REI.12, SMP 2

a. all points in the second quadrant, exclusive of the axes

____

b. all points in the third quadrant, exclusive of the axes

____

c. all points in the fourth quadrant and including the axes

____

d. all points in the first quadrant, including the horizontal axis, but excluding the vertical axis

____

Performance Task

Finding a Break-Even Point

Provide a clear solution to the problem. Be sure to show all of your work, include all relevant drawings, and justify your answers.

Avery is considering whether to buy solar panels for her house. It will cost \$18,000 to buy the panels and have them installed. The panels will produce a maximum of 600 kilowatt-hours (kWh) each month. She predicts that she will save \$0.12 from her power bill for each kilowatt-hour her panels produce. How many months will it be until she saves the amount that she spent on the panels?

Part A

Make a plan to find the answer. Consider these questions:

- Which method of solving systems of equations/inequalities would be best for this problem?
- What will x and y represent?

Part B

Write equations or inequalities to model each of the following.

- The amount of money Avery will spend on the solar panels
- The amount of money Avery will save by using the panels

Record the domain and range for each equation or inequality.

Part C

Graph the system of equations and inequalities that you wrote in **Part B**. Make sure to include a scale and labels.

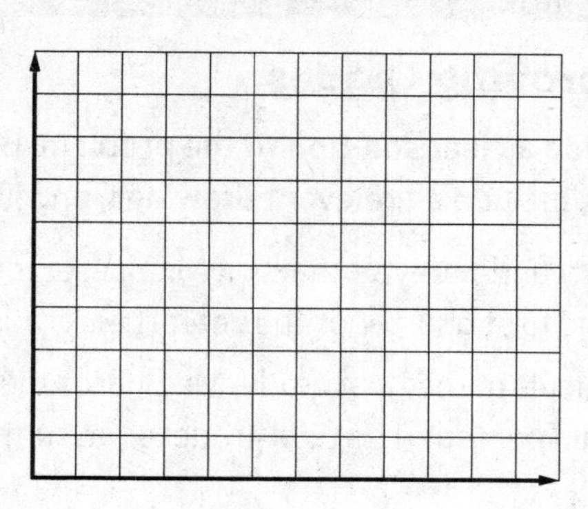

Part D

Interpret your graph to answer the question.

Part E

Suppose that after 10 years of use, the solar panels are less effective. As a result her panels can only produce a maximum of $0.05 per kilowatt-hour. Explain how the equations, inequalities, or graph would be different. How would this change the number of months it takes her to save enough money to cover the cost of the panels? Explain.

Performance Task

Improving Grades

Provide a clear solution to the problem. Be sure to show all of your work, include all relevant drawings, and justify your answers.

Quincy and Shreya's teacher, Mr. Suarez, calculated their math grades for the first quarter of the year. Their grades are shown in the table.

To calculate the final grade, Mr. Suarez multiplies each score by a certain percent. The final grade is the sum of the two products.

	Quincy	Shreya
Average before exam	88	84
Exam score	92	96
Final grade	88.8	86.4

Part A

Write a system of equations to model Quincy and Shreya's grades for the first quarter. Remember to define the variables and list any constraints.

Part B

Solve the system of equations you wrote in **Part A**. By what percent does Mr. Suarez multiply the average before the exam? By what percent does he multiply the exam score?

Part C

The school uses this table to convert percentages to letter grades.

Next quarter, Quincy wants to improve his grade. Write an inequality to show how he could get an A– next quarter. Let x be his grade average before the exam and y be his score on the exam.

Percent	Letter Grade
96 to 100	A+
93 to 95	A
90 to 92	A–
86 to 89	B+
83 to 85	B
80 to 82	B–

Part D

Graph the inequality you wrote in **Part C**. Be sure that your graph reflects all the constraints on the variables.

- What is the minimum average, before the exam, that Quincy needs to achieve his goal?
- If his minimum average is achieved, what must he score on the exam to reach his goal?

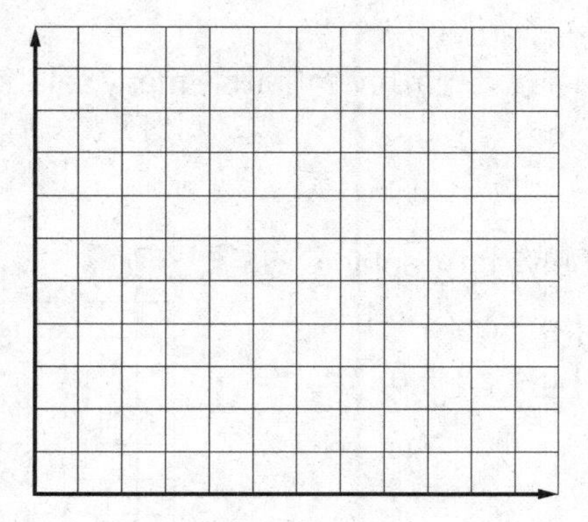

Standardized Test Practice

1. Solve by substitution. CCSS A.CED.6

$$\begin{cases} y = 2x + 3 \\ y = 5x - 6 \end{cases}$$

(☐ , ☐)

2. Select the statements that best describe the following system of linear equations. CCSS A.CED.3

$$\begin{cases} y = \frac{2}{3}x - 1 \\ 4x - 6y = 6 \end{cases}$$

- The lines are parallel.
- The system has exactly one solution.
- The lines are coincident.
- The system has no solutions.
- The system has infinitely many solutions.

3. Solve by graphing. CCSS A.CED.6

$$\begin{cases} y = -2x + 1 \\ y = 3x + 6 \end{cases}$$

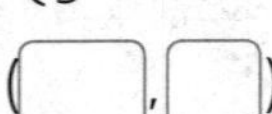

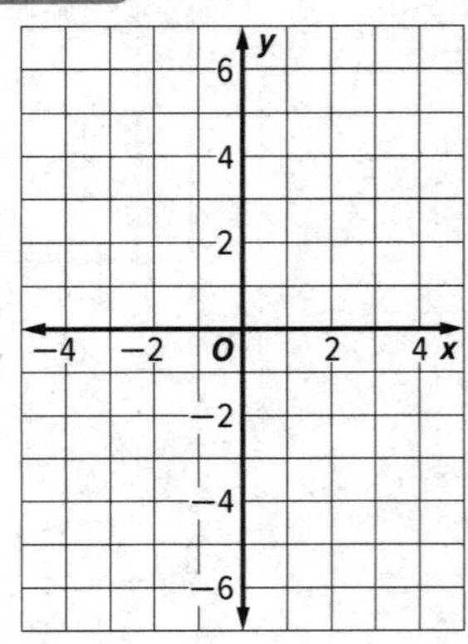

4. Solve by elimination. CCSS A.CED.6

$$\begin{cases} 3x + 2y = -4 \\ -7x - 2y = -4 \end{cases}$$

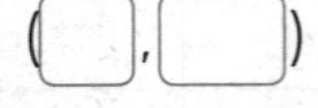

5. The table shows the cost of renting ice skates and the hourly rate for skating at two ice rinks.

	Skate Rental ($)	Hourly Skating Fee ($)
Rink A	5.00	1.50
Rink B	4.25	1.75

Write a system of linear equations that represents the total cost of renting skates and skating for x hours at each rink. CCSS A.CED.2

____

6. Select points that are solutions to the following system of linear inequalities. CCSS A.REI.12

$$\begin{cases} 2y - 5x > -7 \\ y \leq 3x - 1 \end{cases}$$

(3, 1)	(0, −2)
(−1, −4)	(5, 10)
(3, 4)	(−5, −16)

7. Graph the system of linear inequalities. CCSS A.REI.12

$$\begin{cases} 2x - y \geq 5 \\ 4y + 3x > 8 \end{cases}$$

y: 8, 4, −4, −8; x: −8, −4, O, 4, 8

8. Malik works at a hardware store for $12 per hour and as a lifeguard for $18 per hour. He needs to earn at least $280 per week, but he can work no more than 20 hours in total.

a. Write a system of inequalities that represents this situation. CCSS A.CED.2

b. Graph the system of inequalities. CCSS A.REI.12

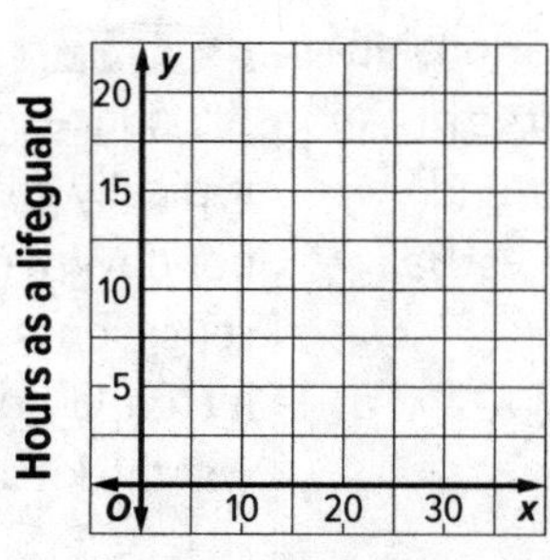

c. Is it possible for Malik to work only at the hardware store and meet his earning goal? Explain how the graph shows the answer to this question. CCSS A.CED.3

9. Solve the system $\begin{cases} 3x - 2y = 5 \\ 4x - 3y = 9 \end{cases}$. Show your work and explain each step. CCSS A.REI.5, A.REI.6

10. Jenna bikes and runs for exercise. The table shows how long it takes her to bike and run one mile and how many calories she burns biking or running for one mile.

	Biking	Running
Time (min)	4	10
Calories burned	45	70

On Saturday, Jenna ran and biked for 57 minutes, burning 535 calories. How many miles did she run, and how many miles did she bike? Explain how you found your answer. CCSS A.CED.3

5 Linear Functions

CHAPTER OVERVIEW Learn about some of the Common Core State Standards that you will explore in this chapter. Answer the preview questions. As you complete each lesson, return to these pages to check your work.

<table>
<tr><th>What You Will Learn</th><th>Preview Question</th></tr>
<tr><td colspan="2">Lesson 5.1: Graphing Linear Functions</td></tr>
<tr><td>CCSS F.IF.4 For a function that models a relationship between two quantities interpret key features of graphs and tables in terms of the quantities, and sketch graphs showing key features given a verbal description of the relationship.

Also addresses: F.IF.7a, F.IF.2, F.IF.5, F.IF.9, NQ.1, A.REI.11, F.LE.5</td><td>CCSS SMP 8 A kitten weighs 3.5 ounces at birth and typically gains about 0.5 ounce per day. Write an equation that represents the weight of the kitten. Graph the equation. What will the kitten weigh in 2 weeks?
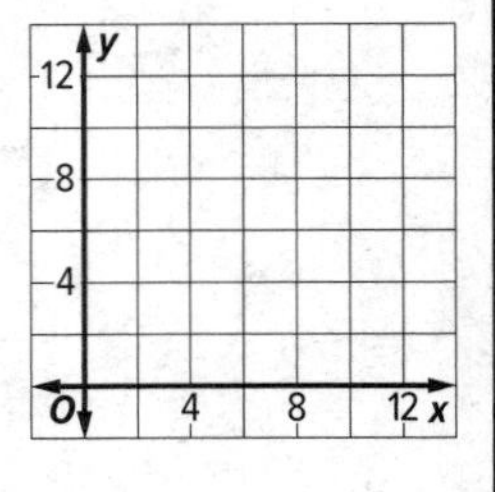
</td></tr>
<tr><td colspan="2">Lesson 5.2: Modeling: Linear Functions</td></tr>
<tr><td>CCSS N.Q.2 Define appropriate quantities for the purpose of descriptive modeling.

Also addresses: N.Q.3, A.SSE.1a, A.SSE.1b, A.CED.1, A.CED.4, F.BF.1a, F.BF.1b, F.LE.5</td><td>CCSS SMP 4 The graph shown is a student's daily wages as a function of the number of hours worked. Find the student's hourly rate of pay.
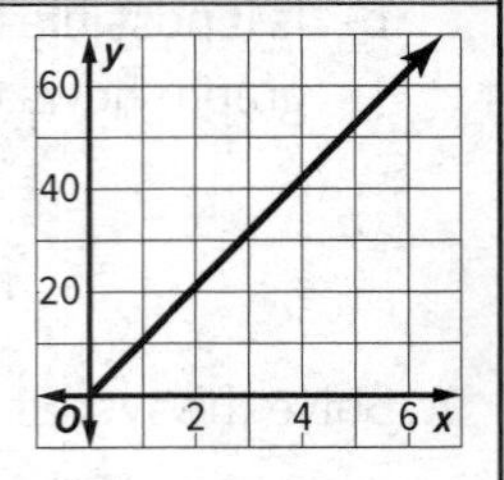
</td></tr>
<tr><td colspan="2">Lesson 5.3: Solving Linear Equations by Graphing</td></tr>
<tr><td>CCSS A.REI.10 Understand that the graph of an equation in two variables is the set of all its solutions plotted in the coordinate plane, often forming a curve (which could be a line).
CCSS F.IF.7a Graph linear and quadratic functions and show intercepts, maxima, and minima.</td><td>CCSS SMP 4 A music service charges $5.50 per month plus $1.50 for each downloaded song. Write and graph the equation. How much will it cost to buy 10 songs?
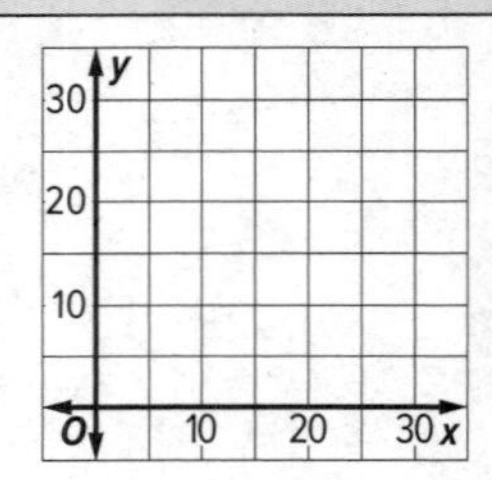
</td></tr>
<tr><td colspan="2">Lesson 5.4: Slope and Rate of Change</td></tr>
<tr><td>CCSS F.IF.6 Calculate and interpret the average rate of change of a function (presented symbolically or as a table) over a specified interval. Estimate the rate of change from a graph.
CCSS F.LE.1a Prove that linear functions grow by equal differences over equal intervals, and that exponential functions grow by equal factors over equal intervals.</td><td>CCSS SMP 2 The table shows the change in elevation of a zip line. Calculate the average rate of change and determine how long it would take to reach the bottom.
<table><tr><th>Time (minutes)</th><th>Elevation (feet)</th></tr><tr><td>1</td><td>258</td></tr><tr><td>2</td><td>225.75</td></tr><tr><td>3</td><td>193.50</td></tr><tr><td>4</td><td>161.25</td></tr></table></td></tr>
</table>

What You Will Learn	Preview Question
Lesson 5.5: Defining Appropriate Quantities	
CCSS N.Q.2 Define appropriate quantities for the purpose of descriptive modeling.	CCSS SMP 6 Describe what the graph shown represents. What would be the coordinate point for 14 inches? 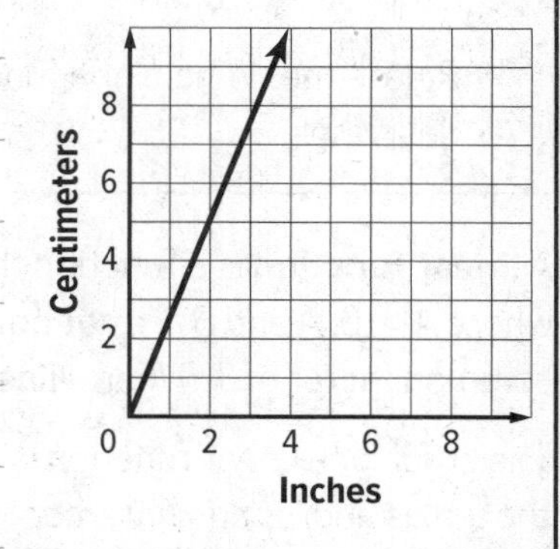
Lesson 5.6: Transforming Linear Functions	
CCSS F.BF.3 Identify the effect on the graph of replacing $f(x)$ by $f(x) + k$, $k\,f(x)$, $f(kx)$, and $f(x + k)$ for specific values of k (both positive and negative); find the value of k given the graphs. Experiment with cases and illustrate an explanation of the effects on the graph using technology. CCSS S.ID.7 Interpret the slope (rate of change) and the intercept (constant term) of a linear model in the context of the data.	CCSS SMP 7 Using the lines shown, find the value of k for each line and explain the difference. 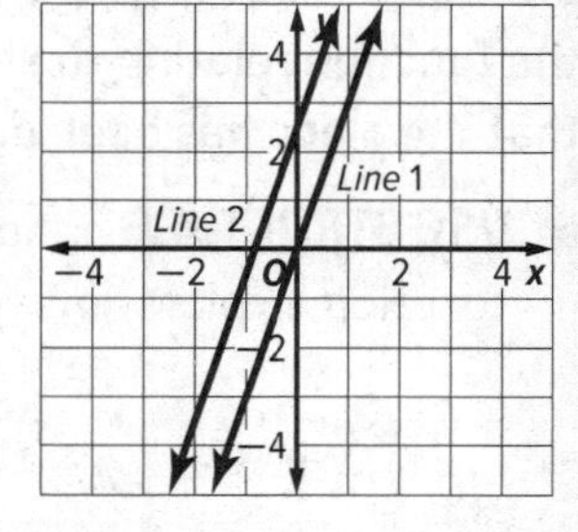
Lesson 5.7: Arithmetic Sequences as Linear Functions	
CCSS F.BF.2 Write arithmetic and geometric sequences both recursively and with an explicit formula, use them to model situations, and translate between the two forms. **Also addresses:** F.LE.2, A.CED.2, F.IF.3, F.BF.1a, A.REI.11	CCSS SMP 8 You borrow \$250 and pay back \$30 every week. Write a formula to show how much you will owe after you have made 6 payments.
Lesson 5.8: Proportional and Nonproportional Relationships	
CCSS F.LE.1b Recognize situations in which one quantity changes at a constant rate per unit interval relative to another. CCSS A.CED.2 Create equations in two or more variables to represent relationships between quantities; graph equations on coordinate axes with labels and scales. **Also addresses:** F.LE.2, F.BF.1a, A.CED.1	CCSS SMP 1 Write an equation to describe the relationship shown in the graph. Is this a proportional relationship? 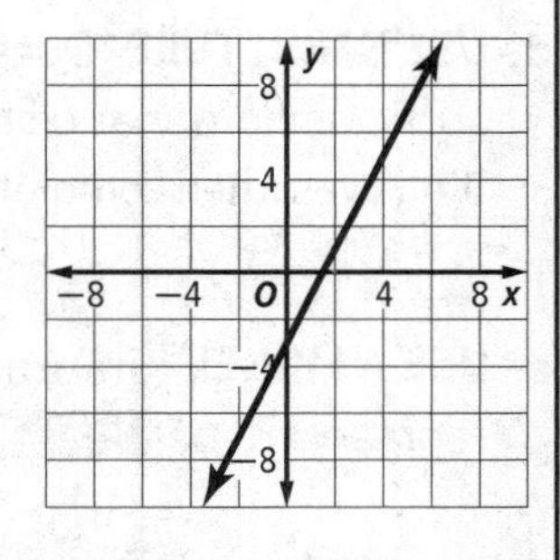

5.1 Graphing Linear Functions

Objectives

- Graph linear functions, and show intercepts.
- Relate the domain of a function to its graph and the relationship it describes.

STANDARDS

Content: F.IF.4, F.IF.7a, F.IF.2, F.IF.5, F.IF.9, N.Q.1, F.LE.5, A.REI.11

Practices: 3, 4, 6, 7

Use with Lesson 3-1

A **linear function** is a function that can be written in the standard form $Ax + By = C$, where $A \geq 0$, A and B are not both zero, and A, B, and C are integers with a greatest common factor of 1. When a linear function is written in standard form, y is a function of x.

The graph of a linear function is a line. The graph has a y-intercept where the graph crosses the y-axis and/or an x-intercept where the graph crosses the x-axis.

Linear models are often used to describe real-world situations.

EXAMPLE 1 Interpret Linear Models

EXPLORE Ana is flying a model airplane on its final descent. The table shows the function relating the height of the plane above the ground and the time that the plane has been descending.

Time(s)	Height (ft)
x	y
0	48
2	36
4	24
6	12
8	0

a. USE STRUCTURE Find the x- and y-intercepts of the graph of the function. Explain how you found each intercept. CCSS F.IF.4, SMP 7

b. USE A MODEL Plot the x-intercept. Interpret what it represents. CCSS N.Q.1, SMP 4

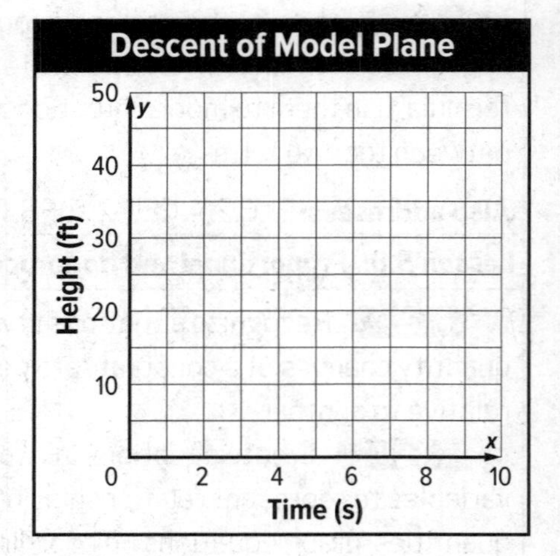

c. USE A MODEL Plot the y-intercept. Interpret what it represents. CCSS N.Q.1, SMP 4

d. USE STRUCTURE Does plotting the x- and y-intercepts give you sufficient information to graph the function? Justify your answer. If it is yes, then complete the graph. CCSS F.IF.4, SMP 7

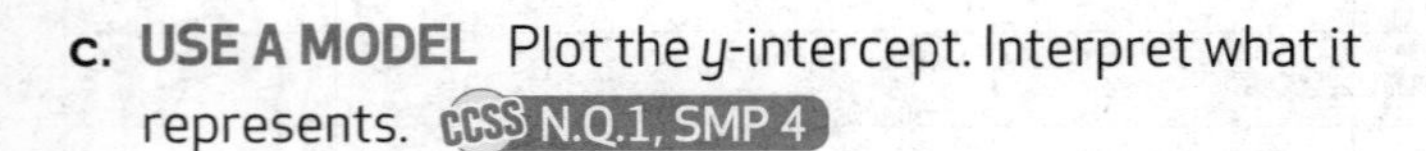

e. USE A MODEL State a reasonable domain for this situation. What does the domain represent? CCSS F.IF.5, SMP 4

By finding the x- and y-intercepts, you have the ordered pairs of two points through which the graph of a linear function passes.

EXAMPLE 2 Graph Linear Functions

USE STRUCTURE Graph each linear function. Identify the intercepts. CCSS F.IF.7a, SMP 7

a. $3x + 2y = 6$

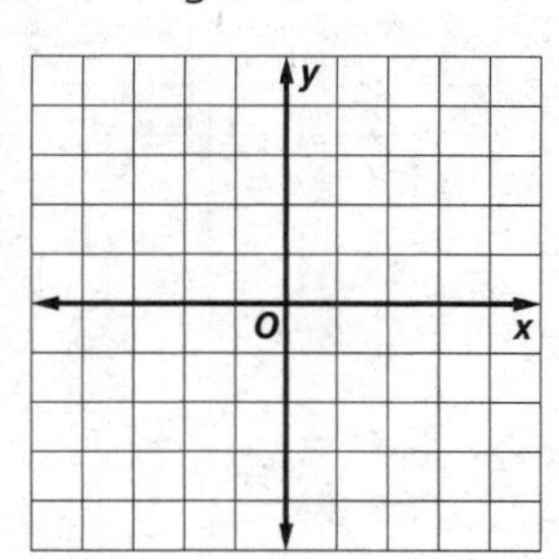

b. $x + 3y = 3$

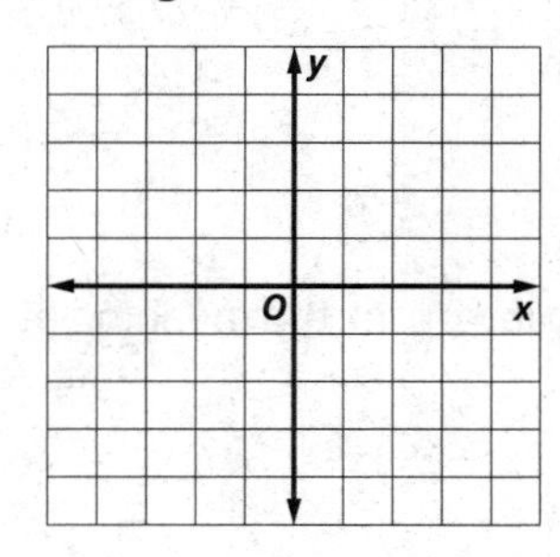

c. $y = 2$

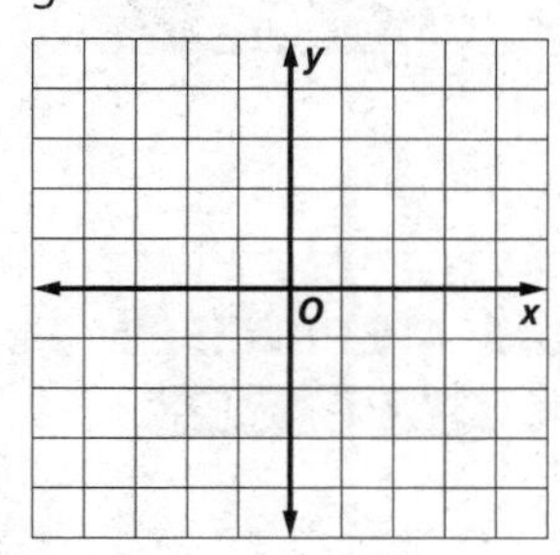

d. USE STRUCTURE The table at the right represents a linear function. Which function(s) in **parts a–c** have graphs with the same x-intercept as the function represented by the table? Which function(s) have graphs with the same y-intercept? Do any of the functions in **parts a–c** have the same graph as the function represented by the table? Explain and sketch a graph to justify your answer. CCSS F.IF.4, F.IF.9, SMP 7

x	y
−4	3
−2	2
0	1
2	0
4	−1

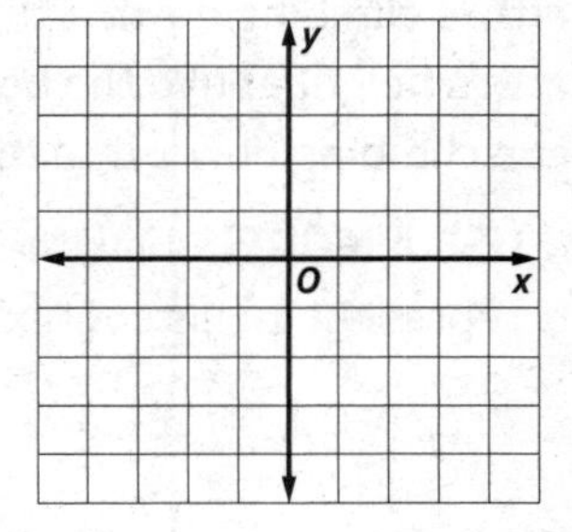

Recall that a function such as $y = 4x - 7$ can also be expressed as $f(x) = 4x - 7$ using function notation.

EXAMPLE 3 Model a Real-World Situation

Ms. Franklin determines that each student's math midterm score $m(x)$ can be represented as a function of the number of hours x the student studied for the midterm by $m(x) = 6x + 40$. The maximum score on the midterm is 100.

a. USE A MODEL What does the domain represent in this context? State a reasonable domain for the situation. Explain your reasoning. CCSS F.IF.4, SMP 4

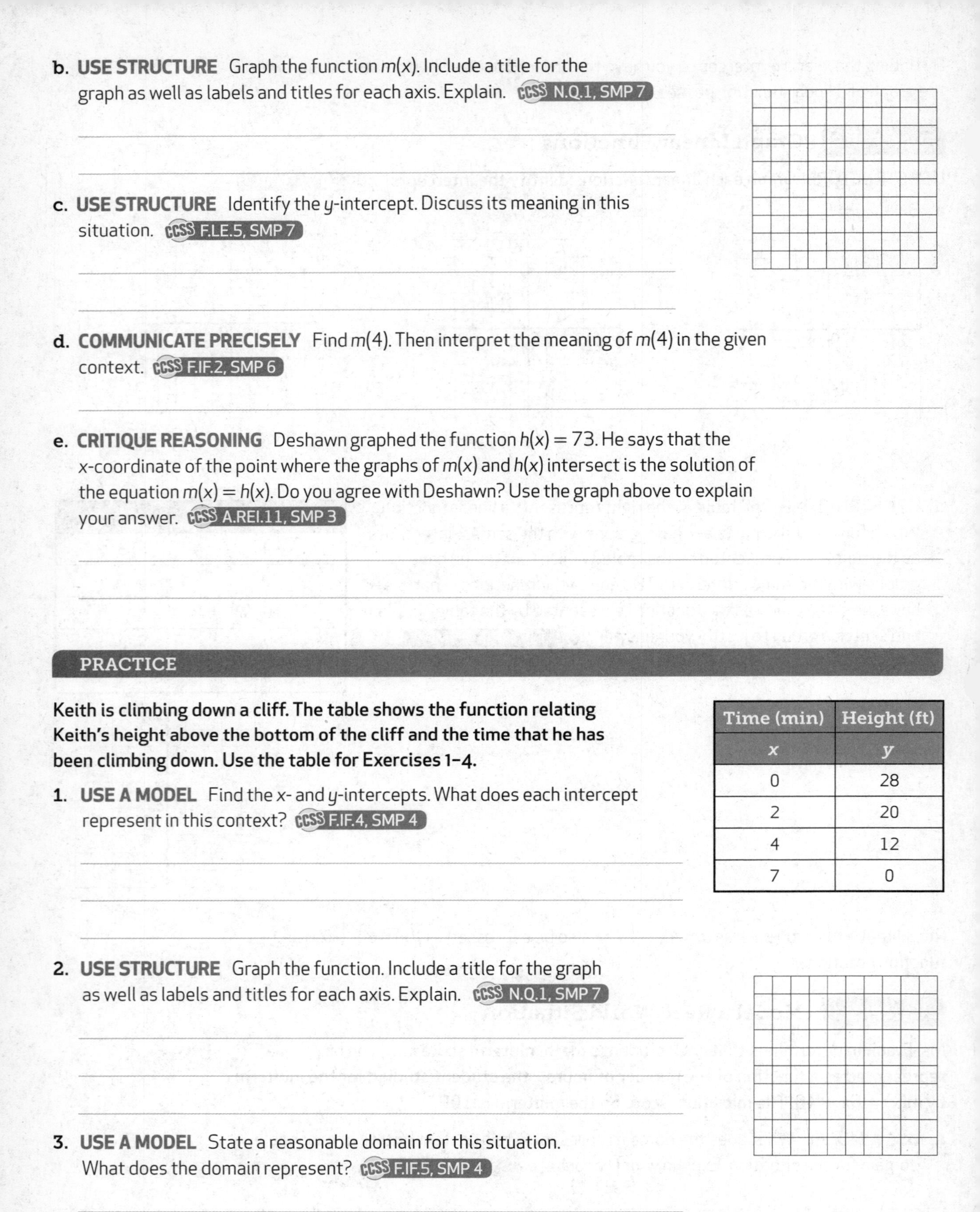

b. USE STRUCTURE Graph the function $m(x)$. Include a title for the graph as well as labels and titles for each axis. Explain. CCSS N.Q.1, SMP 7

c. USE STRUCTURE Identify the y-intercept. Discuss its meaning in this situation. CCSS F.LE.5, SMP 7

d. COMMUNICATE PRECISELY Find $m(4)$. Then interpret the meaning of $m(4)$ in the given context. CCSS F.IF.2, SMP 6

e. CRITIQUE REASONING Deshawn graphed the function $h(x) = 73$. He says that the x-coordinate of the point where the graphs of $m(x)$ and $h(x)$ intersect is the solution of the equation $m(x) = h(x)$. Do you agree with Deshawn? Use the graph above to explain your answer. CCSS A.REI.11, SMP 3

PRACTICE

Keith is climbing down a cliff. The table shows the function relating Keith's height above the bottom of the cliff and the time that he has been climbing down. Use the table for Exercises 1–4.

Time (min)	Height (ft)
x	y
0	28
2	20
4	12
7	0

1. **USE A MODEL** Find the x- and y-intercepts. What does each intercept represent in this context? CCSS F.IF.4, SMP 4

2. **USE STRUCTURE** Graph the function. Include a title for the graph as well as labels and titles for each axis. Explain. CCSS N.Q.1, SMP 7

3. **USE A MODEL** State a reasonable domain for this situation. What does the domain represent? CCSS F.IF.5, SMP 4

4. CONSTRUCT ARGUMENTS Sheila and Rajiv are also each climbing down a cliff. Sheila's height above the ground after x minutes is modeled by the function $h(x) = -5x + 30$. The table shows the function relating Rajiv's height above the bottom of the cliff and the time that he has been climbing down. Of Keith, Sheila, and Rajiv, who starts from the greatest height? Who reaches the ground in the least amount of time? Explain how you can solve the problem by using intercepts. CCSS F.IF.9, SMP 3

Time (min)	Height (ft)
x	y
0	32
2	24
4	16
8	0

5. USE STRUCTURE Graph each linear function. Identify the intercepts. CCSS F.IF.7a, SMP 7

a. $2x - 3y = -6$

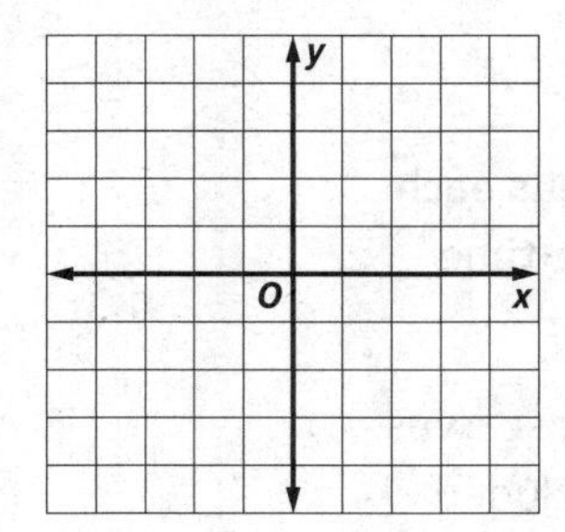

b. $3x + y = -3$

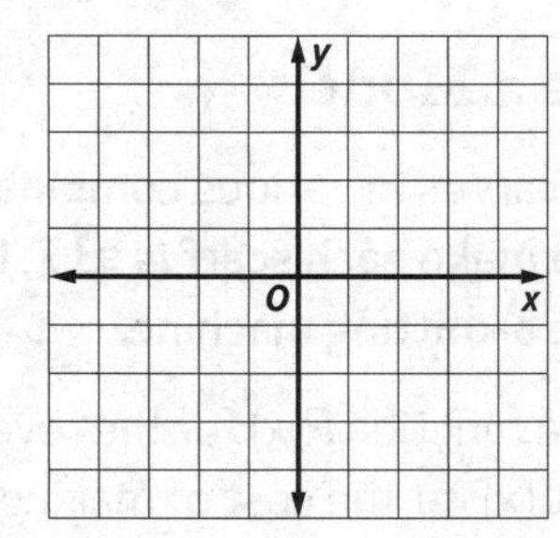

For Exercises 6–7, $f(x) = 2x + 2$ and $g(x) = -4$.

6. COMMUNICATE PRECISELY Paula is solving $f(x) = g(x)$. Graph and label $f(x)$ and $g(x)$. How can Paula use the graphs to find the solution? CCSS A.REI.11, SMP 6

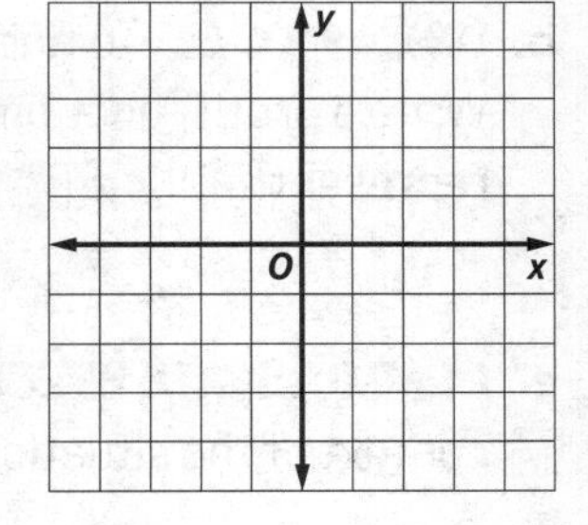

7. USE STRUCTURE Find $f(-3)$. How can you use the graph to find this value? CCSS F.IF.2, SMP 7

8. USE A MODEL Eli has $18 to spend on peanuts and pretzels for a party. If peanuts cost $3 per pound and pretzels cost $2 per pound, then the situation can be modeled by the linear function $3x + 2y = 18$, where x represents pounds of peanuts and y is pounds of pretzels. Graph the function. What does each intercept represent in terms of context? CCSS F.LE.5, SMP 4

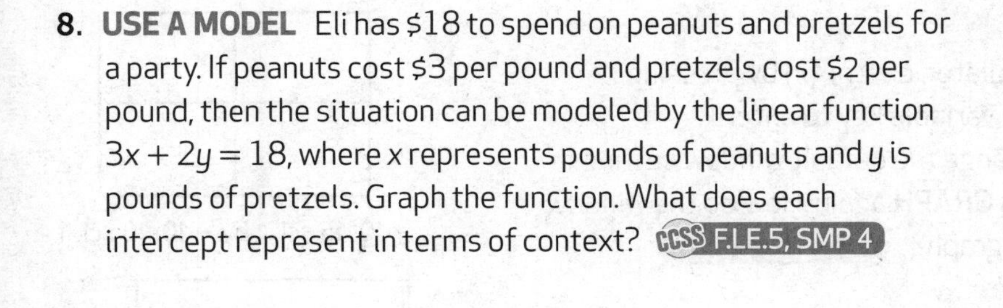

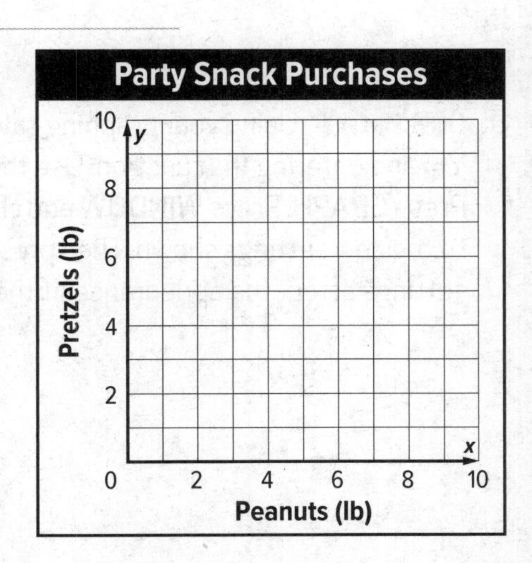

5.2 Modeling: Linear Functions

Objectives

- Build functions to model real-world situations.
- Reason quantitatively when making mathematical models.
- Use context to interpret linear functions.

STANDARDS

Content: N.Q.2, N.Q.3, A.SSE.1a, A.SSE.1b, A.CED.1, A.CED.4, F.BF.1a, F.BF.1b, F.LE.5
Practices: 2, 4, 5, 6, 7
Use with Lesson 3-1

Linear functions can be used to model many different real-world situations. When making mathematical models, technology can be valuable for exploring, making predictions, and asking "What if?" questions about data.

EXAMPLE 1 Create a Model

EXPLORE Sally makes scarves in various combinations of school colors. She sells each scarf for \$25. Her cost to make each scarf is \$17. In addition, she incurred a one-time cost of \$360 to purchase a knitting machine. CCSS N.Q.2

a. **USE A MODEL** Write a function $R(x)$ for the revenue Sally receives by selling x scarves. Then write a function $C(x)$ for the cost of making x scarves. Explain how you included the cost of the knitting machine in $C(x)$. CCSS F.BF.1a, A.SSE.1a, SMP 4

b. **USE A MODEL** To find her profit, Sally uses the formula Profit = Revenue − Cost. Write a profit function $P(x)$ such that $P(x) = R(x) - C(x)$. In words, what does $P(x)$ represent? CCSS F.BF.1b, SMP 4

c. **COMMUNICATE PRECISELY** Find $P(50)$. Then interpret the meaning of $P(50)$ in the context of the situation. CCSS F.BF.1a, SMP 6

d. **USE TOOLS** Using your graphing calculator, enter $P(x)$ by pressing **Y=** and entering the function. Use the variable key to enter x. Press **GRAPH**. Press **WINDOW** and change the default window settings to the new settings shown. Then press **GRAPH** again. How do the window settings affect the appearance of the graph? CCSS F.LE.5, SMP 5

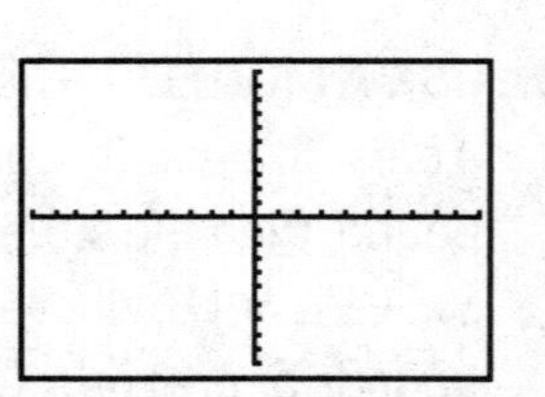

−10, 15 scl: 1 by −10, 10 scl: 1

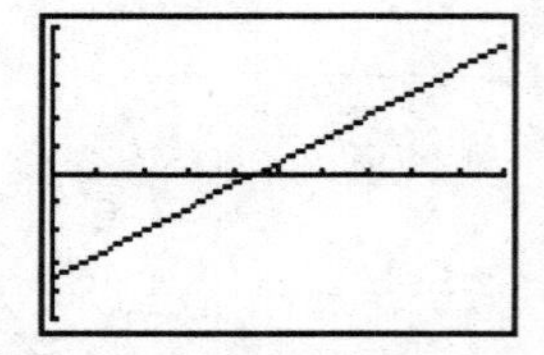

0, 100 scl: 10 by −500, 500 scl: 100

e. **REASON QUANTITATIVELY** Sally wants to find her profit if she makes and sells 75 scarves. Press **TRACE** and use the left and right arrows to move along the line. Explain how to answer Sally's question and discuss the accuracy of the answer that is obtained. CCSS N.Q.3, SMP 2

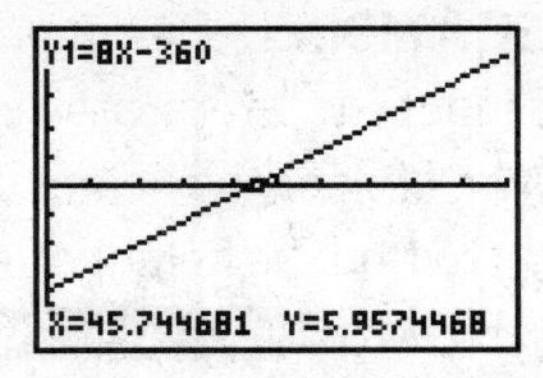

0, 100 scl: 10 by −500, 500 scl: 100

f. **USE TOOLS** Sally wants to find how many scarves she needs to sell to break even (profit = 0). Which key feature of the graph represents the break-even point? Press **2ND** and **TABLE**. Explain how to use the table to find this key feature. What is the break-even point? CCSS F.LE.5, SMP 5

X	Y_1
0	-360
1	-352
2	-344
3	-336
4	-328
5	-320
6	-312

X=0

Formulas can be useful when making a mathematical model of a real-world situation. A formula may need to be rearranged so it will fit the structure of an expression in a function.

EXAMPLE 2 Use a Formula in a Linear Model

A food company is thinking of changing the height of its soup cans. Each can is a cylinder with a label that covers the entire lateral surface. The formula for the lateral area (L) of a cylinder is $L = 2\pi rh$, where r is the radius of its base and h is its height.

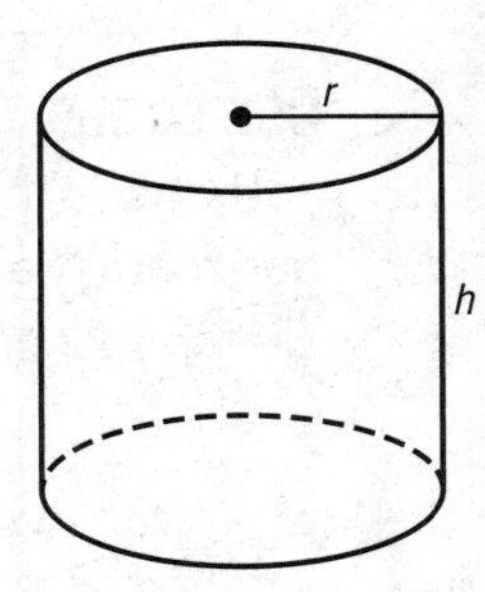

a. **USE STRUCTURE** What does the expression $2\pi r$ represent? How can viewing $2\pi r$ as a single variable simplify your view of the formula? CCSS A.SSE.1b, SMP 7

b. **CALCULATE ACCURATELY** The radius of the can will remain constant at 3.2 centimeters. Using 3.1416 for π, write and simplify an expression for the lateral area of the can where x represents the height. Is the level of precision appropriate to the situation? Explain. CCSS N.Q.3, SMP 6

c. **USE A MODEL** Write a function $L(x)$ for the lateral area of the can as a function of its height x. The height of the soup can was originally 10 centimeters. Explain how to use $L(x)$ to find the area of the label of the original soup can. CCSS A.CED.1, SMP 4

d. USE A MODEL Designers in the packaging department would like an app that takes as input the lateral area in square centimeters and outputs the height of the soup can. Write $h(x)$ for the height of a soup can as a function of lateral area. Explain how to use $h(x)$ to find the height of the new soup can if the area of the original label is decreased by 10%. CCSS A.CED.4, SMP 4

PRACTICE

1. Aidan buys used bicycles, fixes them up, and sells them. His average cost to buy and fix each bicycle is \$47. He also incurred a one-time cost of \$840 to purchase tools and a small shed to use as his workshop. He sells the bikes for \$75 each. CCSS N.Q.2

a. USE A MODEL Write revenue and cost functions $R(x)$ and $C(x)$ for Aidan's situation, where x is the number of bicycles. How did you include the one-time cost in $C(x)$? CCSS F.BF.1a, A.SSE.1a, SMP 4

b. USE A MODEL Write a profit function $P(x)$ such that $P(x) = R(x) - C(x)$. In words, what does $P(x)$ represent? CCSS F.BF.1b, SMP 4

c. USE TOOLS Enter $P(x)$ into your graphing calculator. Press **WINDOW** and enter settings so that the y-axis lies along the left side of the window and the x-axis lies horizontally along the center. Which settings did you use? Then press **GRAPH** and sketch your graph in the space to the right. CCSS F.LE.5, SMP 5

d. USE TOOLS Explain how to use **TRACE** to find Aidan's profit if he sells 120 bikes. Discuss the accuracy of the answer that can be obtained. CCSS N.Q.3, SMP 5

e. USE TOOLS Which key feature of the graph represents Aidan's break-even point (profit = 0)? Explain how to use your graphing calculator to find the most accurate value for this feature. CCSS F.LE.5, SMP 5

2. **The costume department of a theatre company is making cone-shaped hats for a play set in medieval times. Each hat will be covered with satin over its entire lateral surface inside and out. The formula for the lateral area (L) of a cone is $L = \pi r \ell$, where r is the radius of its base and ℓ is its slant height.**

   a. **USE A MODEL** The slant height of each hat will remain constant at 20.0 inches but the radius of the base will vary to accommodate different head sizes. Using 3.1416 for π, write a function $L(x)$ for the lateral area of the hat as a function of the radius of its base, x. Discuss the precision of your calculations. CCSS A.CED.1, A.SSE.1b, SMP 4

   b. **REASON QUANTITATIVELY** Enter $L(x)$ into your graphing calculator. Press **WINDOW** and enter the following settings: Xmin: −10; Xmax: 10; Ymin: −1000; Ymax: 1000. Then press **GRAPH** and sketch your graph in the space to the right. Explain how to use **TRACE** to estimate the amount of satin needed to cover a hat with a head opening of radius 3.5 inches. CCSS N.Q.3, SMP 2

3. **A factory uses a heater in part of its manufacturing process. The product cannot be heated too quickly, nor can it be cooled too quickly after the heating portion of the process is complete.**

   a. **USE A MODEL** The heater is digitally controlled to raise the temperature inside the chamber by 10°F each minute until it reaches the set temperature. Write a function to represent the temperature inside the chamber after x minutes if the starting temperature is 80°F. CCSS A.CED.1, SMP 4

   b. **REASON QUANTITATIVELY** The heating process takes 22 minutes. What is the temperature in the chamber at this point? CCSS A.CED.1, SMP 2

   c. **USE A MODEL** After the heater reaches the temperature determined in **part b**, the temperature is kept constant for 20 minutes before cooling begins. Fans within the heater control the cooling so that the temperature inside the chamber decreases by 5°F each minute. Write a function to represent the temperature inside the chamber x minutes after the cooling begins. CCSS A.CED.1, SMP 4

   d. **REASON QUANTITATIVELY** How long does it take to return the temperature in the chamber to 80°F? CCSS A.CED.1, SMP 2

5.3 Solving Linear Equations by Graphing

Objectives

- Understand how linear equations and linear functions are related.
- Find zeros of linear functions by graphing.

Content: A.REI.10, F.IF.7a
Practices: 1, 2, 3, 4, 5, 6, 7
Use with Lesson 3-2

A **linear equation** has at most one solution, or **root.** You can solve a linear equation by graphing its related **linear function.** To solve $ax + b = 0$, consider the graph of the related linear function $f(x) = ax + b$. Since $f(x) = 0$ at the x-intercept, the x-intercept is the solution of the equation. If the related function has no x-intercept, then the equation has no solution.

KEY CONCEPT

Complete the table by writing the corresponding information for the parent linear function. Then graph this function in the coordinate plane.

Parent Linear Function $f(x) = x$		
Type of graph:		(coordinate plane with axes x, y, origin O)
Domain:		
Range:		

EXAMPLE 1 Investigate Linear Equations

EXPLORE Tiana is solving the linear equation $2.4x - 6 = 0$.

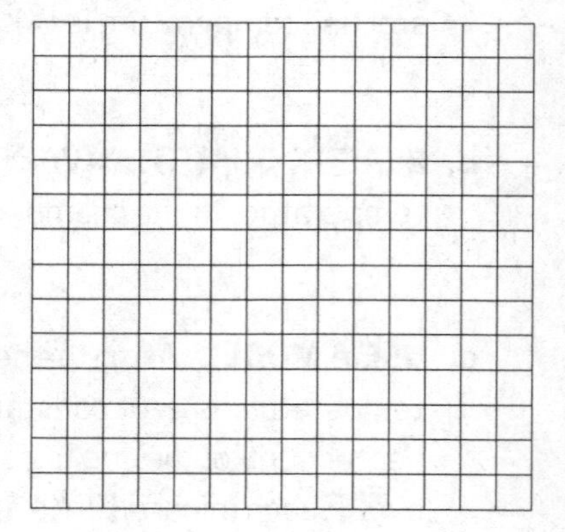

a. INTERPRET PROBLEMS Graph the related function $f(x) = 2.4x - 6$. CCSS F.IF.7a, SMP 1

b. USE STRUCTURE How are the ordered pairs for the points that make up the graph related to the function $f(x) = 2.4x - 6$? CCSS A.REI.10, SMP 7

____

____

____

c. INTERPRET PROBLEMS What is the x-intercept? How is the x-intercept related to Tiana's equation $2.4x - 6 = 0$? Include the terms *root, zero,* and *solution* in your explanation. CCSS F.IF.7a, SMP 1

____

____

____

d. REASON QUANTITATIVELY How can Tiana use the graph of the function to solve the equation $2.4x - 6 = 6$? CCSS A.REI.10, SMP 2

Real-world situations can be modeled with linear equations, which can then be solved by graphing.

EXAMPLE 2 Solve an Equation

Ben bought a new snowboard on credit for a total cost of $290. He'll make payments of $36.25 each week until the remaining balance is paid.

a. USE A MODEL Write an equation that can be solved to determine the number of weeks it will take for the balance to be paid, and write a related function. Then fill in the table of values and use it to graph the function. CCSS F.IF.7a, A.REI.10, SMP 4

x	$y =$	y	(x, y)
0			
4			
8			

b. USE STRUCTURE How is the set of solutions represented by the graph related to the solution of the equation? How is it related to the situation? CCSS A.REI.10, SMP 7

c. CONSTRUCT ARGUMENTS How can you use the graph to tell how many weeks Ben will make payments? Describe how you can be sure that your solution is correct. CCSS F.IF.7a, SMP 3

Technology can be used to solve linear equations by graphing. For example, when a linear function is graphed on a calculator, pressing **TRACE** displays a cursor that moves incrementally along the line as the corresponding x-and y-values for each location are displayed at the bottom of the screen.

EXAMPLE 3 Use Technology to Solve an Equation

Sondra and Trey are each using their graphing calculators to solve the equation $0.875x - 2.45 = 0$. Sondra says that she found an exact solution using TRACE. Trey says that he could not find an exact answer using TRACE and had to estimate a solution. Both students entered the correct function into the calculator. Sondra selected ZDecimal and Trey selected ZStandard from the ZOOM menu, resulting in the window settings shown at the right.

```
WINDOW
 Xmin=-6.6
 Xmax=6.6
 Xscl=1
 Ymin=-4.1
 Ymax=4.1
 Yscl=1
↓Xres=1
```

Sondra's Settings

```
WINDOW
 Xmin=-10
 Xmax=10
 Xscl=1
 Ymin=-10
 Ymax=10
 Yscl=1
 Xres=1■
```

Trey's Settings

a. USE TOOLS Enter the function into your own graphing calculator, press **ZOOM** and select Sondra's option. Use **TRACE** to find the value of points on the graph. How is each pair of x-and y-values related to the equation $y = 0.875x - 2.45$? CCSS A.REI.10, SMP 5

b. CRITIQUE REASONING Describe how Sondra may have used **TRACE** to solve the equation. What is the solution? Why do you think she was confident that her solution was exact? CCSS F.IF.7a, SMP 3

c. USE STRUCTURE Now press **ZOOM** and select Trey's option. Again use **TRACE** to find the value of points on the graph. Describe what could have happened when Trey tried to solve the equation. CCSS F.IF.7a, SMP 7

PRACTICE

1. USE A MODEL Jazmin is participating in a 25.5-kilometer charity walk. She walks at a rate of 4.25 km per hour. If Jazmin walks at the same pace for the entire event, how long will it take her to complete the walk? CCSS SMP 4

a. Complete the table and graph the function that represents the situation. Identify and label the x- and y-intercepts. CCSS F.IF.7a

x	$y = 25.5 -$ _____	y	(x, y)

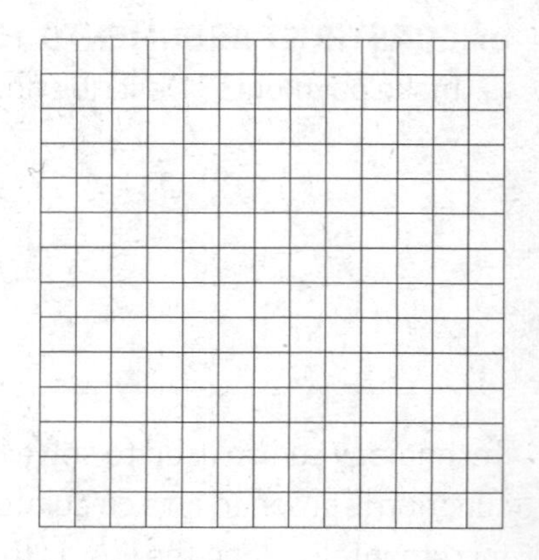

b. What do the x- and y-intercepts represent in this situation? 

c. After Jazmin has walked 17 km, how much longer will it take her to complete the walk? Explain how you can use this graph to answer the question. CCSS A.REI.10

d. Jazmin's friend Reza says you can rewrite the function using 8.5 as the constant in the equation instead of 25.5. Would she have found the solution? Explain. CCSS F.IF.7a

2. REASON QUANTITATIVELY How can you find the solution to $ax + b = c$ by graphing? Identify any restrictions on the values of a, b, and c. CCSS A.REI.10, SMP 2

3. Peggy wants to graph the function $y = -\frac{4}{5}x + \frac{2}{5}$ where $-2 \leq x \leq 5$.

a. COMMUNICATE PRECISELY Graph the function $y = -\frac{4}{5}x + \frac{2}{5}$ over the interval. Identify any values of x where a maximum or minimum value of y occurs. CCSS F.IF.7a, SMP 6

x	$-\frac{4}{5}x + \frac{2}{5}$	y	(x, y)
−2			
5			

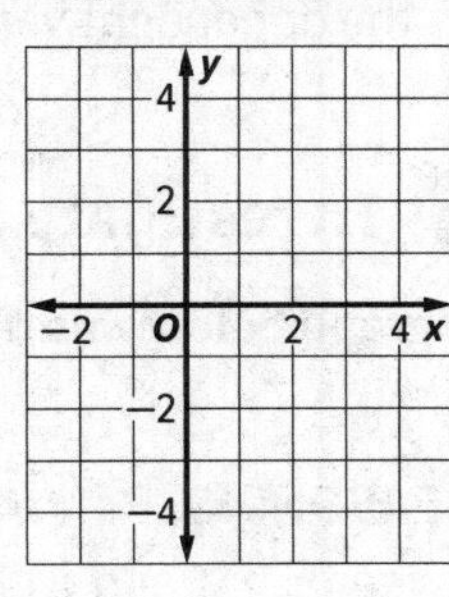

b. USE STRUCTURE Peggy says she can find the solution to $-\frac{4}{5}x + \frac{2}{5} = 0$ using the graph. Explain how she could do this. What is the solution? CCSS A.REI.10, SMP 7

5.4 Slope and Rate of Change

Objectives

- Find the average rate of change of a function over an interval.
- Prove that linear functions change by equal differences over equal intervals.

Content: F.IF.6, F.LE.1a
Practices: 2, 3, 5, 6, 7
Use with Lesson 3-3

A linear function is a function whose graph is a line. One way to describe the graph of a linear function is to find its **slope**, which tells how steep the line is. The slope is the ratio of the change in the y-coordinates **(rise)** to the change in the x-coordinates **(run)**.

KEY CONCEPT

Complete the table.

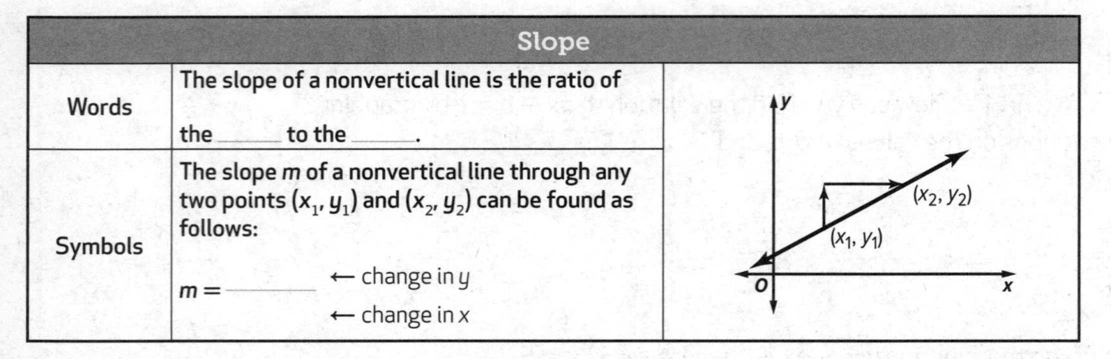

Slope		
Words	The slope of a nonvertical line is the ratio of the ______ to the ______.	
Symbols	The slope m of a nonvertical line through any two points (x_1, y_1) and (x_2, y_2) can be found as follows: $m =$ ______ ← change in y ← change in x	

Slope can be used to describe a **rate of change**, which is a ratio that describes, on average, how much one quantity changes with respect to another quantity. If x is the independent variable and y is the dependent variable, then rate of change $= \frac{\text{change in } y}{\text{change in } x}$.

EXAMPLE 1 Investigate Slope and Rate of Change

EXPLORE The graph below was displayed on a local TV weather report after a cold night.

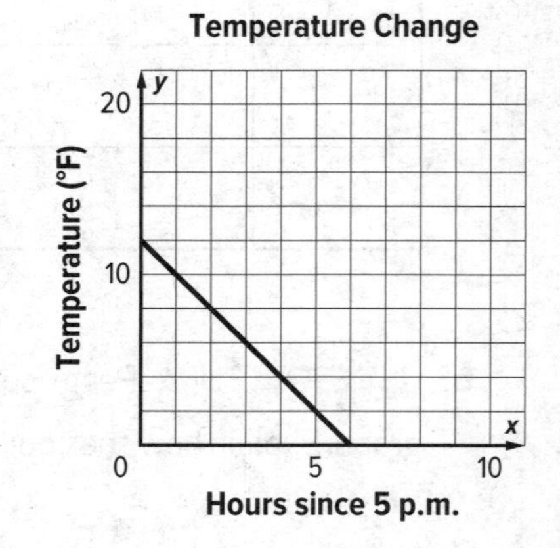

a. CRITIQUE REASONING Referring to the temperature, the reporter stated, "Between 5 P.M. and 11 P.M., the rate of change was minus 2 degrees Fahrenheit each hour." Do you agree? Use the graph to justify your answer. CCSS F.LE.1a, SMP 3

______

______

b. REASONING QUANTITATIVELY Find the x- and y-intercepts and explain how they indicate that the rate of change is negative. CCSS F.LE.1a, SMP 2

______

______

c. **FIND A PATTERN** The table shows the data used to create the graph. Complete the third column. The first calculation has been done for you. How is the change over an interval of 1 hour found? What do you notice? CCSS F.LE.1a, SMP 7

Hours Since 5 p.m.	Temperature (°F)	Change in *y* (°F)
0	12	---
1	10	$10 - 12 = -2$
2	8	
3	6	
4	4	
5	2	
6	0	

d. **MAKE A CONJECTURE** Based on **parts a** and **b**, make a conjecture about how linear functions change over equal intervals. How are the changes in the y values related to the rate of change? CCSS F.LE.1a, SMP 3

e. **USE STRUCTURE** Find the slope of the line on the graph and describe how you found it. How is the slope related to the rate of change? CCSS F.IF.6, SMP 7

You can use technology to investigate whether the rate of change is constant between any two points on the graph of a linear function.

EXAMPLE 2 Use Technology to Investigate Slope and Rate of Change

The function $y = 0.75x + 0.4$ models the number of milliliters of water, y, in a test tube x minutes after a chemical process has been started.

a. **USE TOOLS** Enter the function into your graphing calculator. Set the window size by pressing **ZOOM** and selecting ZDecimal. Then press **GRAPH** and estimate the rate of change for the water in the test tube. Explain your reasoning. CCSS F.IF.6, SMP 5

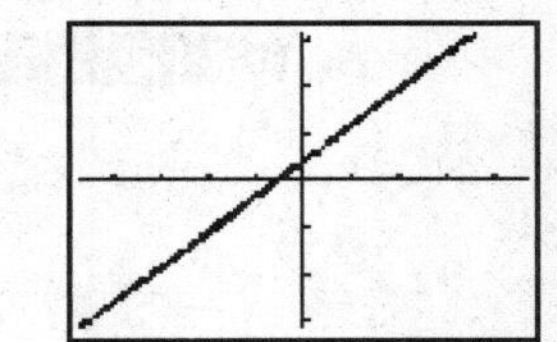

b. **USE TOOLS** Press **TRACE** and use the left and right arrows to explore the graph. What are the values of y for $x = 0$ and $x = 1$? Use the two pairs of coordinates to find the slope of the line. How can finding the slope help you check your estimate of the rate of change in **part a**? CCSS F.IF.6, SMP 5

c. USE TOOLS Explain how to use the table function on your graphing calculator to find the rate of change for $y = 0.75x + 0.4$. CCSS F.IF.6, SMP 5

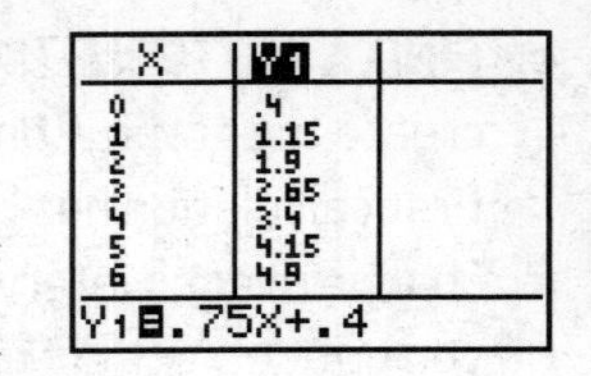

d. MAKE A CONJECTURE Using both **GRAPH** and **TABLE** on your graphing calculator, find the differences in the y-values as x-changes over equal intervals of 0.1, 0.5, 2, and 10, respectively. Based on your results, make a conjecture about how linear functions change over equal intervals and how this is related to the rate of change. CCSS F.LE.1a, SMP 3

e. MAKE A CONJECTURE Using your answer to **part d** make a conjecture about the change in a linear function over an interval of any length. Relate the change to the function equation. CCSS F.LE.1a, SMP 3

It can be proven algebraically that linear functions change by equal differences over equal intervals.

EXAMPLE 3 Construct Arguments about Linear Functions

Elena wants to prove that for any real numbers a, b, and k, a linear function $y = ax + b$ grows by an equal difference over any interval k. She created the graph at the right.

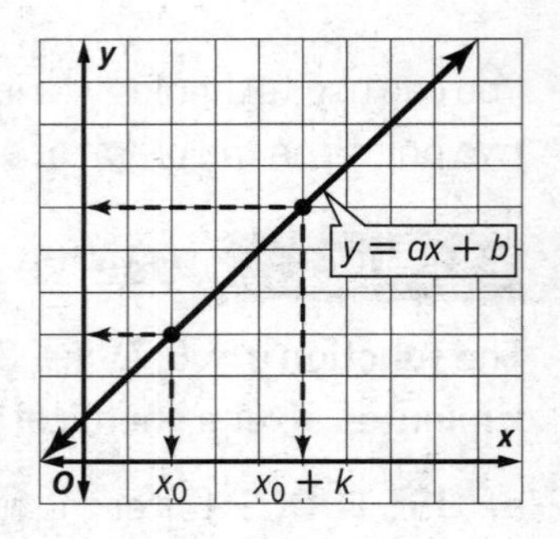

a. USE STRUCTURE Elena forgot to label the y-coordinates for the two points on the line $y = ax + b$. How can she use the x-coordinates of the points to find their y-coordinates? Label each y-coordinate on the y-axis. CCSS F.IF.7a, SMP 7

b. FIND A PATTERN Complete the table below. The first row of calculations has been done for you. What do you notice about the changes in y? CCSS F.LE.1a, SMP 7

x	$y = ax + b$	y	Change in y
x_0	$a(x_0) + b$	$ax_0 + b$	---
$x_0 + k$	$a(x_0 + k) + b$	$ax_0 + ak + b$	ak
$x_0 + 2k$			
$x_0 + 3k$			
$x_0 + 4k$			

c. CONSTRUCT ARGUMENTS Use your answers from **parts a** and **b** to complete Elena's proof that linear functions grow by equal differences over equal intervals. CCSS F.LE.1a, SMP 3

Given: $y = ax + b$ where $a \neq 0$, k where $k \neq 0$

Prove: $y = ax + b$ grows by an equal difference over any interval k.

If $x = x_0$, then $y =$ __________.

If $x = x_0 + k$, then $y =$ __________.

The difference between the y values is

____

So, the linear function $y = ax + b$ grows by an equal difference over any interval k because

____

____

d. CALCULATE ACCURATELY Find the slope using the variable coordinates in Elena's graph for (x_1, y_1) and (x_2, y_2). Based on the slope, predict the rate of change. Explain. CCSS F.IF.6, SMP 6

e. FIND A PATTERN Use the information in the table to find the rate of change for $y = ax + b$. Was your prediction in **part d** correct? CCSS F.LE.1a, SMP 7

f. USE STRUCTURE If you know the rate of change for a linear function and the change in y values, how can you determine the corresponding change in x values? Explain. CCSS F.LE.1a, SMP 7

PRACTICE

The graph shows median prices for small cottages on a lake since 2005. Use the graph for Exercises 1–3.

1. **CRITIQUE REASONING** A real estate agent says that since 2005, the rate of change for house prices is $10,000 each year. Do you agree? Use the graph to justify your answer. CCSS F.IF.6, SMP 3

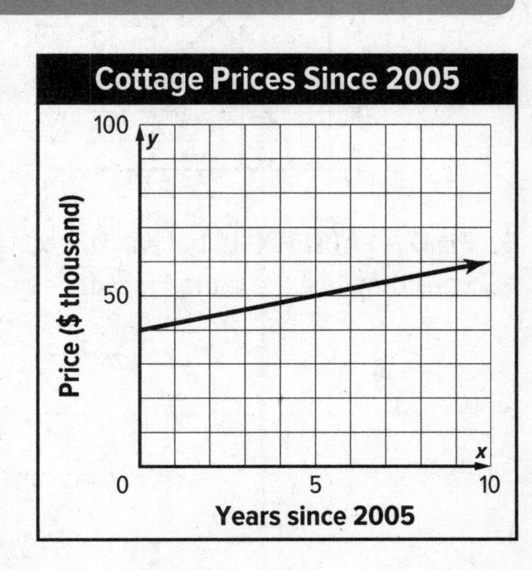

2. **FIND A PATTERN** The table shows the data used to make the graph. Complete the third column. The first calculation has been done for you. Describe the growth of y over each 1-year interval. CCSS F.LE.1a, SMP 7

Years Since 2005 (x)	Cottage Price (\$ thousand) (y)	Change in y (\$ thousand)
0	40	---
1	42	$42 - 40 = 2$
2	44	
3	46	
4	48	
5	50	

3. **USE STRUCTURE** Find the slope. Then find the rate of change. What do you notice? CCSS F.IF.6, SMP 7

The function $y = -0.92x + 2.9$ models the price, y, in dollars of one share of a stock x weeks after Pete purchased 100 shares. Use this information for Exercises 4–5.

4. **USE TOOLS** Using your graphing calculator, press **Y=** and enter the function. Then press **GRAPH** and estimate the rate of change for the stock price by looking at the graph. CCSS F.IF.6, SMP 5

5. **MAKE A CONJECTURE** Using your graphing calculator, find the differences in share prices over equal intervals of 2, 4, and 26 weeks. What do you notice about the rate of change? CCSS F.LE.1a, SMP 3

Eduardo wants to prove that a linear function grows by equal differences over equal intervals even when the values of y are decreasing as the values of x are increasing. He started creating the graph and the table below. Use the graph and the table in Exercises 6–7.

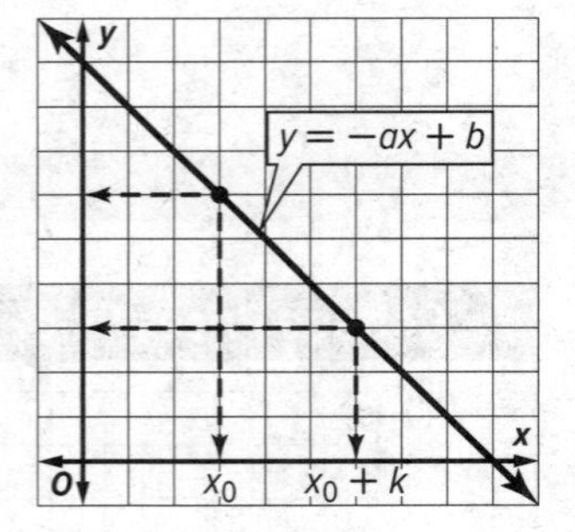

x	$y = -ax + b$	y	Change in y
x_0	$-a(x_0) + b$	$-ax_0 + b$	---
$x_0 + k$	$-a(x_0 + k) + b$	$-ax_0 - ak + b$	$-ak$
$x_0 + 2k$			
$x_0 + 3k$			
$x_0 + 4k$			

6. **FIND A PATTERN** Label each y-coordinate on the y-axis. Then use the graph to help you complete the table. How does the value of y change over each interval k? CCSS F.LE.1a, SMP 7

7. CONSTRUCT ARGUMENTS Use the graph and the table to complete Eduardo's proof. CCSS F.LE.1a, SMP 3

Given: $y = -ax + b$ where $a \neq 0$, k where $k \neq 0$

Prove: $y = -ax + b$ grows by an equal difference over any interval k.

If $x = x_0$, then $y =$ ____________.

If $x = x_0 + k$, then $y =$ ________________.

The difference between the y values is

____

So, the linear function $y = -ax + b$ grows by an equal difference over any interval k because

____

____

8. USE STRUCTURE In this lesson, you proved that linear functions grow by equal differences over equal intervals. Discuss how the differences in the y values over equal intervals are related to the slope and rate of change of the function. CCSS F.LE.1a, SMP 7

____

____

USE STRUCTURE For each graph or equation estimate the rate of change over the indicated intervals. CCSS F.LE.1a, SMP 7

9.

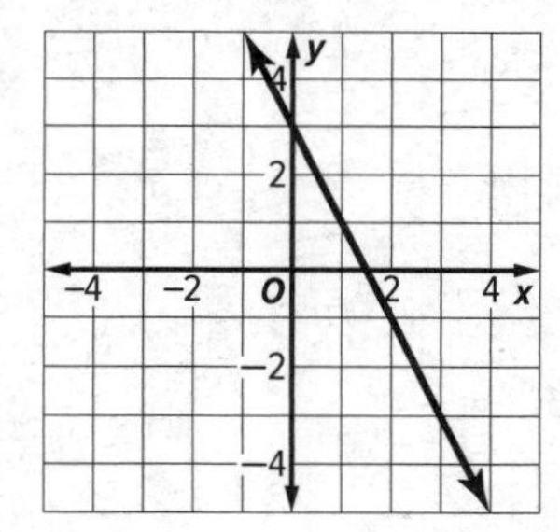

$x = -1$ to $x = 3$: ____________

$x = 0$ to $x = 4$: ____________

10.

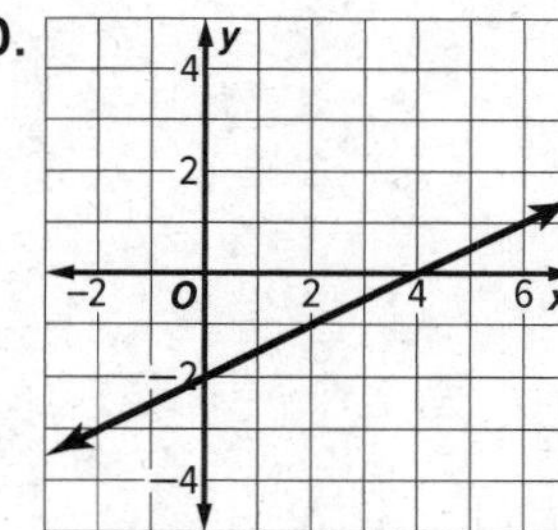

$x = -2$ to $x = 4$: ____________

$x = 0$ to $x = 6$: ____________

11. $y = -0.5x + 4$

$x = -2$ to $x = 2$: ____________

$x = 0$ to $x = 4$: ____________

12. $y = \frac{1}{3}x + 1$

$x = -3$ to $x = 3$: ____________

$x = 0$ to $x = 6$: ____________

13. REASON QUANTITATIVELY Isabel's cell phone bill was \$25 per month three years ago. This year the bill is \$40 per month. If the bill increased by the same amount at the beginning of each year, what is the change in the monthly bill each year? Explain. CCSS F.IF.6, SMP 2

____

5.5 Defining Appropriate Quantities

Objectives

- Define a quantity of interest in a real-world situation.
- Use appropriate quantities when modeling with linear functions.

Content: N.Q.2
Practices: 1, 2, 3, 4, 5, 6, 7
Use with Lesson 2-1

When using linear functions to model real-world situations, one question must constantly be asked: Does the model make sense? For a model to accurately represent a given real-world situation, it is important to define quantities that are appropriate for the situation. Defining appropriate quantities might include determining the parameters relevant to the situation and identifying the appropriate units of measure.

EXAMPLE 1 Define Input and Output

Last year, Simon started a business that makes cupcakes for catered events. As his business grows, Simon wants to develop some mathematical models to help him run his business more efficiently.

a. PLAN A SOLUTION Simon hires pastry chefs on an hourly basis. He estimates that it takes a chef 3 minutes to decorate one cupcake. He wants to write a linear function that models the relationship between the number of cupcakes and how long it takes to decorate them. What quantities would you define for the input and the output? Explain. CCSS N.Q.2, SMP 1

b. USE STRUCTURE Write and graph the function $f(x)$ using the quantities you identified in **part a**. Explain how you wrote the function. CCSS N.Q.2, SMP 7

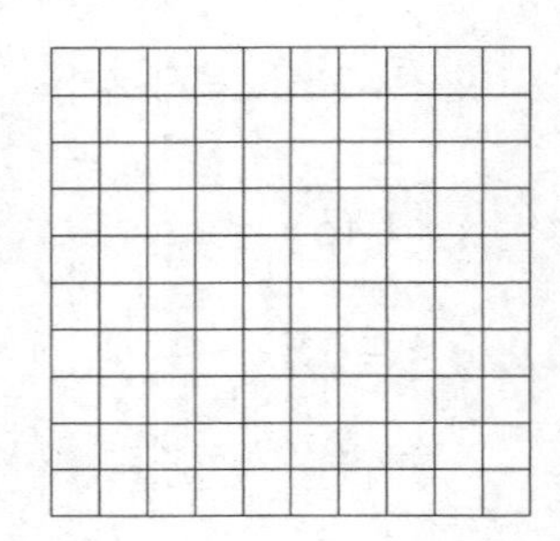

c. FIND A PATTERN Simon pays the pastry chefs \$18 per hour. How could you modify the function you wrote in **part b** to write $g(x)$, the total wages for pastry chefs to decorate x cupcakes. What unit would you choose for the output? Explain. CCSS N.Q.2, SMP 7

In some models, it may be possible to represent quantities in different ways. For example, expressing quantities with larger units may be useful when using the model for some purposes, while smaller units may be more helpful for other purposes.

EXAMPLE 2 Use Different Representations of Quantities

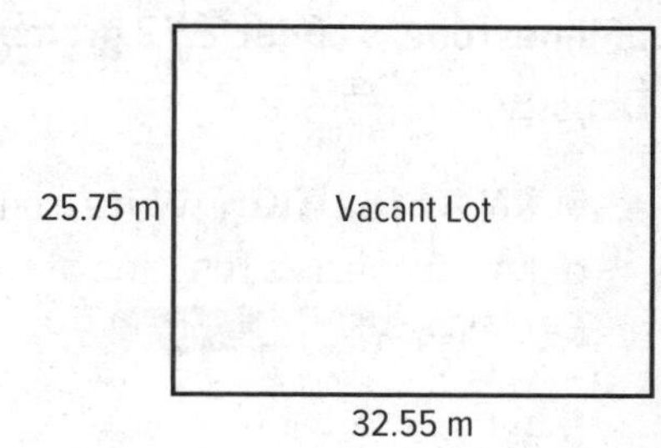

Students in the art club will paint a mural on the 2.5 meter-tall fence that surrounds a rectangular vacant lot near the school. The fence surrounds the entire perimeter of the lot, as shown in the diagram. The mural will extend around the entire perimeter at a constant height.

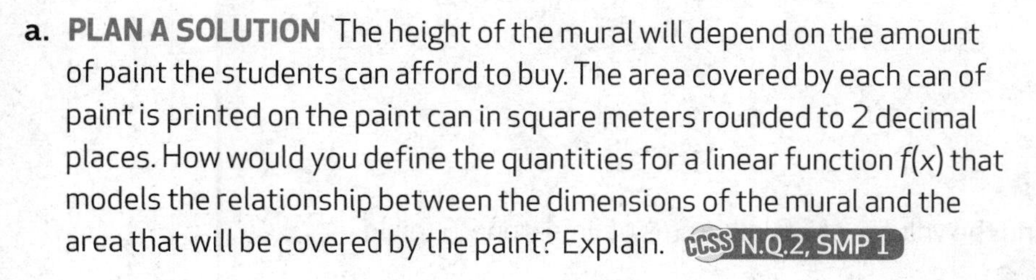

a. **PLAN A SOLUTION** The height of the mural will depend on the amount of paint the students can afford to buy. The area covered by each can of paint is printed on the paint can in square meters rounded to 2 decimal places. How would you define the quantities for a linear function $f(x)$ that models the relationship between the dimensions of the mural and the area that will be covered by the paint? Explain. CCSS N.Q.2, SMP 1

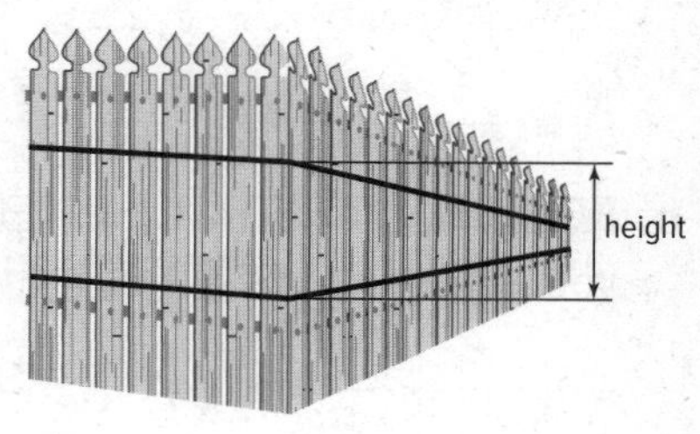

b. **CALCULATE PRECISELY** Write and graph the function $f(x)$ using the quantities you defined in **part a**. Use the function to model a situation where the students buy 8 cans of paint that cover 20.25 square meters each. Describe the results. CCSS N.Q.2, SMP 6

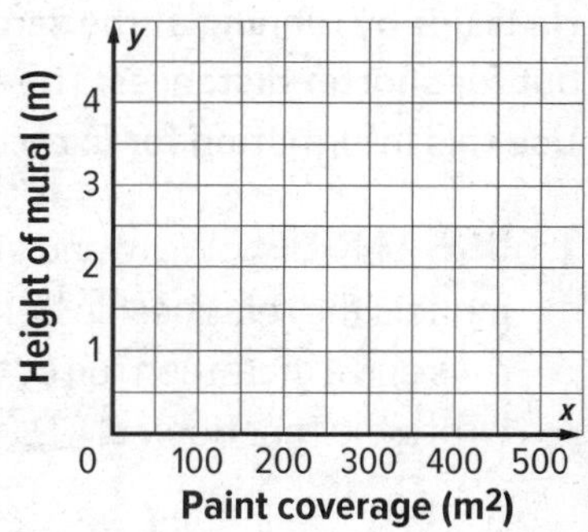

c. **REASON QUANTITATIVELY** Use your graph from **part b** to estimate the amount of paint the art club needs to paint a mural 2 meters tall. If each can of paint covers 20.25 square meters, how many cans of paint will the art club need? Explain. CCSS N.Q.2, SMP 2

d. **CONSTRUCT ARGUMENTS** A student argues that the output quantity should be the height of the mural in centimeters, not meters. Give a reason to support this argument. CCSS N.Q.2, SMP 3

EXAMPLE 3 Defining Appropriate Quantities When Using Technology

Genji is mathematically modeling the relationship between the volume, mass, and density of limestone for a science project. In her research, she found that the density of limestone is about 2.72 grams per cubic centimeter. The formula for density is $\text{Density} = \frac{\text{Mass}}{\text{Volume}}$.

a. **PLAN A SOLUTION** Write a linear function $f(x)$ that models the volume of a quantity of limestone as a function of its mass. Explain how you defined the quantities for the model. CCSS N.Q.2, SMP 1

b. **REASON QUANTITATIVELY** Find the volume of 20 kilograms of limestone. Explain your reasoning. CCSS N.Q.2, SMP 2

PRACTICE

Jeff is training to run a marathon, which is a distance of 26.2 miles. He trains by running at the same pace as he would run in a marathon, but for shorter distances. The table shows his 10 km run times. Use this information for Exercises 1 and 2.

10 km Runs	
Training Day	Time (min)
0	45
5	43
10	42

1. **USE A MODEL** Write and graph a linear function $f(x)$ that models the relationship between the times of Jeff's 10 km runs and a marathon run at the same pace. Explain how you wrote the function. CCSS N.Q.2, SMP 4

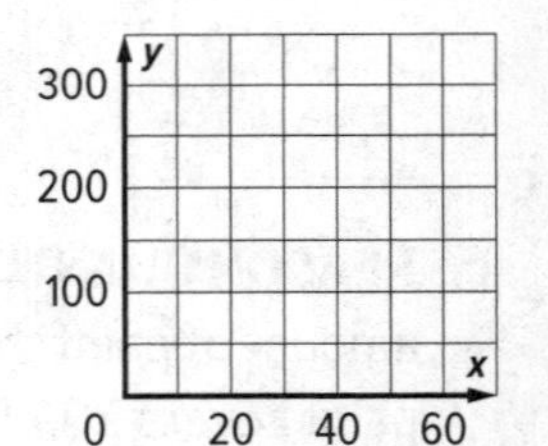

2. **INTERPRET PROBLEMS** How did you define the quantities for a linear function that models the relationship between the times of Jeff's 10 km runs and a marathon? Explain. CCSS N.Q.2, SMP 1

Each employee in Sue's department at work is getting a bonus. The amount of each bonus will be an equal share of a total amount that will range from $0 to $10,000. There are 21 employees in Sue's department. Use this information for Exercises 3–5.

3. **USE A MODEL** Write and graph a linear function $f(x)$ that models the relationship between the total amount to be distributed and the amount of each employee's bonus. Explain how you defined the quantities of interest. CCSS N.Q.2, SMP 4

4. **REASON QUANTITATIVELY** What is the maximum that each employee could receive as a bonus? How much would the total bonus amount have to be in order for each employee to receive a bonus of $250? Explain. CCSS N.Q.2, SMP 2

5. **USE TOOLS** An announcement is made that the total amount will be between $8500 and $8800. Use your graphing calculator to find the range of bonuses. How are the quantities displayed in the calculator related to the quantities you defined? CCSS N.Q.2, SMP 5

6. **COMMUNICATE PRECISELY** Water is leaking from a 100-gallon water tank at a rate of about 30 drops per minute, where 20 drops is about 1 mL. A 10-liter bucket is being used to collect the leaking water. You need to check the bucket from time to time and empty it if it is more than half full. What are the quantities of interest if this situation is modeled by a linear function? How are they represented? Write a function $f(x)$ that models the situation that you can use to predict how often the bucket will need to be emptied. CCSS N.Q.2, SMP 6

7. **PLAN A SOLUTION** Ina is going to begin training for a 60-mile bicycle race. She plans to make several practice rides of 60 miles and wants to know how long the rides will take her at different average speeds. She knows that the relationship between distance, rate, and time is $d = \frac{r}{t}$. Write a linear function $f(x)$ that models the time for a ride as a function of Ina's average speed for the ride. Explain how you defined the quantities for the model. CCSS N.Q.2, SMP 1

5.6 Transforming Linear Functions

Objectives

- Describe the effects of transformations on the graphs of linear functions.
- Interpret the meanings of transformations in real-world situations.

Content: F.BF.3, S.ID.7
Practices: 1, 2, 3, 4, 5, 6
Use with Extend 4–1

A **transformation** is defined as *a movement of a geometric figure*. Since the graph of a linear function is a line, the transformation of a linear function is the movement of a line. When the equation of the line is written in **slope-intercept form**, $y = mx + b$, a transformation may affect the values of m and b.

KEY CONCEPT

Complete the table by writing in the remaining information in the sections Words, Symbols, and Graph.

Slope-Intercept Form		
Words	**The slope-intercept form of a linear equation is $y = mx + b$, where ___ is the slope and ___ is the y-intercept.**	y; (,); $y =$ $x +$; O; x
Symbols	$y = mx + b$ (arrows to two blank boxes)	

EXAMPLE 1 Investigate Linear Transformations

Members of a health club pay a one-time membership fee plus a monthly charge. Recently prices have increased. The new membership cost is modeled by the function $g(x)$. The old membership cost is modeled by the function $f(x) = 25x + 100$, where x is the number of months.

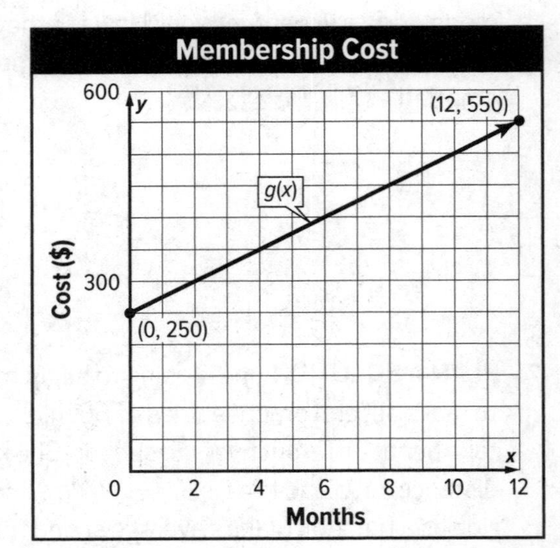

a. **REASON ABSTRACTLY** Graph $f(x)$ in the same coordinate plane as $g(x)$. Describe the transformation that maps $f(x)$ onto $g(x)$. How does the equation for $f(x)$ change and how is it related to the equation for $g(x)$? CCSS F.BF.3, SMP 2

b. INTERPRET PROBLEMS Write the equations for both lines in slope-intercept form. How are the values of m and b affected by transforming $f(x)$ to $g(x)$? Interpret the meaning of the slope and the y-intercept in the context of the problem. CCSS S.ID.7, SMP 1

c. USE A MODEL Suppose the new membership cost could be modeled by the function $h(x) = 50x + 100$. Graph and label $f(x)$ and $h(x)$ in the coordinate plane shown. Describe the transformation from $f(x)$ to $h(x)$. How does this transformation differ from $f(x)$ to $g(x)$? CCSS N.Q.2, SMP 4

Membership Cost

Cost ($): 0, 300, 600 ($y$)

Months: 0, 2, 4, 6, 8, 10, 12 (x)

The starting point for transformations of linear functions is the parent linear function $f(x) = x$. For graphing purposes, the function can also be written as $y = f(x)$ or $y = x$.

EXAMPLE 2 Use Technology to Graph Linear Transformations

You can use your graphing calculator to investigate transformations of linear functions. Start by rewriting the parent linear function $f(x) = x$ as $y = x$ and entering it into your calculator. Graph the function, then sketch and label it in the coordinate plane at the right.

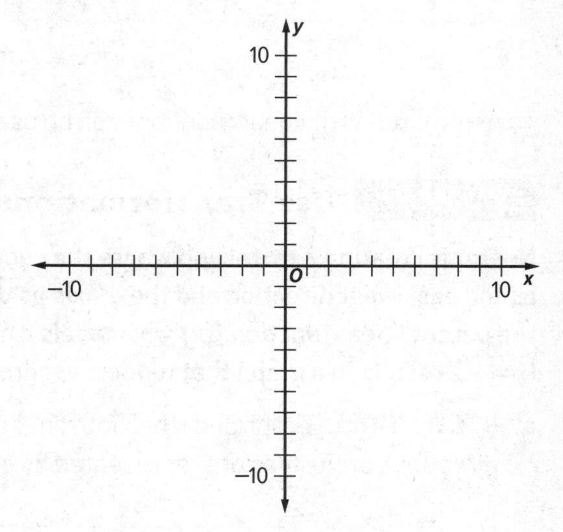

a. PLAN A SOLUTION Enter a second equation into your calculator by replacing $f(x)$ with $f(x) + k$, where $k = 6$, and enter the equation into your calculator. What equation did you write? Explain. CCSS F.BF.3, SMP 1

b. COMMUNICATE PRECISELY Display the graphs of both equations on your calculator. Add and label the graph from **part a** to your sketch above. Describe the transformation of the graph of the parent function. How are the slope and the y-intercept affected? CCSS S.ID.7, SMP 6

c. **USE TOOLS** Enter a third equation into your calculator by replacing $f(x)$ with $f(kx)$, where $k = 2$. Display this graph and the graph of $y = x$ on your graphing calculator. Then sketch and label both graphs in the coordinate plane at the right. Describe the transformation. How is it different from the transformation in **part b**? CCSS F.BF.3, SMP 5

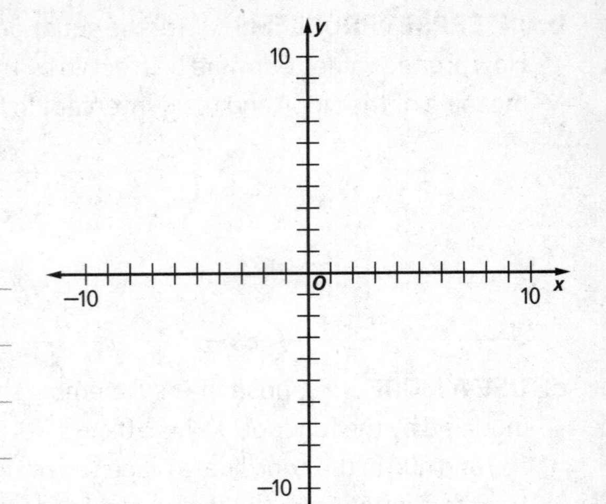

d. **REASON ABSTRACTLY** Formulate a general description for the effect of replacing a linear function $f(x)$ with $f(x) + k$ and $f(kx)$, where k is a real number $\neq 0$. CCSS F.BF.3, SMP 2

e. **REASON ABSTRACTLY** Are there any linear functions for which your conclusions in **part d** do not apply? Explain. CCSS S.ID.7, SMP 2

Transformations of linear functions can be used in various types of models.

EXAMPLE 3 Use Transformations in Models

Nestor is creating a travel app where the coordinate plane is a map, with the x-axis as the east-west direction and the y-axis as the north-south direction. The graph of the parent linear function $f(x) = x$ models a highway. Replacing $f(x)$ with $kf(x)$ where $k = -1$ results in a graph that models a railroad track.

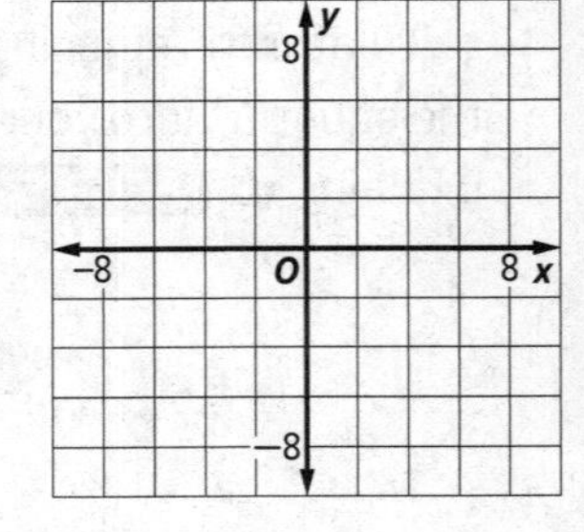

a. **USE A MODEL** Graph and label both lines in the coordinate plane at the right. What directions of the map are represented by each line? CCSS S.ID.7, SMP 4

b. **COMMUNICATE PRECISELY** Describe the transformation of the graph of the parent function. How are the slope and the y-intercept affected? CCSS F.BF.3, SMP 6

PRACTICE

Pilar is planning to ride her bike to the beach and back home again. Use this information for Exercises 1–3.

1. **USE A MODEL** The function $f(x)$, graphed in the coordinate plane, represents Pilar's planned rate of travel for the ride home. Write an equation for the line in slope-intercept form. What do the slope and y-intercept represent in the given context? CCSS S.ID.7, SMP 4

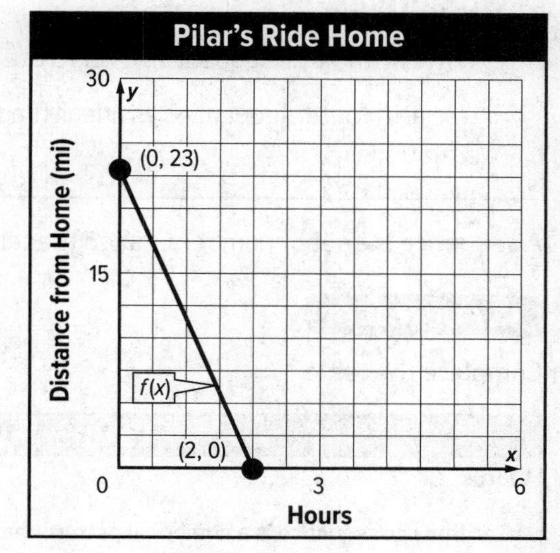

2. **INTERPRET PROBLEMS** Pilar's rate of travel on the ride home can be more accurately modeled by the function $g(x) = -8x + 23$, where x is the time in hours. Graph $g(x)$ in the same coordinate plane as $f(x)$. Describe the transformation that moves the line representing her planned trip to the line representing her actual trip. CCSS F.BF.3, SMP 1

3. **COMMUNICATE PRECISELY** Write the equations for both lines in slope-intercept form. How are the equations related to the transformation described in **Exercise 2**? 

4. **CRITIQUE REASONING** For any linear function, replacing $f(x)$ with $f(x + k)$ results in the graph of $f(x)$ being shifted k units to the right for $k < 0$ and shifted k units to the left for $k > 0$. Ana says that shifting the graph horizontally k units has the same effect as shifting the graph vertically $-k$ units. Do you agree? Create a graph(s) to help you justify your answer. CCSS F.BF.3, SMP 3

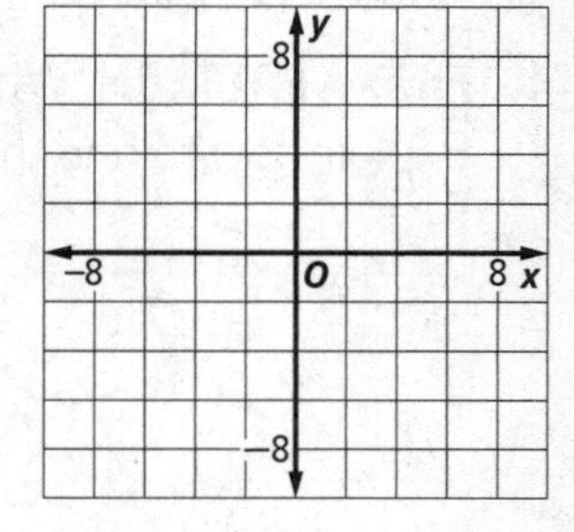

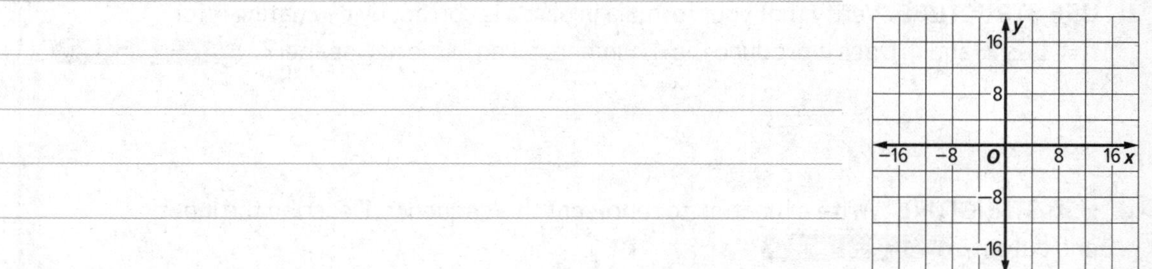

5.7 Arithmetic Sequences as Linear Functions

Objectives

- Write arithmetic sequences as discrete linear functions.
- Use arithmetic sequences as linear functions to model real-world situations.

CCSS STANDARDS

Content: F.BF.2, F.LE.2, A.CED.2, F.IF.3, F.BF.1a, A.REI.11
Practices: 1, 3, 4, 5, 7
Use with Lesson 3–5

A **sequence** is a set of numbers, called the terms of the sequence, in a specific order.

KEY CONCEPT

Complete the table.

Arithmetic Sequence		
Words An arithmetic sequence is a numerical pattern that increases or decreases at a constant rate called the ________________.		
Examples	2, 4, 6, 8, 10, … ____ ____ ____ ____ $d =$ ____	41, 38, 35, 32, 29, … ____ ____ ____ ____ $d =$ ____

To the find the nth term of an arithmetic sequence with the first term a_1, you can use the formula $a_n = a_1 + (n - 1)d$, where d is the common difference and n is a positive integer.

EXAMPLE 1 Investigate Arithmetic Sequences

EXPLORE Jorge noticed a contest advertised in the local newspaper, as shown. He thinks that he can solve the puzzle by using what he knows about arithmetic sequences.

Solve this Puzzle! Win a Prize!
What is the 100th number in the list below?
0.2, 0.35, 0.5, 0.65, …

a. FIND A PATTERN What do you notice about the numbers in the list? Explain how to find the nth term in the list. CCSS F.BF.2, SMP 7

____

____

____

b. USE STRUCTURE Verify that your formula in **part a** is correct by evaluating it for $n = 1, 2, 3$, and 4. Does it produce the 4 numbers given in the newspaper? CCSS F.IF.3, SMP 7

____

____

c. USE STRUCTURE Write a function to represent this sequence. Describe the input and output. CCSS F.IF.3, SMP 7

____

d. USE STRUCTURE Complete the table and graph the function. Determine the domain and the range of $a(n)$. CCSS F.IF.3, A.CED.2, SMP 7

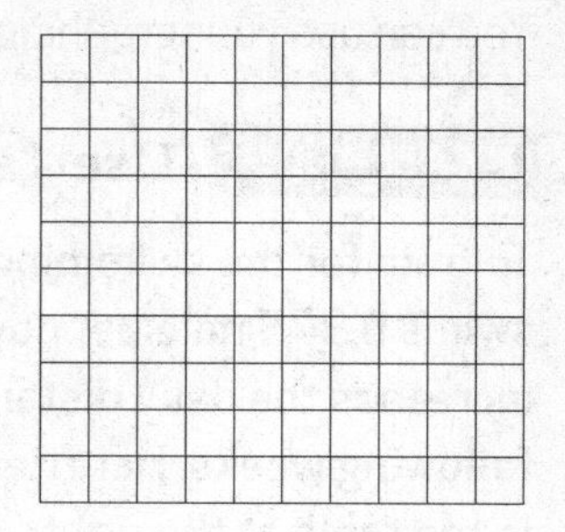

n						
$a(n)$						

e. REASON QUANTITATIVELY Solve the puzzle. Show your work. CCSS F.BF.2, SMP 2

You can use arithmetic sequences as linear functions to model real-world situations. While the domain of most linear functions is the set of all real numbers, the domain of an arithmetic sequence as a linear function is the set of counting numbers, {1, 2, 3, 4, ...}.

EXAMPLE 2 Create a Model Using an Arithmetic Sequences

An elevator stops at each floor as it descends from the top floor of a tall building to the lowest level of the parking garage 4 floors below ground level. The graph represents the feet above ground level as the elevator makes its first five stops.

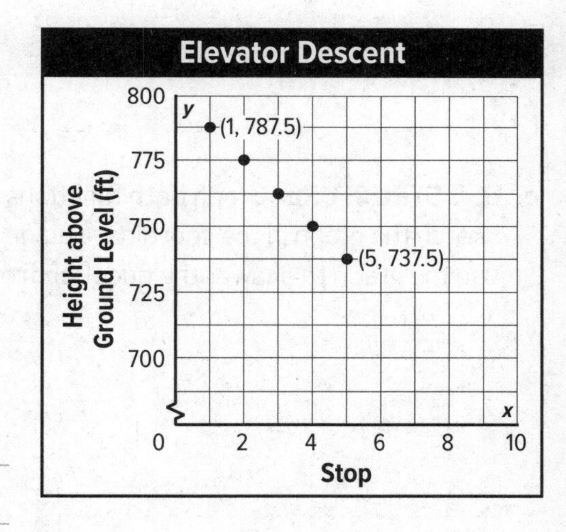

a. PLAN A SOLUTION Write an arithmetic sequence to represent the feet above ground level at each stop as the elevator descends. Then, write the function for the nth term. CCSS F.LE.2, F.BF.1a, SMP 1

b. USE STRUCTURE Determine the domain and range of the function you wrote in **part a** in the given context. Explain your reasoning. CCSS F.IF.3, SMP 7

You can use your graphing calculator to investigate arithmetic sequences as linear functions.

EXAMPLE 3 Use Technology with Arithmetic Sequences

To train for the swimming component of a triathlon, Marta swims 0.375 mile per day for the first week and then increases the daily distance by 0.125 mile each of the following weeks. Her friend Luz's training schedule is shown in the table at the right.

Week	Daily Distance
1	0.5
2	0.6
3	0.7
4	0.8

a. USE A MODEL Write arithmetic sequences to model each girl's weekly swimming distances. Then write the function for the *n*th term of each sequence. CCSS F.BF.2, F.BF.1a, SMP 4

b. USE TOOLS Enter both functions from **part a** into your calculator. Use the table feature to investigate how the functions are related. Do both girls ever swim the same daily distance during the same week? If so, during which week? Explain how you used the calculator to help you solve the problem. CCSS A.REI.11, F.IF.3, SMP 5

c. USE STRUCTURE Graph both functions on your calculator. Then sketch the graph in the coordinate plane at the right. How can you use the graph to answer the question from **part b**? CCSS A.CED.2, SMP 7

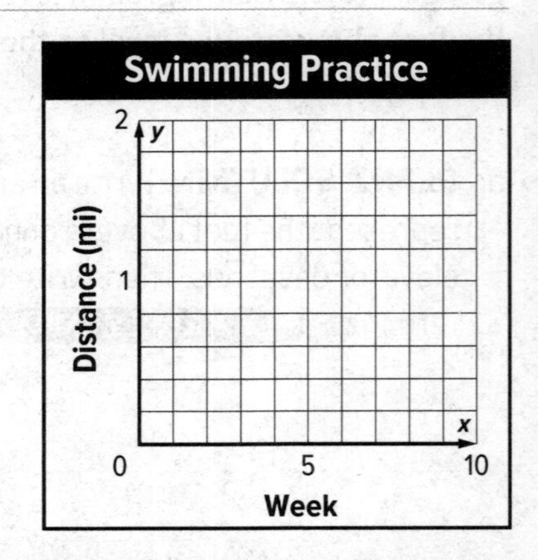

PRACTICE

1. **CRITIQUE REASONING** To represent the arithmetic sequence −0.8, −0.5, −0.2, 0.1, …, Dewayne wrote the linear function $f(n) = 0.3n - 0.5$. Is his function correct? If not, explain how to correct the error. CCSS F.BF.2, SMP 3

Kate is making a necklace with beads. The necklace is 8 cm long when she starts. Each bead she adds increases the length of the necklace by 2 mm. Use this information for Exercises 2 and 3.

2. USE A MODEL Write an arithmetic sequence to represent the length in centimeters of the necklace as Kate adds each bead. Then, write the formula for the *n*th term. CCSS F.BF.2, SMP 4

3. USE STRUCTURE Determine the domain and range if the greatest possible length of the necklace is 60 cm. Explain. CCSS F.IF.3, SMP 7

Andre and Sam are both reading the same novel. Andre reads 30 pages each day. Sam created the table at the right. Use this information for Exercises 4–6.

Sam's Reading Progress	
Day	**Pages Left to Read**
1	430
2	410
3	390
4	370

4. USE A MODEL Write arithmetic sequences to represent each boy's daily progress. Then write the function for the *n*th term of each sequence. CCSS F.BF.2, F.BF.1a, F.LE.2, SMP 4

5. USE TOOLS Enter both functions from **Exercise 4** into your calculator. Use the table to determine if there is a day when the number of pages Andre has read is equal to the number of pages Sam has left to read. If so, which day is it? Explain how you used the table feature to help you solve the problem. CCSS A.REI.11, SMP 5

6. USE STRUCTURE Graph both functions on your calculator, then sketch the graph in the coordinate plane at the right. How can you use the graph to answer the question from **Exercise 5**? CCSS A.CED.2, SMP 7

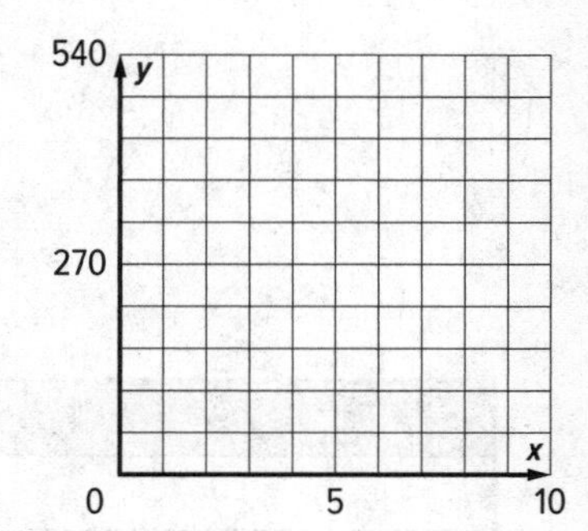

7. USE STRUCTURE For each arithmetic sequence described, write a formula for the *n*th term of a sequence that satisfies the description. CCSS F.BF.2, SMP 7

a. first term is negative, common difference is negative

b. second term is -5, common difference is 7

c. $a_2 = 8, a_3 = 6$

 5.8 Proportional and Nonproportional Relationships

Objectives

- Identify and describe proportional relationships in real world contexts.
- Construct linear functions that describe proportional relationships given a graph, table, or verbal description of a relationship.

Content: F.LE.1b, F.LE.2, A.CED.2, F.BF.1a
Practices: 1, 2, 3, 4, 5, 6
Use with Lesson 3-6

In a **proportional relationship**, two related quantities increase or decrease together. For example, if one value doubles, then the other value doubles as well. A proportional relationship between x and y can be described by an equation of the form $y = kx$.
A proportional relationship can be represented by a linear equation. However, not every linear equation represents a proportional relationship. If the graph of the linear equation does not pass through (0, 0), then the equation describes a **nonproportional relationship**.

EXAMPLE 1 Identifying Proportional Relationships CCSS F.LE.1b, SMP 6

EXPLORE Determine whether each relationship is *proportional*, *nonproportional*, or *nonlinear*. Explain your reasoning.

a. Caleb walks dogs in his neighborhood to earn money. He earns $2 for each hour he spends with a dog. Analyze the relationship between time spent and money earned.

b. A function containing the following points: {(−2,−5), (1, 4), (5, 16) (8, 25), (12, 37)}

c.

x	y
0	0
1	1
2	4
3	9
4	16

d.

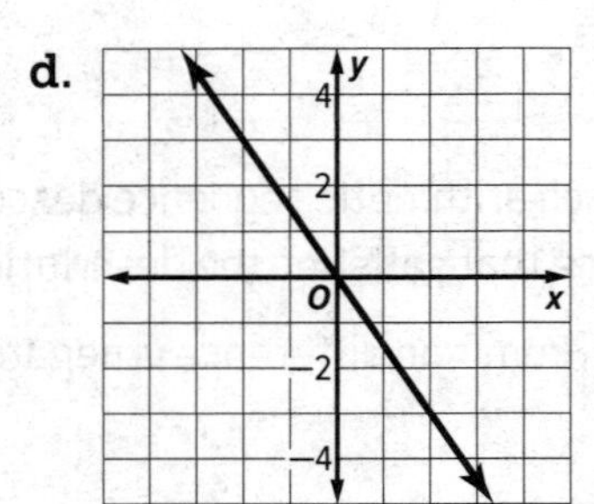

EXAMPLE 2 Write an Algebraic Expression CCSS F.LE.2, A.CED.2, F.BF.1a

INTERPRET PROBLEMS For each linear relationship in Example 1, write an equation that represents the relationship. If the relationship is not linear, write *not linear.* CCSS SMP 1

a. ______________________ b. ______________________

c. ______________________ d. ______________________

EXAMPLE 3 Model a Relationship

The table shows the relationship between different sizes and prices for one brand of snack crackers.

Ounces	Price
4	$0.95
10	$2.25
16	$4.00
28	$6.80

a. **COMMUNICATE PRECISELY** Describe the relationship represented by the table. Is it linear? Is it proportional? Explain your reasoning. CCSS F.LE.1b, SMP 6

____

____

b. **USE TOOLS** Real-world relationships are not always perfectly proportional or linear. Sometimes, a *regression line* is used to approximate these real world relationships. Plot the data from the table. Identify a proportional relationship that closely models this relationship of the data. Sketch your line. CCSS A.CED.2, SMP 5

____

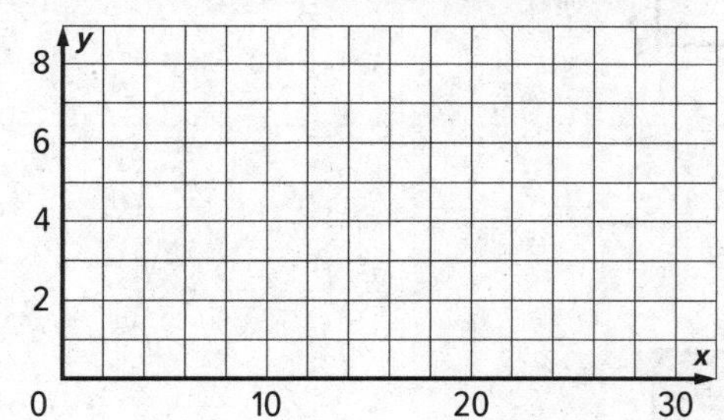

c. **USE A MODEL** Use your equation to predict a reasonable price for a 48-ounce box of crackers. CCSS F.BF.1a, SMP 4

____

d. **REASON QUANTITATIVELY** Determine the price per ounce for each of the sizes. Which size represents the best value? CCSS F.BF.1a, SMP 2

____

____

e. **USE A MODEL** If a 22-ounce box of snack crackers was placed on sale for $4.75, would this represent a good value? Why or why not? CCSS F.BF.1a, SMP 4

____

____

f. **REASON QUANTITATIVELY** Find the cost of a 32-ounce box of crackers if it has the same price per ounce as the 10-ounce box. CCSS A.CED.1, SMP 2

____

PRACTICE

1. **CRITIQUE REASONING** Avery claims that, if she is given a single point on the graph of a proportional relationship, she can write a function that defines the relationship. Is Avery correct? CCSS F.LE.1b, SMP 3

2. **CONSTRUCT ARGUMENTS** Determine whether each relationship is *proportional*, *nonproportional*, or *nonlinear*. Explain your reasoning. CCSS F.LE.1b, SMP 3

 a. $3(y - x) + 7 = 2(2x + 1) - 6x + 5$

 b.

x	y
−1	−4
0	−1
1	2
2	5

 c.

x	y
−2	−8
−1	−1
0	0
1	1
2	8

3. **CRITIQUE REASONING** Seung claims that the data in the table represents a proportional relationship. He says the values vary directly because each increase of 4 in the x value corresponds to an increase of 4 in the y value. Is he correct? Explain why or why not. CCSS F.LE.1b, SMP 3

x	y
−6	−9
−2	−5
2	−1
6	3

4. **REASON QUANTITATIVELY** As a real-estate agent, Mateo earns a commission of \$2975 on the sale of an \$85,000 house. CCSS SMP 2

 a. Write an equation that represents his commission rate. CCSS F.LE.2

 b. How much would Mateo earn on the sale of a \$140,000 house? CCSS F.BF.1a

 c. How much would a house have to sell for in order for Mateo to earn a commission of \$7875? CCSS F.BF.1a

 d. Use your equation from **part a** to determine the commission Mateo earns on the sale of a \$170,000 house. CCSS F.BF.1a

 e. Compare your answer to **part d** to the commission that Mateo earns on the sale of an \$85,000 house. Does it make sense? Explain. CCSS F.LE.1b

5. Antwan is comparing cell phone plans. Use the plan summaries to help you answer the following questions.

	Plan A	Plan B
Monthly Fee	\$0	\$25
Data Charge per GB	\$35	\$15

 a. **INTERPRET PROBLEMS** Write a function that describes the relationship between the number of gigabytes (GB) of data *d* and the cost of service *c* for each plan. CCSS F.BF.1a, F.LE.2, SMP 1

 b. **COMMUNICATE PRECISELY** Describe the relationship between number of GB of data and the total cost of service for each plan. Are they proportional, non-proportional, or neither? Explain your reasoning. CCSS F.LE.1b, SMP 6

 c. **USE A MODEL** Sketch the graph of each function and compare the two plans. Does Plan A always represent a better value than Plan B? Explain your reasoning. CCSS A.CED.2, SMP 4

Improving Cycling Speed

Provide a clear solution to the problem. Be sure to show all of your work, include all relevant drawings, and justify your answers.

A cyclist wants to improve his average speed for a race. The table shows the distance he has covered for different times. The data in the table can be used to calculate his speed over four 20-minute intervals. The goal is to cut the cyclist's overall time by 5 minutes.

Time (mins)	Distance (miles)
0	0
20	5
40	10
60	17
80	25

Part A

Sketch a graph of the data. Connect each pair of adjacent points with a line segment.

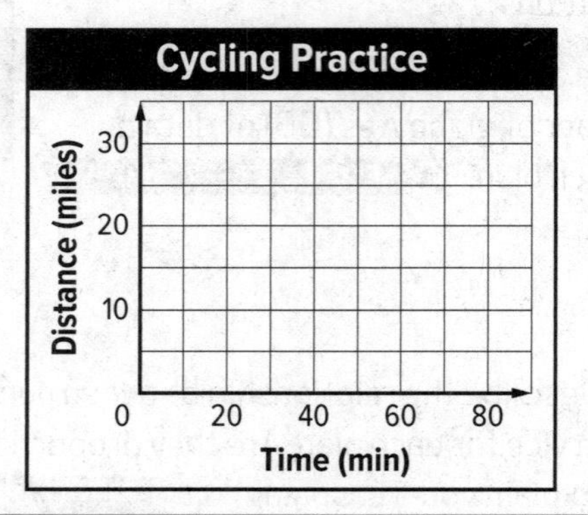

Part B

Calculate the average speed during each interval using the slope formula. Complete the table. Connect the first and last points and then find the average speed for the entire interval (0–80 mins).

	Average speed (miles/min)
Interval 1 (0–20 min)	
Interval 2 (20–40 min)	
Interval 3 (40–60 min)	
Interval 4 (60–80 min)	
Entire Interval (0–80)	

Part C

During which interval is the cyclist going the fastest? The slowest?

Part D

What recommendations would you make to cut his 25-mile time by 5 minutes? What would be the new overall average speed?

Performance Task

Estimating Income

Provide a clear solution to the problem. Be sure to show all of your work, include all relevant drawings, and justify your answers.

Jean has two part-time jobs, one that pays $7.50/hr. and another that pays $8.25/hr. The number of hours from one week to another varies, depending on the demand from each job. Help her find a way to earn $500 over a 4-week period. How many hours will she work each week?

Part A

Write an expression that shows how much money she earns from her first job.
Write an expression that shows how much money she earns from her second job.
Write an equation that shows how she can earn $500.

Part B

Write the equation from **Part A** in slope-intercept form. Sketch a graph of the function below.

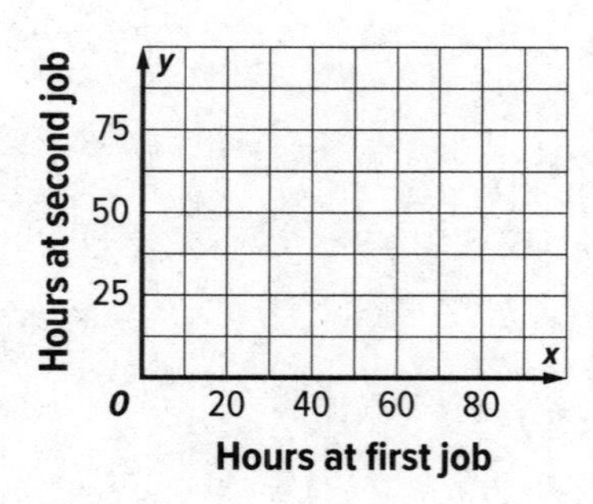

Part C

What do the *x*-*y* coordinates represent for each point along the line that represents the graph of the function?

Part D

Make a recommendation to Jean for a combination of hours at each job that will allow her to make $500. Determine the total number of hours she needs to work and divide it evenly among the four weeks.

Standardized Test Practice

1. Isadorra is walking home from school. In the graph below, y represents her distance from home in yards, and x represents the number of minutes since she left school.

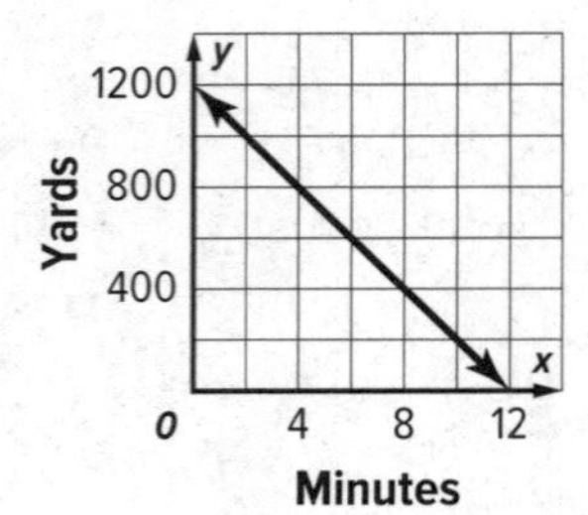

Use the graph to complete the following statements. CCSS A.REI.10

Isadorra's school is [ ] yards from her home.

After 3 minutes of walking, Isadorra was [ ] yards from home.

After [ ] minutes of walking, Isadorra was 200 yards from home.

It took Isadorra [ ] minutes to walk home.

2. Write a linear function that corresponds to the sequence 7, 10, 13, 16,…. CCSS F.BF.2

[ ]

3. The table shows the weight of Simon's puppy at several ages.

Age (weeks)	6	9	11	14
Weight (lbs)	3.5	6.2	7.8	12.7

Use the information to complete the following statements. CCSS F.IF.6

The average rate of change between weeks 6 and 14 is [ ] pounds per week.

The average rate of change between weeks 9 and 11 is [ ] pounds per week.

4. Ayashe pays a monthly fee for a gym membership and an additional $5 for each class he takes. If he takes 3 classes, his total bill is $30. Graph his monthly gym fee based on the number of classes he takes that month. CCSS F.IF.4

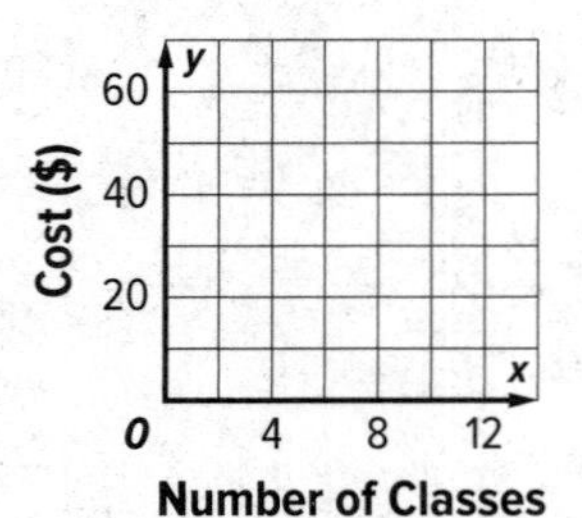

5. The function $y = 5.23x - 1.4$ models the height in millimeters of a tomato plant x days after a seed is planted. Select the true statements below. CCSS N.Q.2, N.Q.3

The plant grows about 36 mm per week.

The plant was exactly 19.52 mm tall 4 days after planting.

The plant is about 5 cm tall 10 days after planting.

The plant grows 5.23 mm every day.

6. The temperature inside an oven is 75°F and increases by a certain amount each minute after it is turned on. After 3 minutes, the temperature is 111°F. After 9 minutes, the temperature is 183°F. CCSS F.LE.2

The function that represents this situation, where x is the number of minutes after the oven is turned on, is [ ].

The first four terms of the arithmetic sequence that corresponds to the oven temperature each minute are [ ].

7. Sam and Aditi are both saving money for college. Sam's savings is represented by the function $y = 30x + 250$, and Aditi's savings is represented by the function $y = 30x + 180$, where x is the number of months they have been saving.

a. Explain the meaning of the 30 and the 180 in Aditi's function. 

b. Describe how the graphs of the two functions are different. 

8. Tell whether each equation, table, or description represents a proportional or nonproportional relationship. Explain your reasoning. CCSS F.LE.1b

a. $y = -2x + 7$

b.

x	3	7	22
y	1.5	3.5	11

c. Jason earns $9.50 per hour at his after school job.

9. A wading pool holds 160 gallons of water. When the plug is pulled, water drains from the pool at a rate of 5 gallons per minute.

a. What function represents this situation? 

b. What do x and y represent, and what are the units? 

c. Graph the function. CCSS F.IF.7a

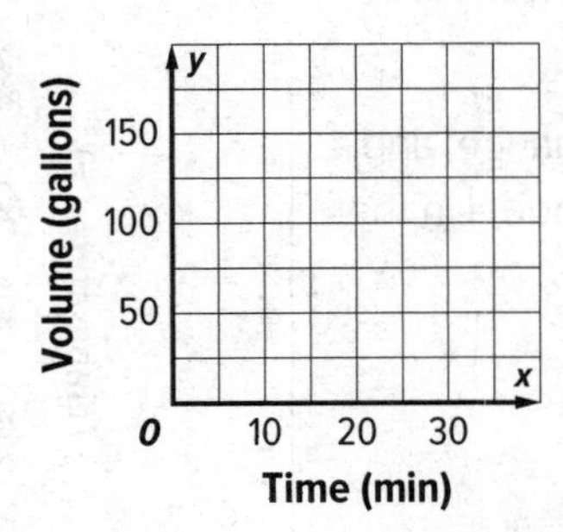

d. Use your graph to determine the number of minutes it takes for all of the water to drain from the pool. Verify your answer using your function from **part a**.

6 Equations of Linear Functions

CHAPTER FOCUS Learn about some of the Common Core State Standards that you will explore in this chapter. Answer the preview questions. As you complete each lesson, return to these pages to check your work.

What You Will Learn	Preview Question
Lesson 6.1: Equations in Slope-Intercept Form	
CCSS A.CED.2 Create equations in two or more variables to represent relationships between quantities; graph equations on coordinate axes with labels and scales. CCSS S.ID.7 Interpret the slope (rate of change) and the intercept (constant term) of a linear model in the context of the data. **Also addresses:** F.IF.7a, F.LE.2	CCSS SMP 2 A hang glider begins his trip at 400 feet above sea level. He descends at a rate of 40 feet per minute. Write an equation in slope-intercept form for the change in elevation. What do the x and y-intercepts represent?
Lesson 6.2: Equations in Point-Slope Form	
CCSS F.LE.2 Construct linear and exponential functions, including arithmetic and geometric sequences, given a graph, a description of a relationship, or two input-output pairs. **Also addresses:** F.IF.7a, A.CED.1, A.CED.2	CCSS SMP 2 Write an equation in point-slope form for the line that passes through the points, (1, 4) and (–1, 1). Then write the equation in slope-intercept form.
Lesson 6.3: Scatter Plots and Lines of Fit	
CCSS S.ID.6a Fit a function to the data; use functions fitted to data to solve problems in the context of the data. Use given functions or choose a function suggested by the context. Emphasize linear quadratic, and exponential models. CCSS S.ID.6c Fit a linear function for a scatter plot that suggests a linear association. CCSS S.ID.7 Interpret the slope (rate of change) and the intercept (constant term) of a linear model in the context of the data.	CCSS SMP 1 Temperature is related to latitude. Make a scatter plot using the data in the table. What type of relationship is shown?

Latitude	35	33	30	25	40
Temp. (°F)	46	52	67	76	37

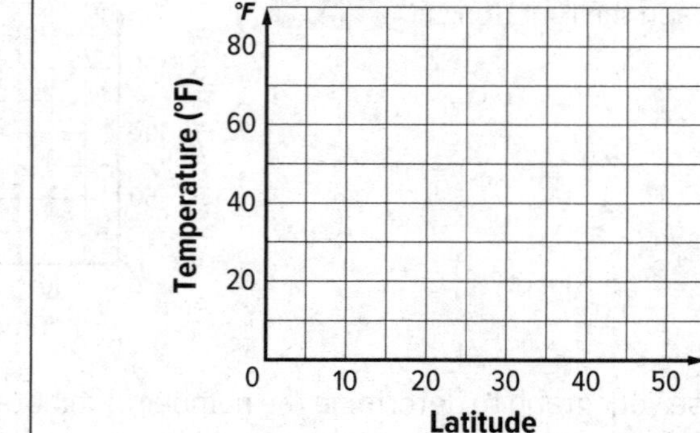

What You Will Learn	Preview Question
Lesson 6.4: Regression and Median-Fit Lines	
CCSS S.ID.6b Informally assess the fit of a function by plotting and analyzing residuals. CCSS S.ID.8 Compute (using technology) and interpret the correlation coefficient of a linear fit.	CCSS SMP 5 Use the data from the Preview Question for Lesson 6.3 to calculate the equation for the median-fit line. Use a graphing calculator and round to the nearest tenth. Predict the average temperature for the latitude 45°N.
Lesson 6.5: Correlation and Causation	
CCSS S.ID.9 Distinguish between correlation and causation.	CCSS SMP 2 For the situation, describe if a correlation is likely and if there is a causal relationship. Explain your reasoning. *the time spent at soccer practice and number of goals saved by the goalie*
Lesson 6.6: Inverse Linear Functions	
CCSS A.CED.2 Create equations in two or more variables to represent relationships between quantities; graph equations on coordinate axes with labels and scales. CCSS F.BF.4a Solve an equation of the form $f(x) = c$ for a simple function f that has an inverse and write an expression for the inverse.	CCSS SMP 6 Find the inverse of the function $y = -2x + 4$ and graph both functions. y; 4; 2; −4; −2; O; 2; 4; x; −2; −4 CCSS SMP 7 The point (0, 4) lies on the graph of the given function. Based on this, what point must lie on the graph of the inverse function? Show that the coordinates of this point satisfy the equation you wrote for the inverse.

6.1 Equations in Slope-Intercept Form

Objectives

- Write and graph equations using the slope-intercept form.
- Model and solve problems using equations in the slope-intercept form and their graphs.

CCSS STANDARDS

Content: F.IF.7a, F.LE.2, A.CED.2, S.ID.7
Practices: 1, 2, 4, 7, 8
Use with Lessons 4-1, 4-2, and 4-4

The **slope** of a line is the ratio of the change in the y-coordinates (rise) to the corresponding change in the x-coordinates (run) as you move from one point to another along a line. The ***y*-intercept** of a line is the y-coordinate of the point where the graph crosses the y-axis. The **slope-intercept** form of an equation is $\boldsymbol{y = mx + b}$, where m is the slope and b is the y-intercept.

EXAMPLE 1 Investigate Graphs of Linear Equations CCSS S.ID.7, A.CED.2

EXPLORE The tables show how the heights of two candles changed in relation to how long they burned.

a. **USE A MODEL** Graph the data for each candle. Connect the points with a straight line. CCSS SMP 4

Blue Candle	
Hours Burned	Height (in.)
0	9
4	6
8	3

Red Candle	
Hours Burned	Height (in.)
0	8
4	6
8	4

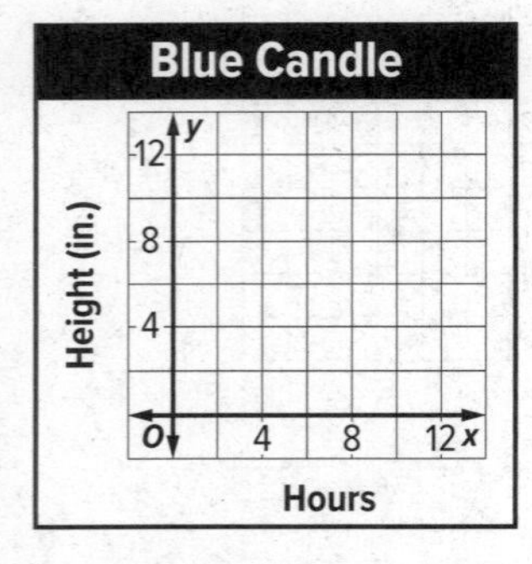

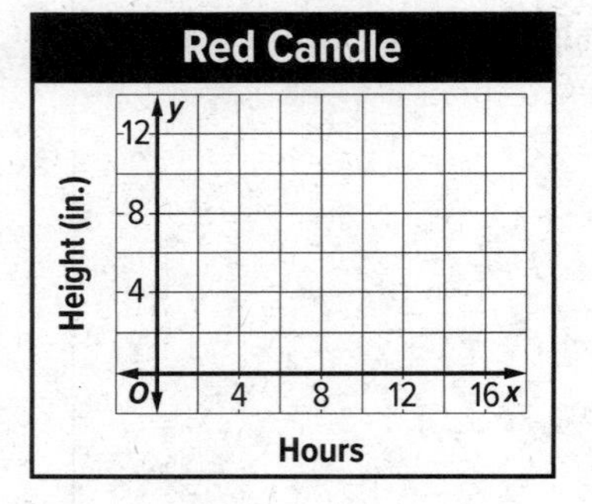

b. **INTERPRET PROBLEMS** What is the y-intercept of each graph? What do the y-intercepts mean in the context of the situation? CCSS SMP 1

c. **INTERPRET PROBLEMS** What is the x-intercept of each graph? What do the x-intercepts mean in the context of the situation? CCSS SMP 1

d. **REASON ABSTRACTLY** How can you tell, just by looking at the graphs, which candle burns at a faster rate? CCSS SMP 2

e. REASON QUANTITATIVELY What is the slope of each line? What does the slope tell you about the candles? CCSS SMP 2

f. REASON QUANTITATIVELY The slope in each equation is negative. Why does this make sense in the context of the situation? CCSS SMP 2

g. USE STRUCTURE What is the slope-intercept form of the equation for each graph? Explain your solution process. CCSS SMP 7

EXAMPLE 2 Write an Equation in Slope-Intercept Form CCSS F.LE.2, A.CED.2

USE STRUCTURE Write the equation of each line in slope-intercept form, $y = mx + b$. CCSS SMP 7

a.

x	y
−4	3
0	−5
4	−13

b.

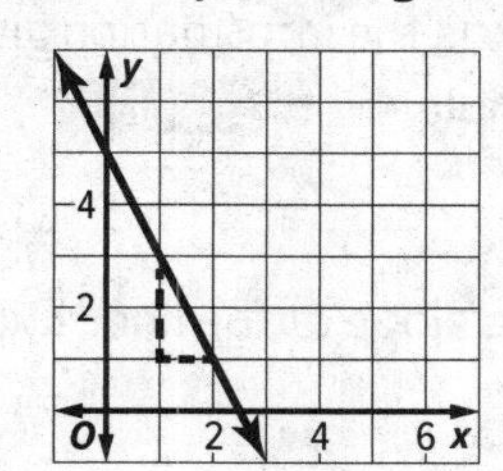

c. USE STRUCTURE Determine whether the lines are *parallel*, *perpendicular*, or *neither*. Explain your reasoning. CCSS SMP 7

EXAMPLE 3 Graph an Equation in Slope-Intercept Form

USE STRUCTURE Identify the slope and y-intercept of each equation. Then graph it. CCSS F.IF.7a, SMP 7

a. $y = \frac{3}{2}x - 2$

slope: ________ y-intercept ________

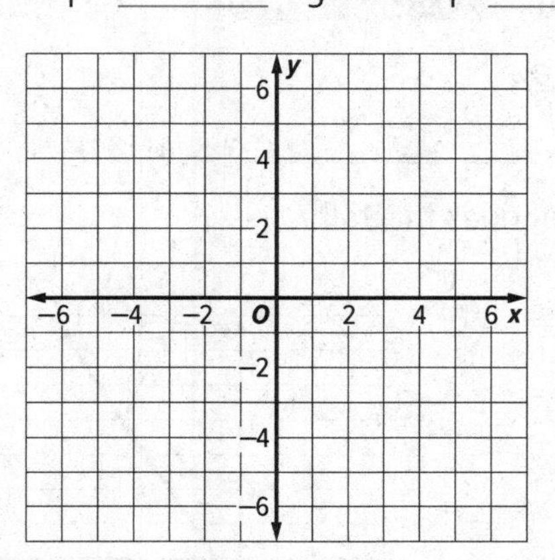

b. $2x + 3y = 12$

slope: ________ y-intercept ________

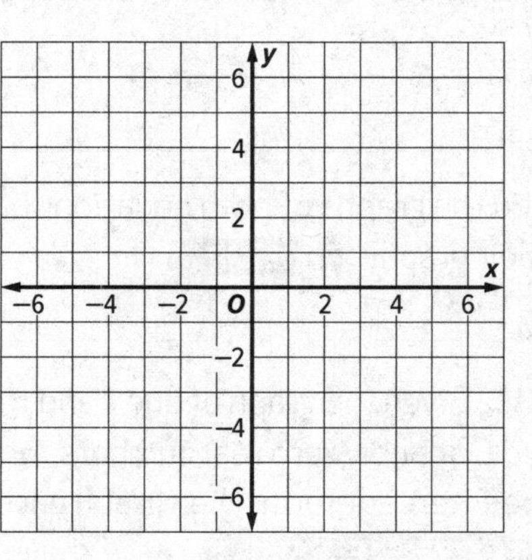

c. USE STRUCTURE Determine whether the lines are *parallel*, *perpendicular*, or *neither*. Explain your reasoning. CCSS SMP 7

EXAMPLE 4 Write the Equation of a Line Given a Point and the Slope CCSS F.LE.2, A.CED.2

Write the equation of a line that passes through $(-12, 6)$ and has a slope of $\frac{5}{6}$.

a. PLAN A SOLUTION Explain how to find the y-intercept CCSS SMP 1

____

b. USE STRUCTURE Find the y-intercept. 

____

c. USE STRUCTURE What is the equation in slope-intercept form? CCSS SMP 7

____

EXAMPLE 5 Write an Equation Given Two Points CCSS F.LE.2, A.CED.2

Write the equation of a line that passes through $(-3, 3)$ and $(3, -1)$.

a. INTERPRET PROBLEMS How is the information given in this example different from the information given in **Example 4**? CCSS SMP 1

____

b. USE STRUCTURE What is the slope of the line? Explain your solution process. 

____

c. DESCRIBE A METHOD Explain how to write the equation in slope-intercept form. 

____

____

d. USE STRUCTURE What is the equation of the line in slope-intercept form? How can you check that the equation is correct? CCSS SMP 7

____

PRACTICE

1. USE STRUCTURE Write the equation of each line in slope-intercept form. CCSS F.LE.2, SMP 7

a. slope: $\frac{4}{5}$

y-intercept: -8 ______________________

b. $8x + 3y = -24$

____

c. What point do the graphs of both equations have in common and what does this tell you about their graph? CCSS SMP 7

____

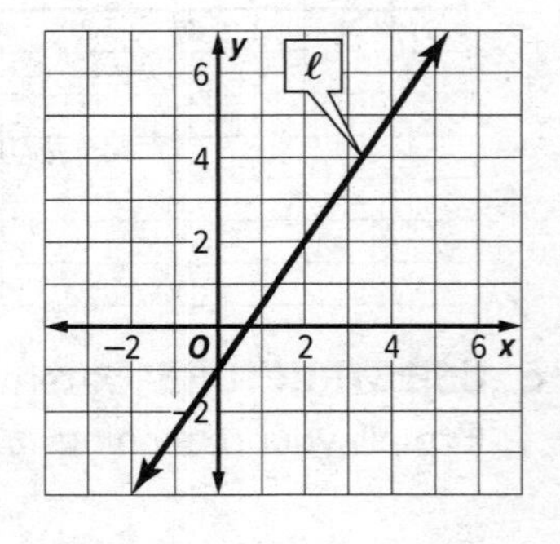

2. USE STRUCTURE Given the graph of line ℓ and the equation $Ax + By = C$, in which A, B, and C are non-zero real numbers, write the equation of each line in slope-intercept form. Then graph each line and label its intercepts. CCSS F.IF.7a, F.LE.2, SMP 7

a. Line h: $A = 6$, $C = 10$, and $h \parallel \ell$

____

b. Line $\mathcal{J}$: $B = 1$, $\mathcal{J} \perp \ell$, and has the same y-intercept as ℓ

3. REASON QUANTITATIVELY Koby tracked the weight of his puppy for 6 months. Her growth can be modeled with the equation $p = 7.5m + 1$ with $m =$ age in months and $p =$ weight in pounds. CCSS S.ID.7, SMP 2

a. What is the y-intercept? What does the y-intercept mean in the context of the problem?

b. What is the slope? What does the slope mean in the context of the problem?

4. a. USE STRUCTURE Levy claims that the line through $(-6, -2)$ and $(2, 10)$ is perpendicular to the graph of $3x - 2y = 10$? Do you agree? Explain why or why not. CCSS S.ID.7, SMP 7

b. USE STRUCTURE Jamal says that the line through $(7, -10)$ and $(3, -2)$ is parallel to $2x - y = -5$. Do you agree? Explain why or why not. CCSS S.ID.7, SMP 7

c. REASON QUANTITAVELY Alonae says that the line through $(1, -4)$ and $(5, -6)$ is parallel to the line through $(2, -7)$ and $(5, -6)$. How can you tell she is mistaken without determining the slope? CCSS S.ID.7, SMP 2

5. INTERPRET PROBLEMS Sierra borrowed money from her brother and paid back a set amount each week. The table shows how much she owed in a given week. CCSS S.ID.7, SMP 1

a. Graph the data and connect the points with a straight line.

Week	3	6	8	10	13
Amount Owed	\$32.50	\$25.00	\$20.00	\$15.00	\$7.50

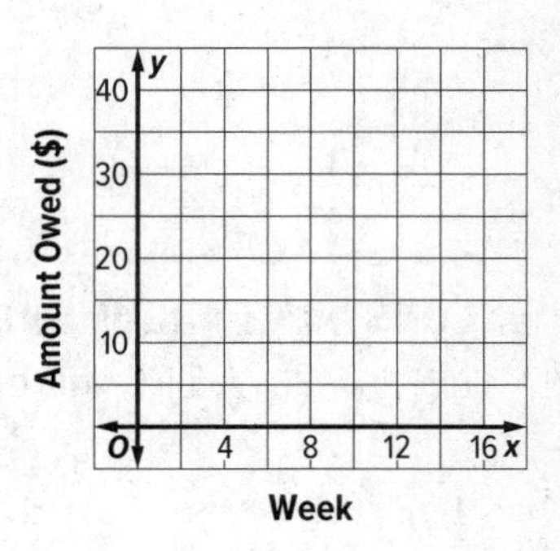

b. What does the y-intercept tell you about the problem?

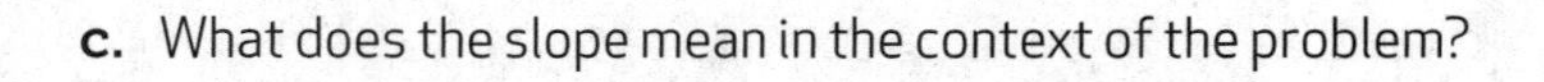

c. What does the slope mean in the context of the problem?

d. What does the x-intercept mean in the context of this situation?

e. What does the fact that the graph is a straight line tell you about the problem?

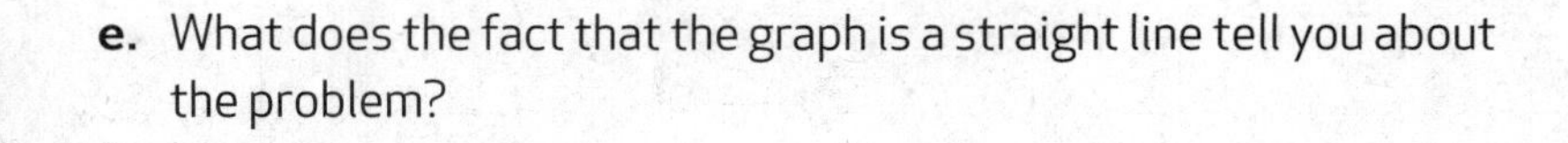

6.2 Equations in Point-Slope Form

Objectives

- Write the equation of a line in the point-slope form.
- Graph the equation of a line using the point-slope form.

CCSS STANDARDS

Content: F.IF.7a, F.LE.2, A.CED.1, A.CED.2
Practices: 1, 2, 3, 4, 7
Use with Lesson 4-3

The linear equation $y - y_1 = m(x - x_1)$ is written in **point-slope form** when (x_1, y_1) is a point on the line and m is the slope of the line.

EXAMPLE 1 Investigate the Point Slope Formula

EXPLORE **Consider the graph at the right.** CCSS F.LE.2

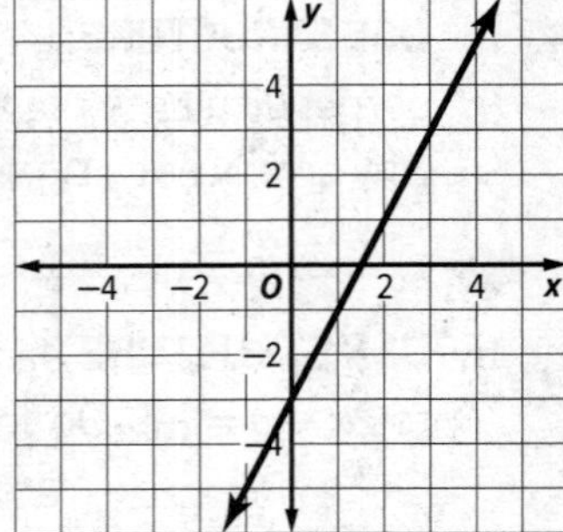

a. **USE STRUCTURE** Find the slope of the line. Substitute the slope, the coordinates of one point, and an arbitrary point (x, y), into the slope formula. CCSS SMP 7

b. **CONSTRUCT ARGUMENTS** Write the equation from **part a** in point-slope form. Compare your equation with other students' equations. Explain any differences or similarities. CCSS SMP 3

c. **CRITIQUE REASONING** Write this equation in slope-intercept form. Compare your equation with other students' equations. Explain any differences or similarities. CCSS SMP 3

d. **CONSTRUCT ARGUMENTS** What advantage does the slope-intercept form of a line have over the point-slope form? What advantage does the point-slope form have? CCSS SMP 3

e. **CONSTRUCT ARGUMENTS** Brianna writes the point-slope equation as $y + 1 = 2(x - 1)$. Is this equation correct? Explain. CCSS SMP 3

f. **REASON ABSTRACTLY** Manipulate the general forms of the point-slope and slope-intercept equations to explain how these forms are related. CCSS SMP 2

EXAMPLE 2 Graph a Line Using the Point-Slope Form

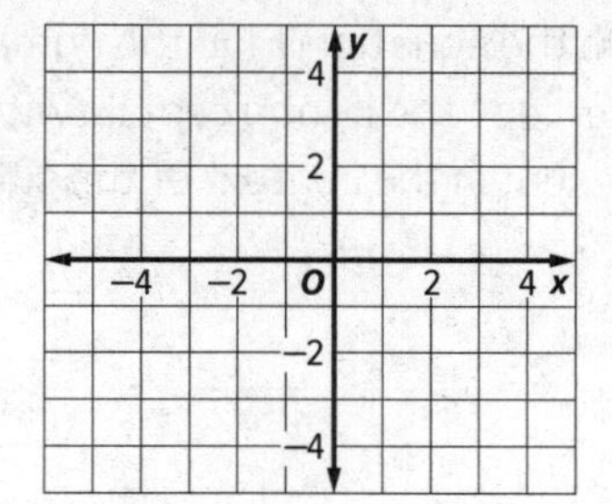

USE STRUCTURE Use the equation to find the slope and the coordinates of a point on the line. Then graph the line and show and label the intercepts. CCSS F.IF.7a, SMP 7

a. $y - 3 = -\frac{3}{4}(x + 5)$

point: ______________ slope: ______________

x-intercept: ______________ y-intercept: ______________

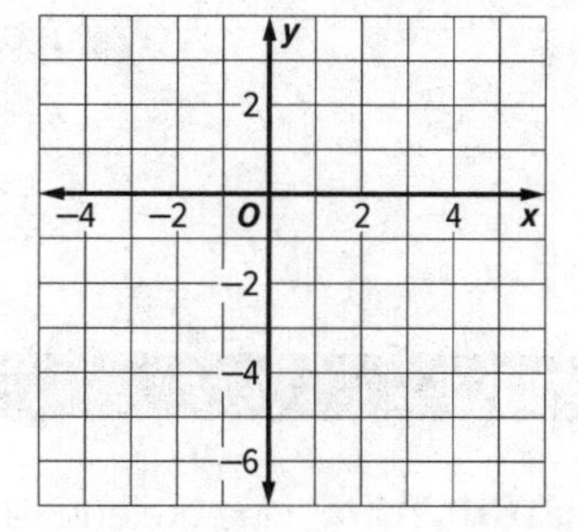

b. $y + 5 = 2(x + 1)$

point: ______________ slope: ______________

x-intercept: ______________ y-intercept: ______________

EXAMPLE 3 Write the Equation of a Line in Point-Slope Form CCSS F.LE.2

USE STRUCTURE Write an equation in point-slope form using the given information. CCSS SMP 7

a. parallel to $6x - 5y = 5$ and passes through $(-4, 6)$

____

b. passes through $(3, -2)$ and $(5, 4)$

____

c.

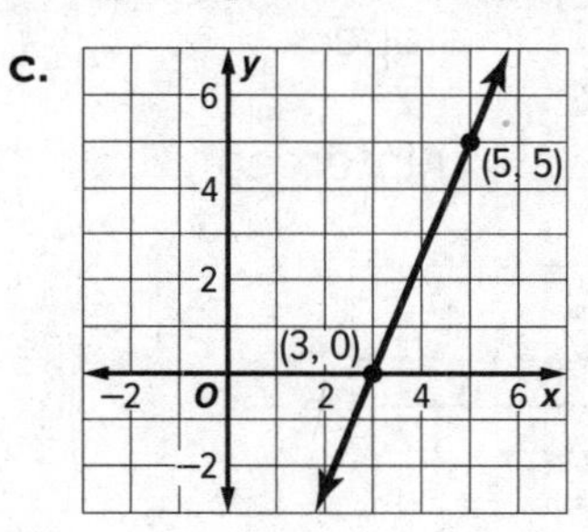

____

EXAMPLE 4 Solve a Real World Problem

INTERPRET PROBLEMS Sofia pays $20 per month for a gym membership, after paying an initial fee. After 15 months, she has paid a total of $450. CCSS SMP 1

a. Write an equation in point-slope form that represents the total amount she has paid, y, after a given number of months, x. Explain the meaning of each constant. CCSS A.CED.2, F.LE.2

____

____

____

b. Rewrite the equation in slope-intercept form where y represents the total amount she has paid after a given number of months, x. Explain the meaning of each constant. CCSS A.CED.2, F.LE.2

____

____

c. Graph the equation. Find the domain and range. Determine whether the x-intercept and y-intercept lie within the domain and range. What do they represent in the context of the situation? CCSS F.IF.7a

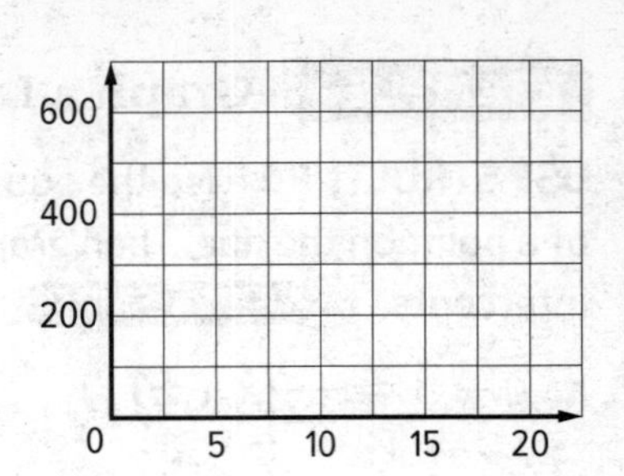

PRACTICE

1. USE STRUCTURE Use the equation to find the slope and the coordinates of a point on the line. Then graph the line. Show and label the intercepts. CCSS F.IF.7a, SMP 7

a. $y + 4 = \frac{1}{3}(x - 5)$

point: ______ slope: ______

x-intercept: ______ y-intercept: ______

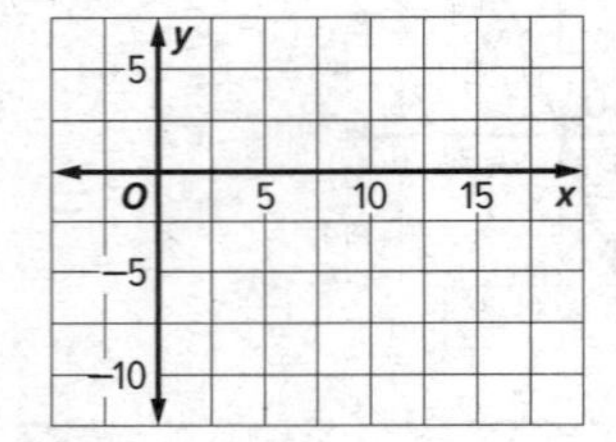

b. $y - 2 = -2.5(x + 8)$

point: ______ slope: ______

x-intercept: ______ y-intercept: ______

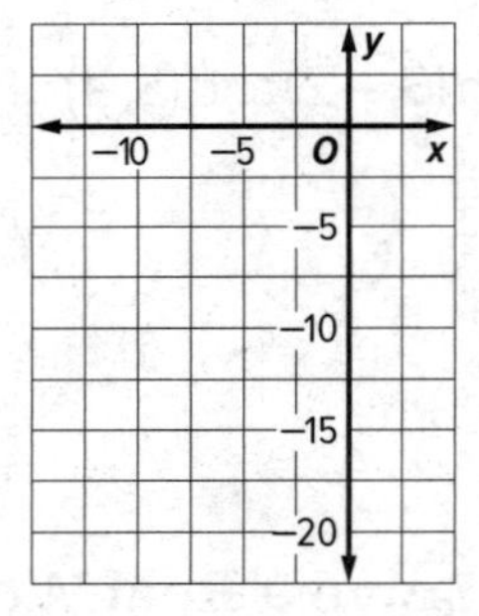

2. USE STRUCTURE Write an equation in point-slope form using the given information. CCSS F.LE.2, SMP 7

a. passes through $(2, -1)$ and perpendicular to $5y - x = 5$

b. The following are solutions:

x	−5	−2	1	4
y	7	5	3	1

c.

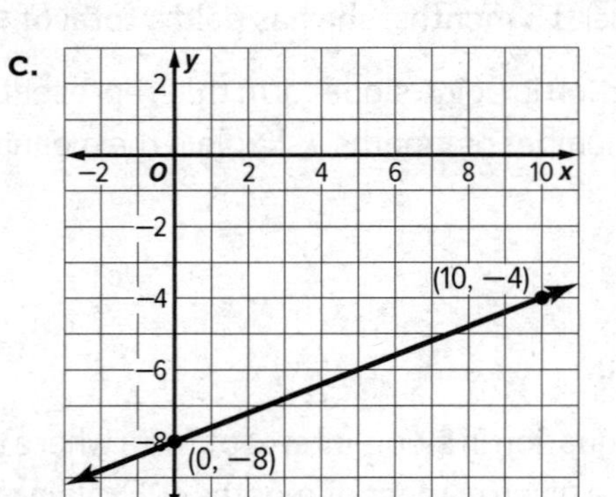

3. INTERPRET PROBLEMS When Santo was born, his uncle put money in a bank account to help pay for a car when Santo became a teenager. Each birthday, his uncle adds \$200 to the account. When Santo was 10, there was \$4000 in the account. CCSS SMP 1

a. Write an equation in point-slope form that represents the amount y after x years. How much will be in the account when Santo is 16? Justify your answer. CCSS A.CED.2, F.LE.2

b. How much was Santo's uncle's initial deposit? Explain your reasoning. CCSS A.CED.1

4. The equation $y - 750 = 60(x - 5)$ represents the height in feet of a hot-air balloon x seconds after students in a science class began to track it. CCSS F.IF.7a

a. USE A MODEL Graph this function. Label the x- and y-intercepts. CCSS SMP 4

b. REASON ABSTRACTLY The teacher asks the students different questions about the hot-air balloon. Jermaine's answer is "the y-intercept, 450 feet;" Suri's answer is "the x-intercept, -7.5 seconds;" and Cid's answer is "60 feet per second." What question did each student answer? CCSS SMP 2

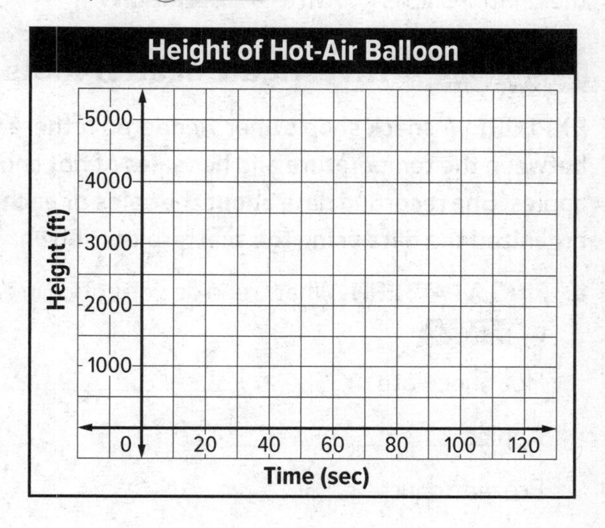

5. Sonia is ordering custom baseball caps online. The table shows the price for having the caps printed and shipped.

a. USE A MODEL Write an equation in point-slope form to represent the relationship between the price and the number of caps. CCSS F.LE.2, A.CED.2, SMP 4

Number of Caps	Total Price
5	55
10	95
15	135
20	175

b. INTERPRET PROBLEMS How many caps can Sonia order with a budget of \$200? Justify your answer. CCSS A.CED.1, SMP 1

c. USE STRUCTURE Use the equation you wrote in **part a** to draw a graph for the relationship between the price and the number of caps. Label the intercepts. CCSS F.IF.7a, SMP 7

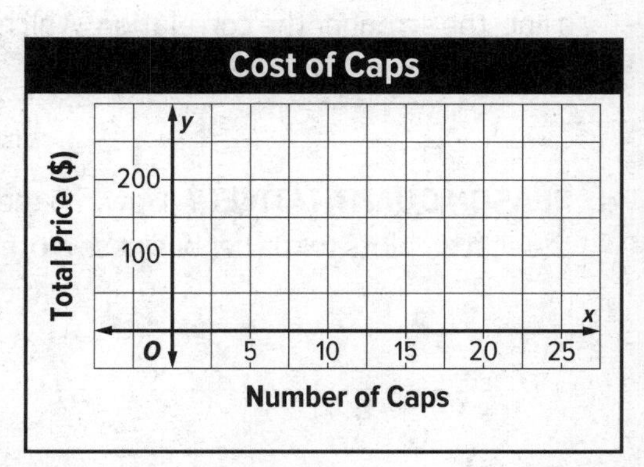

6.3 Scatter Plots and Lines of Fit

Objectives

- Interpret scatter plots.
- Draw a line of fit for a scatter plot.
- Use lines of fit to solve problems.

CCSS STANDARDS

Content: S.ID.6a, S.ID.6c, S.ID.7
Practices: 1, 2, 4, 7, 8
Use with Lesson 4–5

Data involving two variables is called **bivariate data**. A **scatter plot** shows the relation between a set of bivariate data. If the data clusters in a linear pattern, the relationship is known as a **correlation**.

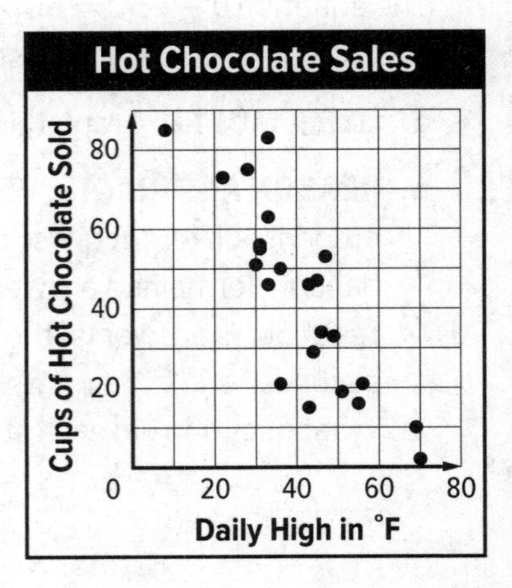

EXAMPLE 1 Investigate Scatter Plots CCSS S.ID.6c

EXPLORE A snack shop owner wondered if there was a relationship between the temperature and her sales of hot chocolate, frozen yogurt, and apples. She recorded data about the sales of each and the temperature. She organized the data using the scatterplots shown.

a. FIND A PATTERN What trend do you notice in each scatter plot? CCSS SMP 7

Hot Chocolate ________________________________

Apples ________________________________

Frozen Yogurt ________________________________

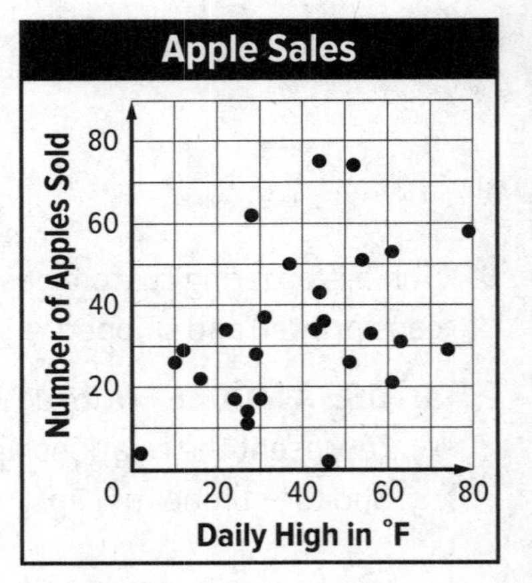

b. INTERPRET PROBLEMS When the dots appear to follow a line it, the data has a correlation. When one value increases as the other increases, the data has a **positive correlation**. When one value decreases as the other increases, the data has a **negative correlation**. Describe the type of correlation, if any, shown by each scatter plot. CCSS SMP 1

Hot Chocolate ________________________________

Apples ________________________________

Frozen Yogurt ________________________________

c. INTERPRET PROBLEMS The more closely the data points cluster along a line, the stronger the correlation. Which data shows the strongest correlation? CCSS SMP 1

____

d. REASON QUANTITATIVELY How can the owner of the snack shop use the scatter plots to run her business more efficiently? CCSS SMP 2

____

____

EXAMPLE 2 Lines of Fit CCSS S.ID.6a

Victoria's basketball coach kept a record of the number of baskets each player attempted compared to the number made. She recorded the data in a scatter plot. How can she write an equation that models the relationship?

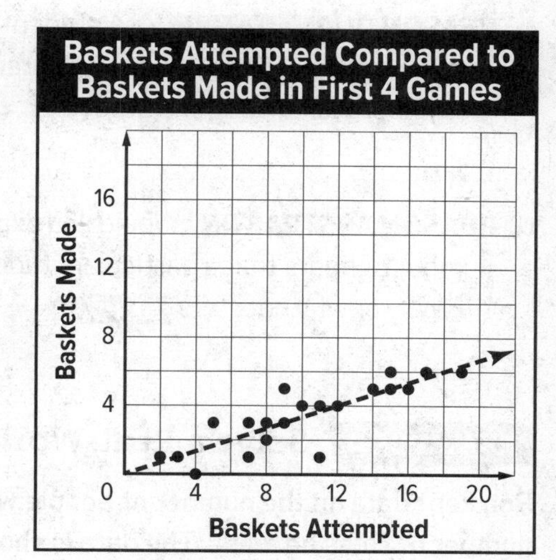

a. DESCRIBE A METHOD The coach drew a line that showed the trend of the data. The line is called a **trend line** or a **line of fit**. About half the points are above it and half the points are below it. How can the coach write an equation of the line? CCSS SMP 8

b. USE STRUCTURE What are two points on the line of fit? CCSS SMP 7

c. USE STRUCTURE Does a line of fit *have to* include any points from the collected data? CCSS SMP 7

d. USE STRUCTURE What is the slope of the line of fit? CCSS SMP 7

e. USE STRUCTURE What is the equation of the line of fit? CCSS SMP 7

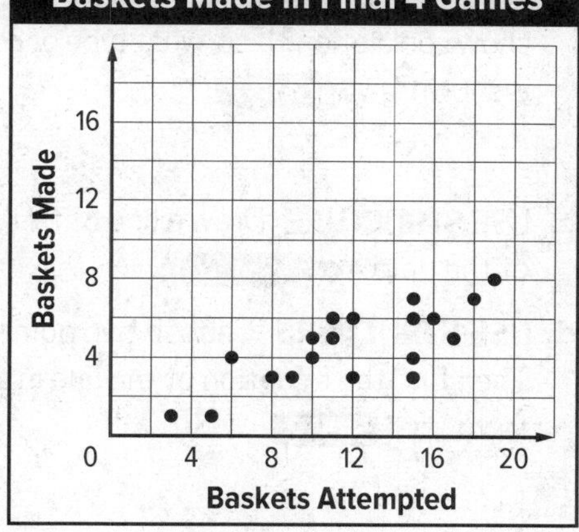

f. USE STRUCTURE The coach collected and recorded the data for the final 4 games of the season. Draw a line of fit for the data. Then name two points on the line. CCSS SMP 7

g. USE STRUCTURE What is the slope of the line of fit? CCSS SMP 7

h. USE STRUCTURE What is the equation of the line of fit? CCSS SMP 7

EXAMPLE 3 Make Predictions Using a Line of Fit CCSS S.ID.6a, S.ID.7

The scatter plot compares the hours students spend studying with the hours they spend watching television. The equation of the line of fit is $y = -\frac{4}{3}x + 25$ in which y = hours of television watched and x = hours spent studying.

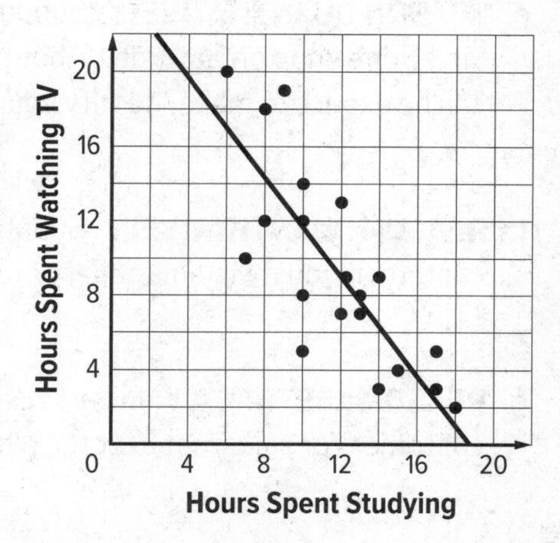

a. USE STRUCTURE Is the correlation positive or negative? Explain your reasoning CCSS SMP 7

b. REASON QUANTITATIVELY A student says she studies 15 hours a week. Explain how to use the line of fit to predict the number of hours she watches television. How many hours does she spend watching television? CCSS SMP 2

c. REASON QUANTITATIVELY A student says he watches television 15 hours a week. Explain how to use the line of fit to predict the number hours he spends studying. How many hours does he spend studying? CCSS SMP 2

d. REASON ABSTRACTLY Would you expect to find the data point that represents the number of hours that a particular student spent studying and watching TV on the line of fit? Why or why not? CCSS SMP 2

EXAMPLE 4 Solve a Real-World Problem CCSS S.ID.6a, S.ID.6c

Ron kept data on the number of people who visited his photography gallery and the number of sales he made. The data is shown in the table.

Customers	39	34	28	20	43	42	40	53	44	64	12	28	38	24
Sales	2	2	3	1	2	4	3	5	3	6	1	0	3	2

a. USE A MODEL Display the data on the scatter plot. CCSS SMP 4

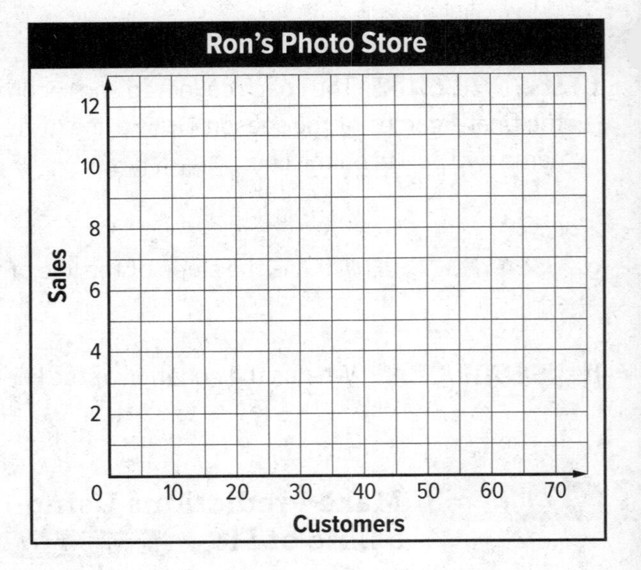

b. INTERPRET PROBLEMS Does the scatter plot show correlation? If so, what type of correlation is shown? CCSS SMP 1

c. USE STRUCTURE Draw a line of fit to show the trend of the data. CCSS SMP 7

d. USE STRUCTURE Choose two points on the line. Then find the equation of the line in slope-intercept form. CCSS SMP 7

e. REASON QUANTITATIVELY Suppose 50 customers visit Ron's shop on Saturday. About how many sales can he expect to make? Justify your answer. CCSS SMP 2

f. REASON QUANTITATIVELY Suppose Ron wants to make 10 sales. About how many customers must visit his gallery? Justify your answer. CCSS SMP 2

g. REASON ABSTRACTLY In parts *e* and *f* you were able to tell Ron *about* how many customers or sales, not *exactly*. Why could you not provide exact information? CCSS SMP 2

1. **USE STRUCTURE** Decide if the data in the scatter plot shows positive correlation, negative correlation, or no correlation. Then draw a line of fit that shows the trend of the data. Finally, write the equation of your line of fit. CCSS S.ID.6c, SMP 7

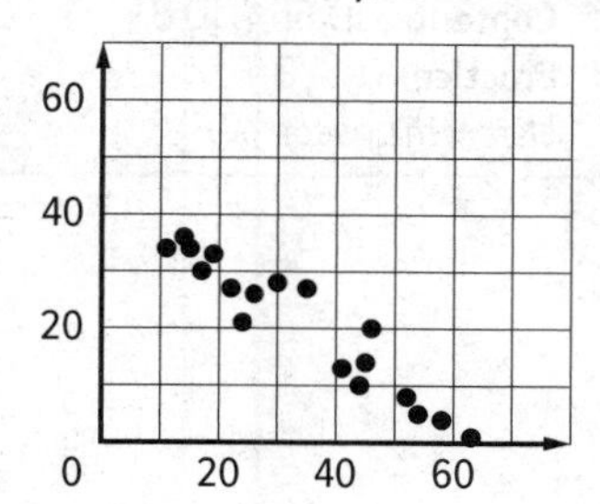

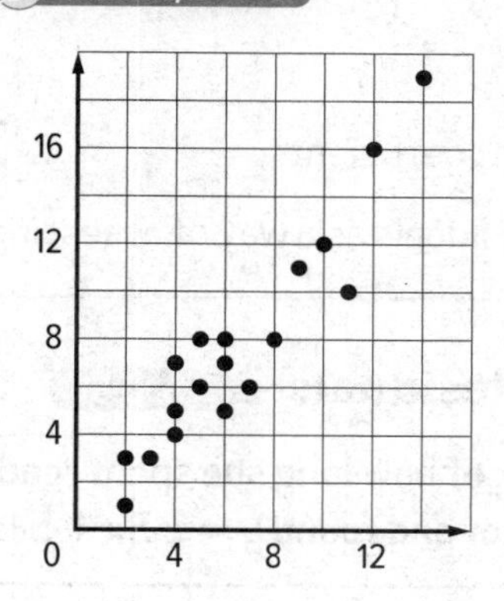

Correlation: ____________ Correlation: ____________

Equation of Line: ____________ Equation of Line: ____________

2. Several groups volunteered to each clean up litter along a mile of the highway near their town. The table shows how many people were in each group and how long it took each group to finish the job. CCSS S.ID.6a, S.ID.6c

Workers	9	16	18	8	15	11	9	17	9	15	11	12
Minutes	80	40	35	90	60	60	70	30	70	50	80	70

a. **USE A MODEL** Display the data on the scatter plot. 

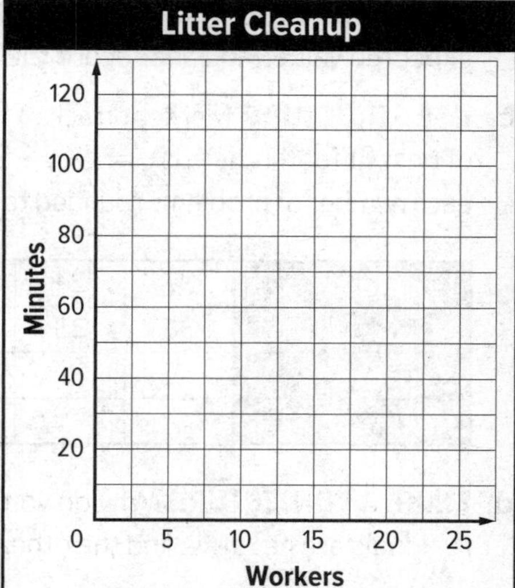

b. **USE STRUCTURE** Draw a line of fit to show the trend of the data. CCSS SMP 7

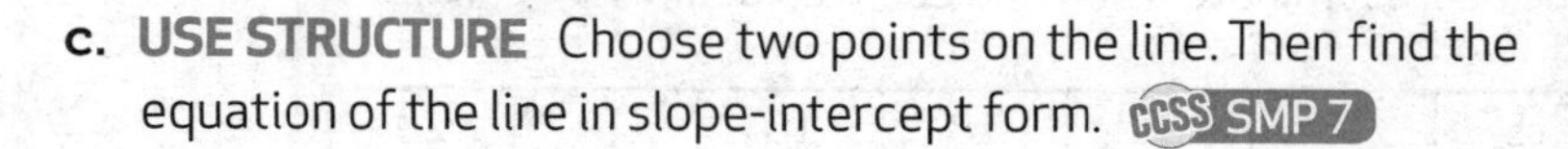

c. **USE STRUCTURE** Choose two points on the line. Then find the equation of the line in slope-intercept form. CCSS SMP 7

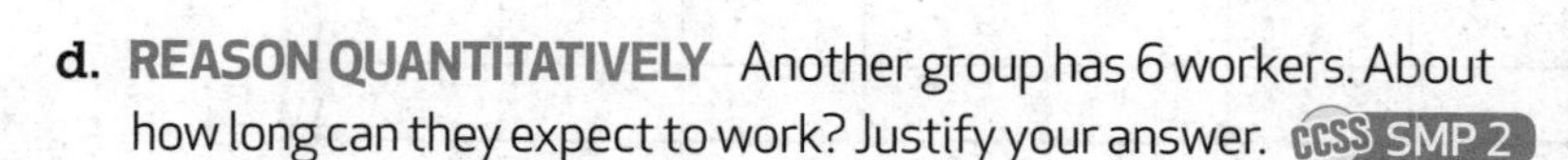

d. **REASON QUANTITATIVELY** Another group has 6 workers. About how long can they expect to work? Justify your answer. CCSS SMP 2

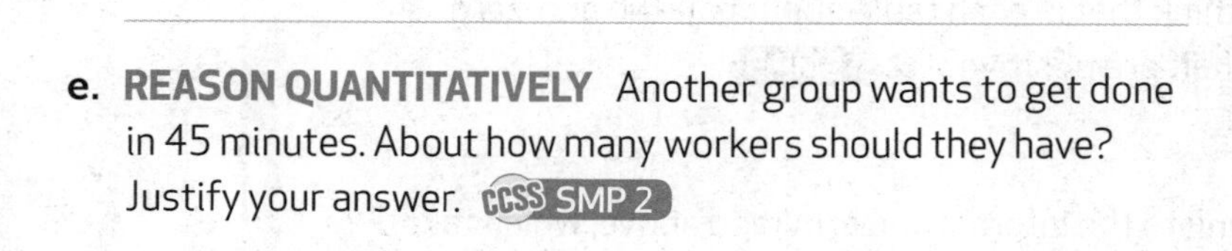

e. **REASON QUANTITATIVELY** Another group wants to get done in 45 minutes. About how many workers should they have? Justify your answer. CCSS SMP 2

f. **INTERPRET PROBLEMS** Find the y-intercept of your equations and discuss the problem with its meaning in this situation. CCSS SMP 1

6.4 Regression and Median-Fit Lines

Objectives

- Find residuals
- Interpret correlation coefficients
- Plot and interpret residuals as a way of assessing a line of fit

CCSS STANDARDS

Content: S.ID.6b, S.ID.8
Practices: 1, 2, 3, 6, 7
Use with Lesson 4-6

EXAMPLE 1 Find Residuals CCSS S.ID.6b

EXPLORE Mai kept track of how long she spent reading her history book. She entered the data into her calculator and found $y = 1.3x + 5$ is an equation for the line of best fit.

Pages (x)	12	10	8	11	17	10	26	18	12
Minutes (y)	22	18	14	19	27	20	40	29	20
Expected y	$y = 1.3(12) + 5 \approx \mathbf{21}$								
Residual	$22 - 21 = 1$								

a. **USE STRUCTURE** The y-values shown in the table are the actual times it took her to read the given number of pages. The **expected y** is the estimated time given by substituting each number of pages (x) into the equation for the best fit line. Use the equation to find and record the expected time for each number of pages. Round to the nearest minute. CCSS SMP 7

b. **USE STRUCTURE** The **residual** is the difference between the actual value and the expected value. Find and record the residuals for the data in the table. CCSS SMP 7

c. **USE STRUCTURE** Mai kept track of how long she spent working math problems. The line of best fit for this data is $y = 1.4x + 2.5$. Use the equation to find the expected time for each number of problems rounded to the nearest minute. Then find each residual. CCSS SMP 7

Problems (x)	27	24	15	16	25	18	16	24	20
Minutes (y)	38	33	33	19	49	21	24	43	34
Expected y									
Residual									

d. **MAKE A CONJECTURE** Why do you think that in each table about half the non-zero residuals are negative and the other half are positive? CCSS SMP 3

e. **REASON QUANTITATIVELY** According to the information provided above, which takes Mai longer: reading a page from her history book or completing a math problem? How much longer? How do you know? CCSS SMP 2

EXAMPLE 2 Use Residuals to Assess the Fit of a Function

USE STRUCTURE **To see if the linear function used to model the data is a good fit, it is possible to use the graph of the residuals. If the graph of (x, residual for the x) shows randomly placed points, the line of fit is accurate. If the graph shows a pattern, the function does not accurately reflect the data. Often it happens because the data is not linear in nature.** CCSS SMP 7

a. **USE STRUCTURE** Plot the residuals of Mai's data about her math homework. CCSS SMP 7

Pages (x)									
Minutes (y)	38	33	33	19	49	21	24	43	34
Expected y	40	36	24	25	38	28	25	36	31
Residual									

b. **INTERPRET PROBLEMS** What does the shape of the data points tell about the line of fit? CCSS SMP 1

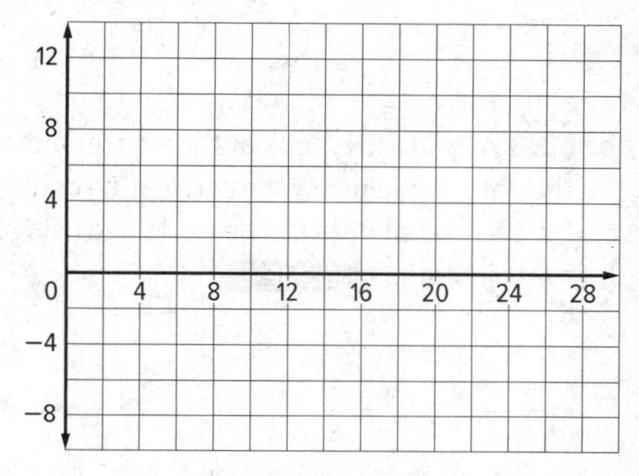

c. **CALCULATE ACCURATELY** Mai kept track of how long she spent on science homework. The line of best fit for this data is $y = 0.19x + 23$. Complete the table. Then plot the residuals. CCSS SMP 6

Pages (x)	4	6	8	10	14	18	20	24	16
Minutes (y)	33	27	21	24	18	24	27	39	18
Expected y									
Residual									

d. **CONSTRUCT ARGUMENTS** Why does the plot show that Mai's line of best fit is not accurate? CCSS SMP 3

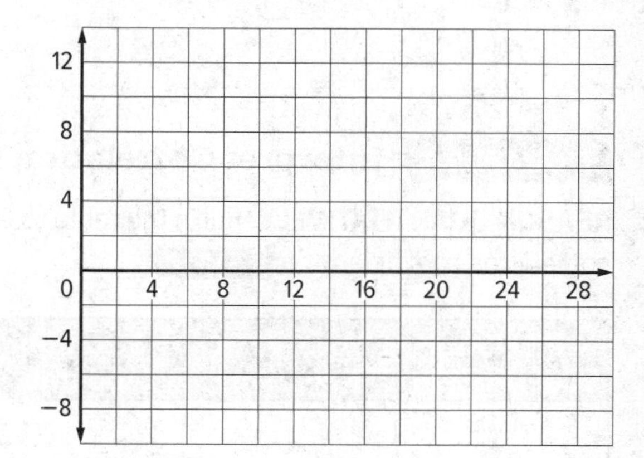

e. **PLAN A SOLUTION** What should Mai look for when reviewing her work? CCSS SMP 1

EXAMPLE 3 Correlation Coefficients

A correlation coefficient tells the strength of the relationship between the two sets of data. If there is no correlation, the correlation coefficient is zero. A perfect positive correlation has a coefficient of +1. The closer a positive correlation is to 1, the stronger the correlation is. A perfect negative correlation has a coefficient of −1. The closer a negative correlation is to −1, the stronger the correlation is.

a. INTERPRET PROBLEMS Mai used her calculator to find the correlation coefficients for her data. The correlation coefficient of the history homework data is +0.99. The correlation coefficient for the math homework is +0.75. What do the correlation coefficients mean in terms of the data? CCSS SMP 1

b. FIND A PATTERN Look at the scatter plots of the data. Explain how the correlation coefficient for each set of data can be explained using the scatter plots. CCSS SMP 7

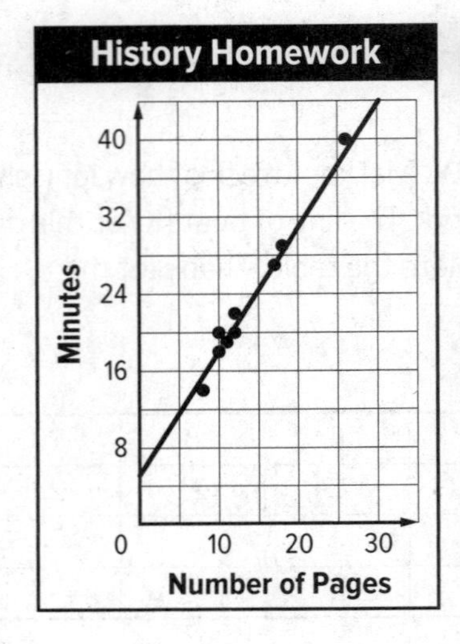

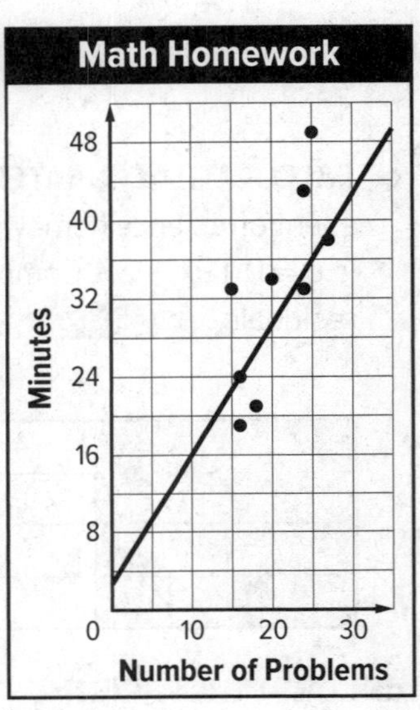

EXAMPLE 4 Interpret Correlation Coefficients CCSS S.ID.8

REASON QUANTITATIVELY Fill in the table to explain each correlation coefficient. CCSS SMP 2

Correlation Coefficient	Positive or Negative?	Strong or Moderate?	Graph
−0.88			
−0.48			
+0.59			
+0.83			

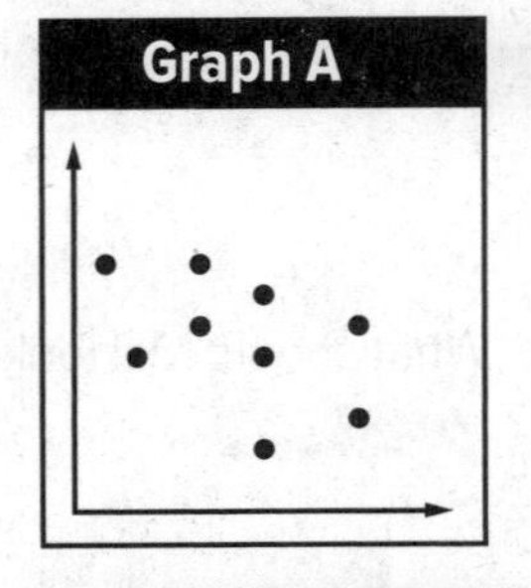

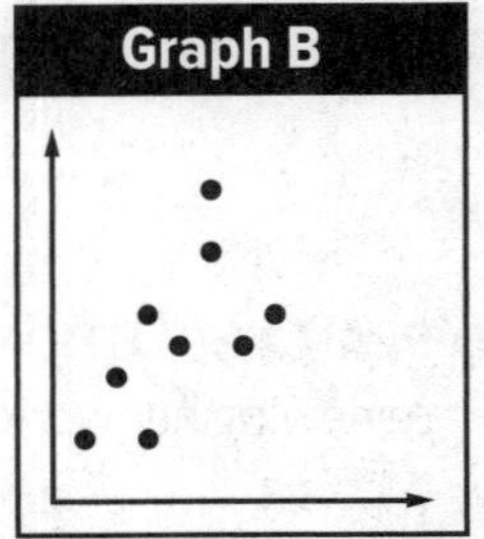

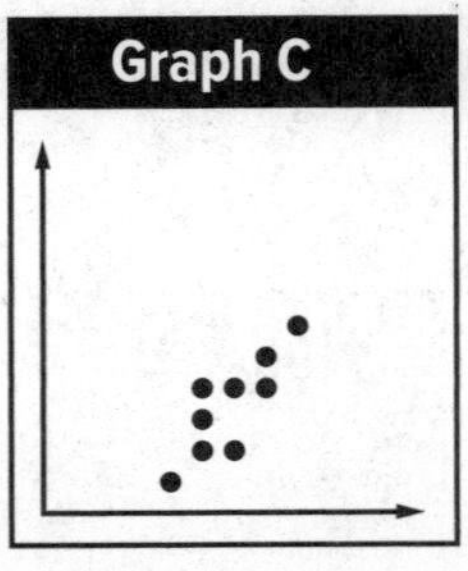

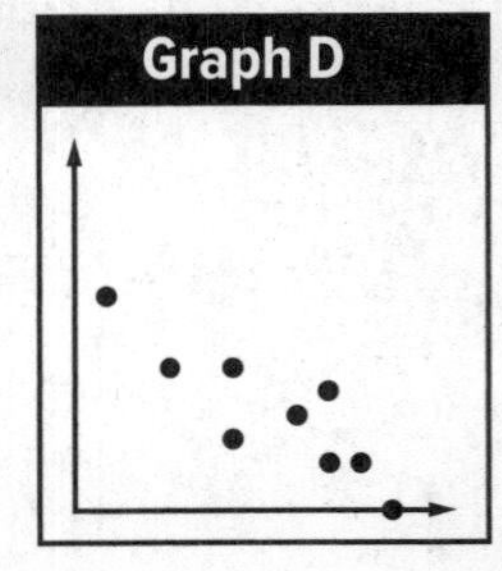

PRACTICE

1. USE STRUCTURE For a science project Noah measured the effect of light on plant growth. At the end of 3 weeks, he recorded the height of each plant and how many hours of light it received each day. CCSS S.ID.6b, SMP 7

a. He used his calculator to find that the equation of a line of best fit was $y = 0.5x + 2.4$. Use the equation to complete the table. Round values to the nearest whole number.

Hours Sunlight (x)	0	3	6	10	4	9	7	8	12	11	5
Height in Inches (y)	1	3	4	8	4	6	7	8	9	6	5
Expected y											
Residual											

b. Plot the residuals. Then use the plot to assess the line of best fit.

4
0 4 8 12 16
−4

2. USE STRUCTURE For her project, Madison measured the effect of fertilizer on plant growth. At the end of 3 weeks, she recorded the height of each plant and how many drops of fertilizer it received each day. CCSS S.ID.6b, SMP 7

a. She used her calculator to find that the equation of a line of best fit was $y = -0.6x + 8.5$. Use the equation to complete the table. Round values to the nearest whole number.

Hours Sunlight (x)	0	3	6	10	4	9	7	8	2	1	5
Height in Inches (y)	5	8	9	0	8	0	9	0	7	6	9
Expected y											
Residual											

b. Plot the residuals. Then use the plot to assess the line of best fit.

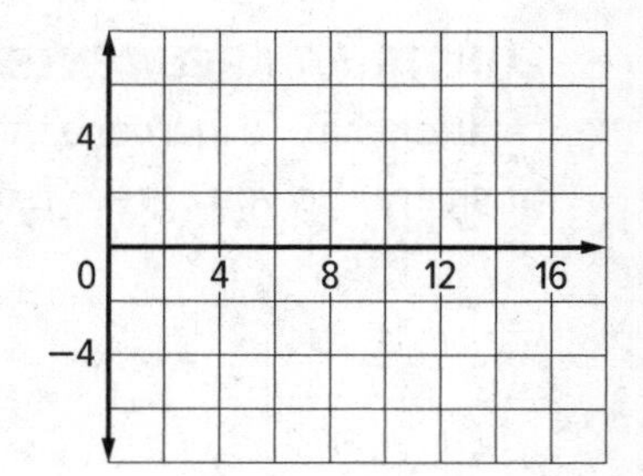

3. REASON QUANTITATIVELY A fertilizer company researched how effective each potential new product was on the yield of potatoes per acre. The correlation coefficient for each product is given. Explain what each means for a potato farmer's crop. CCSS S.ID.8, SMP 2

a. +0.39

b. −0.10

c. +0.80

6.5 Correlation and Causation

Objectives

- Explain the difference between correlation and causation.

CCSS STANDARDS

Content: S.ID.9
Practices: 1, 2, 3, 8
Use with Extend 4–5

EXAMPLE 1 Examine Correlation and Causation CCSS S.ID.9

EXPLORE The main source of income in the town of Buenavida is revenue from tourists who ski in the winter and boat in the summer. A group of business owners was trying to determine how to increase the number of summer tourists. They studied data from both seasons.

a. DESCRIBE A METHOD One business owner noticed that when there were a lot of skiers in the winter, there were a lot of boaters the following summer. He wondered if there is a relationship between the two. How could he investigate the relationship mathematically? CCSS SMP 8

b. INTERPRET PROBLEMS The business owners asked the high school math teacher to look into the matter. Some of her students made a scatter plot. Others found that the correlation coefficient is $+0.91$. How could they interpret each in terms of the problem and explain their reasoning so that the business owners would understand? CCSS SMP 1

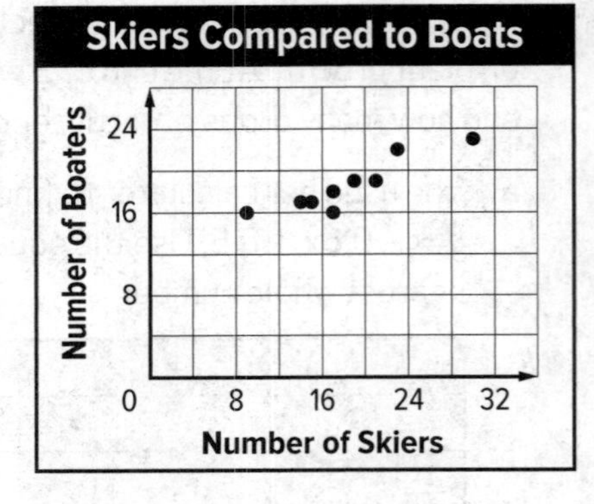

Scatter plot:

Correlation Coefficient:

c. CONSTRUCT ARGUMENTS The business owner's conclusion was, "There is a relationship. To increase the number of boaters, we just need to increase the number of skiers." Do you agree? Explain. CCSS SMP 3

d. MAKE A CONJECTURE Another business owner said, "They are related, but I don't think one causes the other. Something else must be causing both." What else might cause the correlation? CCSS SMP 3

EXAMPLE 2 Evaluate Correlation and Causation CCSS S.ID.9

CRITIQUE REASONING Sometimes another factor affects both elements being examined. Although it may appear that one of the elements affects the other, the third factor affects them both. The group of business owners studied other reasons for a correlation between the number of skiers and the number of boaters. They had the math class analyze each set of data. Tell what conclusions can be reached about each set of data. CCSS SMP 3

a. One business owner said, "The amount of snowfall affects the number of skiers and boaters since it affects how deep the snow is and how much water is in the lake for boating and fishing. Therefore the amount of snowfall causes an increase in the number or skiers and boaters." Do you agree? Explain. CCSS SMP 3

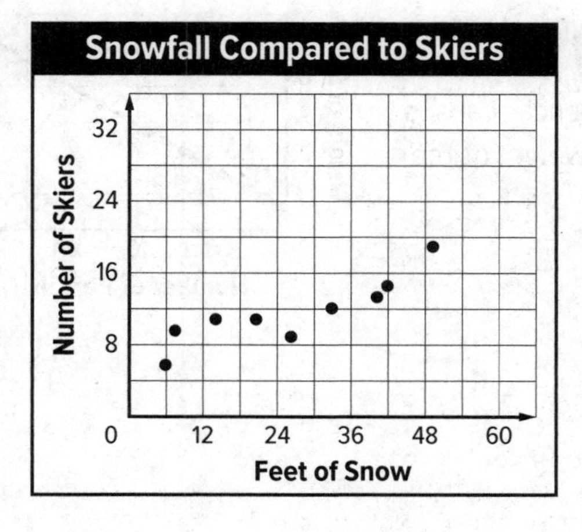

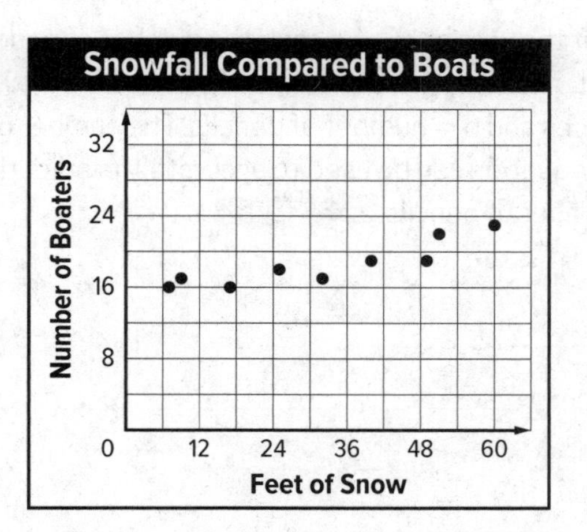

b. Another business owner contends that the economy affects how many skiers and boaters there are because when people have more money, they will spend it on recreation. He shows that the correlation coefficient between the economy and skiers is 0.92 and between the economy and boaters is 0.32. CCSS SMP 3

c. Another business owner contends that the attendance of blockbuster movies affects how many skiers and boaters visit. The correlation coefficient for movie attendance and skiers is +0.05 and for movie attendance and boaters is −0.12. CCSS SMP 3

EXAMPLE 3 Think about Correlation and Causation 

CRITIQUE REASONING Study each example of correlation and causation. Decide if the conclusion drawn is valid. Explain your reasoning. CCSS SMP 3

a. The principal of an elementary school noticed that there was a correlation factor of +0.89 between students' shoe sizes and their reading scores. She concluded that instead of spending time and money giving reading tests, she could just use shoe size to evaluate students' reading ability. CCSS SMP 3

b. The relationship of the weight of a box and the number of pencils in it is shown by the scatter plot. The equation of the line of best fit is $w = 0.3p + 1$ where w = weight in ounces and p = number of pencils. The number of pencils causes the weight so the equation can be used to accurately predict the weight of the box for a given number of pencils. CCSS SMP 3

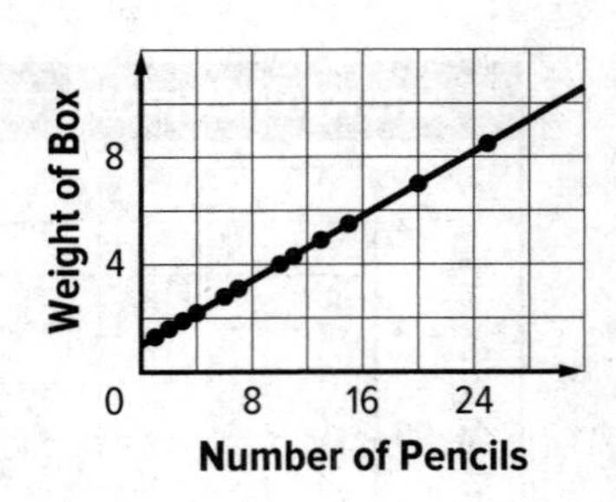

c. Armando's mother claims that the number of hours spent playing video games brings down his grades. Armando collected data from his friends and displayed it in the scatter plot shown. He says it does not matter how much time he spends playing video games. His mother claimed that a published research project showed a correlation factor of −0.73 between hours spent playing video games and grade point averages. CCSS SMP 3

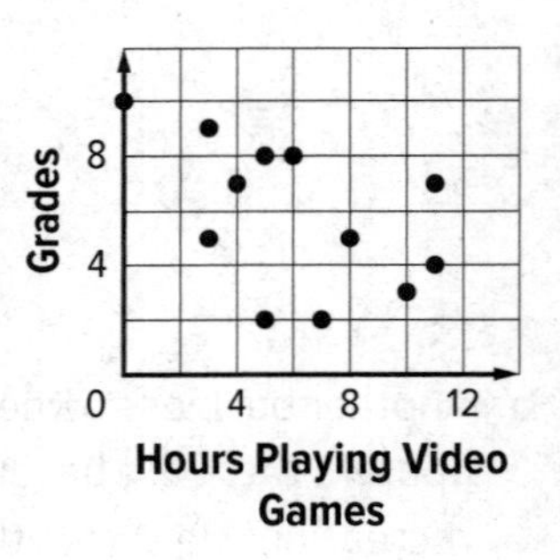

d. In general, why can't you use correlation to prove causation? CCSS SMP 3

e. What would be true of the scatterplot of a data set if the two factors are perfectly correlated? Is your answer valid for both negative and positive correlations? CCSS SMP 3

PRACTICE

A study compared the average monthly amount spent on swimsuits with the average monthly amount spent on air conditioning for several months in Sunnyside. The data is shown in the scatter plot. The correlation coefficient for the data is + 0.90.

1. **INTERPRET PROBLEMS** Explain what is shown by the scatter plot and the correlation coefficient. CCSS S.ID.9, SMP 1

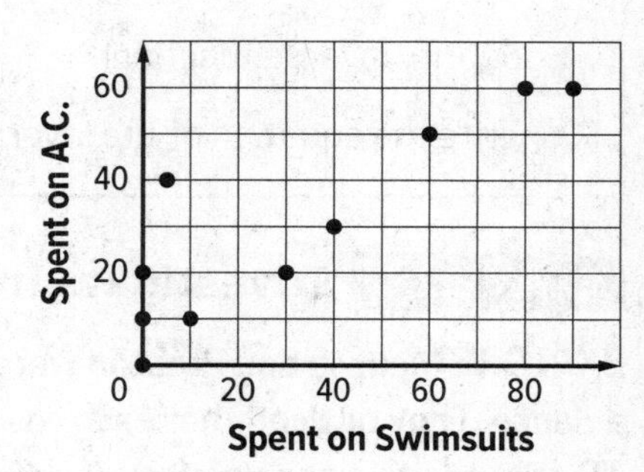

2. **CRITIQUE REASONING** Explain whether or not this statement is accurate. "There is a strong correlation between spending money on swimsuits and spending money on air conditioning. Therefore, to cut down the amount of electricity used in Sunnyside, people should buy fewer swimsuits." CCSS S.ID.9, SMP 3

3. **REASON ABSTRACTLY** What could affect both swimsuit sales and air conditioner use? Explain. CCSS S.ID.9, SMP 2

4. **CONSTRUCT ARGUMENTS** What is meant by this statement: *Correlation does not imply causation.* CCSS S.ID.9, SMP 3

5. **CRITIQUE REASONING** Kami claims that the greater the correlation between two factors, the greater the correlation coefficient will be. Is she correct? Explain. CCSS S.ID.9, SMP 3

6.6 Inverse Linear Functions

Objectives

- Identify inverse functions
- Write the equation of the inverse of a linear function

Content: A.CED.2, F.BF.4a
Practices: 1, 2, 3, 6, 7, 8
Use with Lesson 4–7

EXAMPLE 1 Investigate Inverse Functions CCSS A.CED.2

EXPLORE Melissa and Preston were making paper roses to decorate the gym for a dance. They noticed that each rose took 2 minutes to make. In addition, it took 10 minutes to clean up after each work session. Their teacher asked them to write an equation to model the situation, let *r* represent the number of roses completed and *t* represent time.

a. INTERPRET PROBLEMS Melissa thought, "I want to figure out how long it takes to make *r* roses. How can I write an equation to show how long it will take?" CCSS SMP 1

b. INTERPRET PROBLEMS Preston thought, "I want to figure out how many roses we can make in *t* minutes. How can I write an equation to show how many roses we will make after taking into consideration the clean up time?" CCSS SMP 1

c. CONSTRUCT ARGUMENTS Which equation is correct? Explain your thinking. CCSS SMP 3

d. CALCULATE ACCURATELY Use the students' equation to complete the tables. CCSS SMP 6

Melissa: $t = 2r + 10$	Roses	5	10	15	20	25	30	35	40
	Minutes								

Preston: $r = \frac{t - 10}{2}$	Minutes	20	30	40	50	60	70	80	90
	Roses								

e. FIND A PATTERN How are the values in the tables similar? CCSS SMP 7

f. USE STRUCTURE Solve Melissa's equation for *r*. Solve Preston's for *t*. What do you notice? CCSS SMP 7

EXAMPLE 2 Recognize Inverse Functions CCSS F.BF.4a

An inverse function generates ordered pairs in the reverse order of the original function.

a. **CALCULATE ACCURATELY** Complete each table. CCSS SMP 6

$f(x) = 2x + 2$	
x	$f(x)$
−2	
−1	
0	
1	
2	

$g(x) = 2(x + 2)$	
x	$g(x)$
−2	
−1	
0	
1	
2	

$h(x) = \frac{1}{2}x - 2$	
x	$h(x)$
0	
2	
4	
6	
8	

$k(x) = \frac{x-2}{2}$	
x	$k(x)$
0	
2	
4	
6	
8	

b. **FIND A PATTERN** Which functions are inverses of each other? How do you know? CCSS SMP 7

EXAMPLE 3 Write Inverse Functions 

USE STRUCTURE Write the inverse of $f(x) = \frac{1}{3}x + 5$ CCSS SMP 7

a. Rewrite $f(x)$ as y and then switch the variables in the equation.

b. Solve the new equation for y.

c. Rewrite y as $f^{-1}(x)$. Remember, $f^{-1}(x)$ means "the inverse of $f(x)$."

d. Compare the structure of the two functions. CCSS SMP 2

EXAMPLE 4 Use Inverse Functions CCSS A.CED.2, F.BF.4a

Beto and Carrie attend the same yoga studio. The monthly fee is \$20 plus \$4 for each class.

a. **REASON QUANTITATIVELY** Write a function $f(x)$ which gives the fee for attending x classes during the month. CCSS SMP 2

b. **USE STRUCTURE** Write the inverse, $f^{-1}(x)$, of $f(x)$. CCSS SMP 7

c. **CONSTRUCT ARGUMENTS** What does the inverse mean in terms of the problem? CCSS SMP 3

d. **USE STRUCTURE** Graph $f(x)$ and $f^{-1}(x)$ on the same coordinate plane. What do you notice about the graphs? Graph the line $f(x) = x$ on the same coordinate plane. What do you notice about the relationship between the first two graphs and $f(x) = x$? CCSS SMP 7

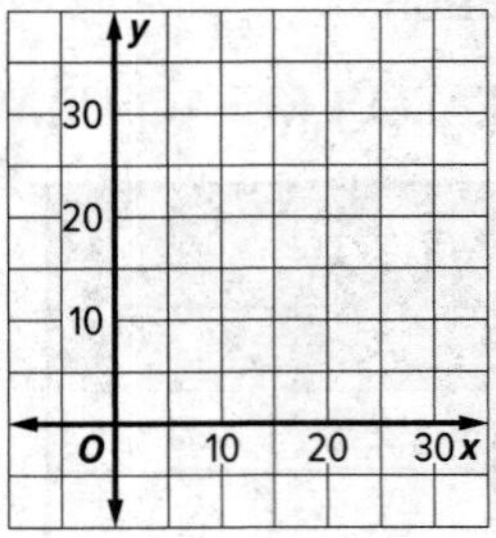

e. **INTERPRET PROBLEMS** Beto budgeted $60 for yoga this month. Which function allows him to find the number of classes he can afford most efficiently? How many classes can he take? CCSS SMP 1

f. **INTERPRET PROBLEMS** Carrie wants to take 12 classes this month. Which function allows her to find the monthly fee most efficiently? How many classes can she take? CCSS SMP 1

g. **DESCRIBE A METHOD** How can Beto and Carrie use each other's functions to check their answers? CCSS SMP 8

h. **REASON QUANTITATIVELY** Suppose you substitute a number into the equation in **part a**, and then substitute the result into the equation in **part b**. Describe the results. CCSS SMP 2

PRACTICE

1. **COMMUNICATE PRECISELY** What are inverse functions? CCSS F.BF.4a, SMP 6

2. **DESCRIBE A METHOD** How can you use ordered pairs to check the inverse of a function is correct? CCSS F.BF.4a, SMP 8

3. a. **REASON QUANTITATIVELY** What is the relationship between the slopes of two lines that are inverse functions of one another? Give an example. CCSS F.BF.4a, SMP 2

b. **REASON QUANTITATIVELY** What is the relationship between the x- and y-intercepts of two lines that are inverse functions of one another? Give an example. CCSS F.BF.4a, SMP 2

4. **USE STRUCTURE** Write the inverse of each function. CCSS F.BF.4a, SMP 7

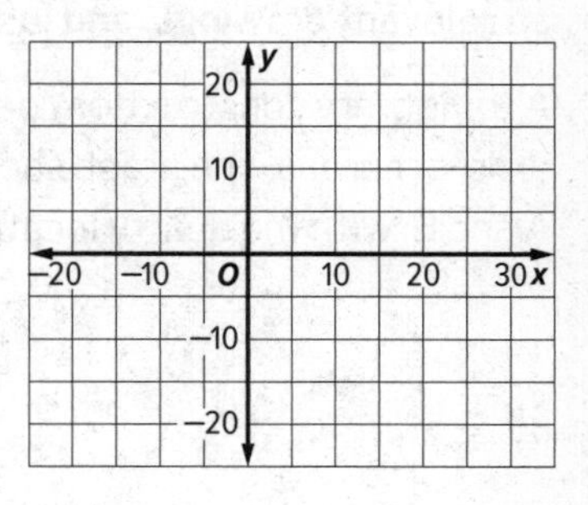

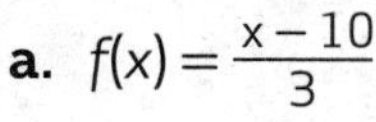
a. $f(x) = \frac{x-10}{3}$

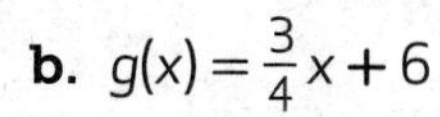
b. $g(x) = \frac{3}{4}x + 6$

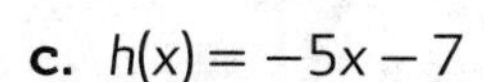
c. $h(x) = -5x - 7$

d. Graph $h(x)$ and $h^{-1}(x)$ on the same coordinate plane to check your answer.

5. Four friends had a small business creating personalized cards and signs. At the end of each month they put \$100 of their profit in the bank to buy equipment to expand their business. They split the remaining profit 4 ways. CCSS A.CED.2, F.BF.4a

a. **INTERPRET PROBLEMS** Write a function $f(x)$ which gives the amount each friend receives when the business has x dollars in profit. CCSS SMP 1

b. **USE STRUCTURE** Write the inverse, $f^{-1}(x)$, of $f(x)$. CCSS SMP 7

c. **INTERPRET PROBLEMS** What does the inverse mean in terms of the problem? CCSS SMP 1

d. **REASON QUANTITATIVELY** The business made a total profit of \$250 last month. Which function is most efficient for the friends to find out how much each should receive? How much should each receive? CCSS SMP 2

e. **REASON QUANTITATIVELY** One of the business partners wants to earn \$75 next month. Which function allows him to most efficiently find the amount of profit the business must have? What amount of profit will result in each partner getting paid \$75? CCSS SMP 2

6. **CRITIQUE REASONING** Jonathan claims that finding the inverse of the function $f(x)$ is much easier than we have been making it. To find the inverse, $f^{-1}(x)$, we need only remember that raising something to the power of -1 is the same as taking its reciprocal. So the inverse of $f(x)$ is simply $\frac{1}{f(x)}$. Is Jonathan correct? Explain by giving an example. CCSS F.BF.4a, SMP 3

Performance Task

Planning a Road

Provide a clear solution to the problem. Be sure to show all of your work, include all relevant drawings, and justify your answers.

A temporary construction road must be built from the quarry and pass by two drop-off areas. The graph shows the proposed road and the two drop-off areas A and B, with the quarry located at (0, 4).

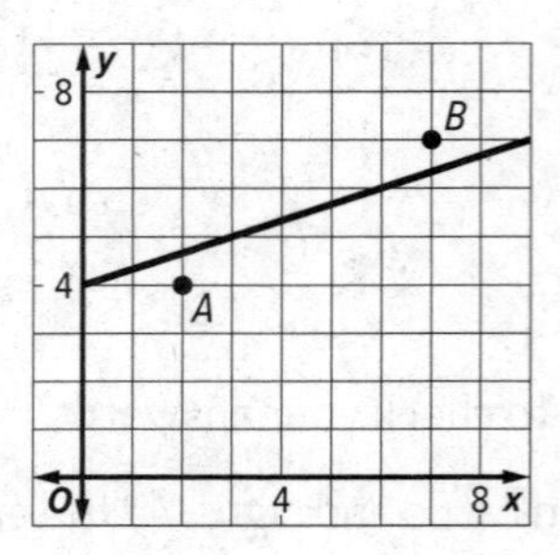

Part A

The road must be built so that the distance from each drop-off area to the road is minimized. Define a function in slope-intercept form to represent the line. What are the possible values for the slope of the line? Explain.

Part B

If the vertical distance from each drop-off area to the line is the same, then the distances are minimized. Write and solve an equation to find the slope, and then write the equation of the line that represents the construction road.

Part C

Use a calculator to find the best-fit line that passes through the quarry and both drop-off points. Does this satisfy the conditions for the proposed road? Explain.

Part D

Yolanda says, "Let's just have the construction road pass through both drop-off areas. This will be easier." If by "easier" Yolanda means that the vertical distance from the quarry will be less than the combined vertical distances to each drop-off site, do you agree with Yolanda's assessment? Explain.

Performance Task

Do You Measure Up?

Provide a clear solution to the problem. Be sure to show all of your work, include all relevant drawings, and justify your answers.

Anthropologists use relationships between bones of the human body to predict the height of a person from skeletal remains. Follow the task below to look for such a relationship.

Part A

Collect data from 10 individuals. Have each person measure, to the nearest centimeter, the length of the femur bone. Record this data and the height of each person on the table. Create a scatterplot of the data on the coordinate grid. Include labels and the scale of the axes.

Name	Femur Length (cm)	Height (cm)

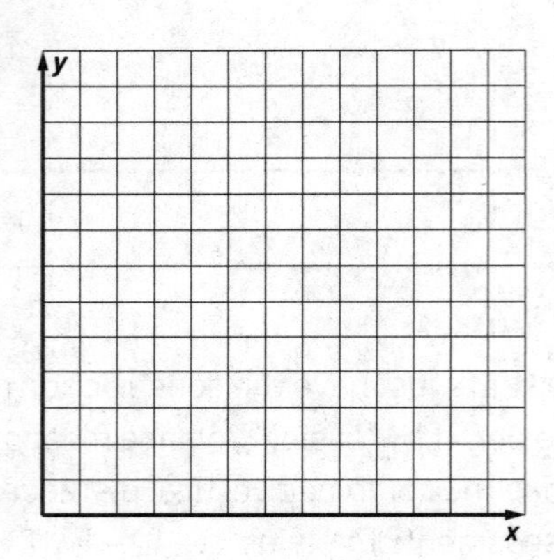

Part B

Use technology to find the best-fit line for the data. Determine the correlation coefficient, and interpret what it means in the context of the situation.

Part C

A Web site claims that the length of the femur represents 26% of a person's height. Write a linear function to represent this fact. Compare your best-fit line from **Part B** to your linear function. Why might they be different?

Part D

In general, your data should show a linear association. Describe the correlation that the data exhibits. Would you classify the association as causation? Justify your answer.

Standardized Test Practice

1. A line passes through $(8, -1)$ and is perpendicular to $y = 2x - 9$. What is the equation of the line in point-slope form? CCSS F.LE.2

____

What is the equation of the line in slope-intercept form? CCSS F.LE.2

____

2. Consider the linear functions shown in the graph below.

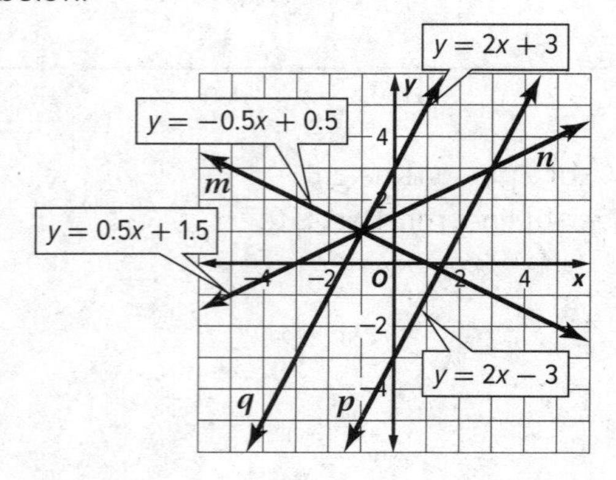

Which lines represent functions that are inverses of each other? CCSS F.BF.4a

line ☐ and line ☐

3. If the correlation coefficient between x and y is 0.86, which of the following statements are true? CCSS S.ID.9

As x increases, y decreases.

An increase in x causes an increase in y.

x and y are positively correlated.

x and y are negatively correlated.

x and y are not correlated.

As x increases, y increases.

4. Find the inverse function of $y = -\frac{2}{7}x + 5$. Show your work. CCSS F.BF.4a

____

5. A caterer charges a setup fee and a flat amount per person served. If 45 people are served, the total bill will be \$654. If 68 people are served, the total bill will be \$953.

a. Write an equation in slope-intercept form that gives the cost of serving x people. CCSS A.CED.3

____

b. What are the slope and y-intercept, and what do they mean? CCSS S.ID.7

____

c. How many people could be served for \$1100? Explain how you found your answer. CCSS A.CED.2

____

6. The table shows the number of people who entered a store and the total sales each hour.

Number of People	42	17	26	50	57	42	39	44	64	31
Total Sales ($)	155	93	109	240	237	129	136	189	271	105

a. Use your calculator to find a linear regression model that fits the data. CCSS S.ID.6a

b. What is the *r*-value for your regression model? What does this *r*-value mean? CCSS S.ID.8

c. Use your graphing calculator to find a median-fit line for the data. CCSS S.ID.6a

d. For what number of people do the models predict about the same amount of sales? Explain how you found your answer. CCSS S.ID.6a

7. The table below shows the number of absences and test scores for a group of students.

Number of Absences	1	5	0	8	3	2	0	3
Test Score	98	67	93	52	88	86	90	83

a. Create a scatter plot for the data and draw a line of fit. CCSS S.ID.6c

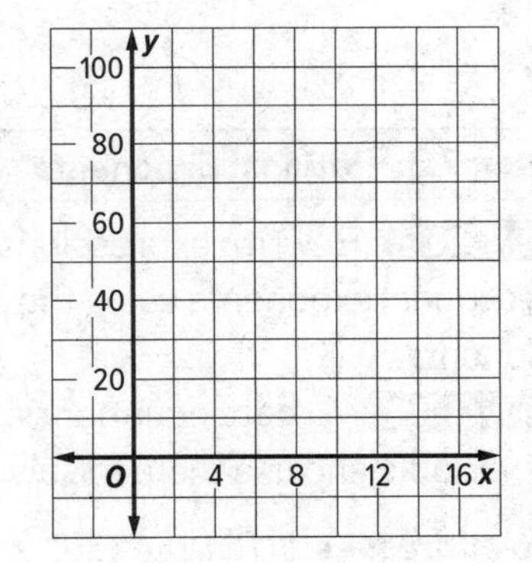

b. Using your line of fit, what do you predict the test score would be for a student with 10 absences? Is this an example of interpolation or extrapolation? CCSS S.ID.6a

c. Is the data correlated? If yes, give the type and explain whether this is a causal relationship. If no, explain why not. CCSS S.ID.9

8. What is the equation of the inverse of the linear function that whose graph contains the points $(2, -1)$ and $(-1. -7)$. Explain your solution. CCSS F.BF.4a

7 Exponential Functions

CHAPTER FOCUS Learn about some of the Common Core State Standards that you will explore in this chapter. Answer the preview questions. As you complete each lesson, return to these pages to check your work.

What You Will Learn	Preview Question
Lesson 7.1: Multiplication Properties of Exponents	
CCSS A.SSE.3c Use the properties of exponents to transform expressions for exponential functions. CCSS F.IF.8b Use properties of exponents to interpret expressions for exponential functions.	CCSS SMP 7 Simplify the expression $(-4x^3y^4z)^4(2x^4y)^3$
Lesson 7.2: Division Properties of Exponents	
CCSS F.IF.8b Use the properties of exponents to interpret expressions for exponential functions. CCSS A.SSE.3c Use properties of exponents to transform expressions for exponential functions.	CCSS SMP 2 The formula for density is *mass* ÷ *volume*. Earth's mass is approximately 5.972×10^{24} kg and its volume is approximately 1.083×10^{12} km^3. Write and simplify an expression for Earth's density in scientific notation. Round to the nearest thousandth.
Lesson 7.3: Rational Exponents	
CCSS N.RN.2 Rewrite expressions involving radicals and rational exponents using the properties of exponents. CCSS A.CED.1 Create equations and inequalities in one variable and use them to solve problems. **Also addresses:** N.RN.1, A.CED.2, F.IF.8b	CCSS SMP 7 Find the area of the figure. Write your answer in radical form. $(4x)^{\frac{3}{2}}$ $8x^{\frac{3}{4}}$
Lesson 7.4: Exponential Function	
CCSS F.IF.7e Graph exponential and logarithmic functions, showing intercepts and end behavior, and trigonometric functions, showing period, midline, and amplitude. CCSS F.IF.8b Use properties of exponents to interpret expressions for exponential functions. CCSS F.LE.2 Construct linear and exponential functions, including arithmetic and geometric sequences, given a graph, a description of a relationship, or two input-output pairs (include reading these from a table). **Also addresses:** F.LE.1a, F.LE.5, A.REI.11	CCSS SMP 4 Graph the function $y = 4^x - 5$. Find the y-intercept and state the domain and range. y 4 2 −4 −2 O 2 4 x −2 −4

What You Will Learn	Preview Question
Lesson 7.5: Transforming Exponential Functions	
CCSS F.BF.3 Identify the effect on the graph of replacing $f(x)$ by $f(x) + k$, $kf(x)$, $f(kx)$, and $f(x + k)$ for specific values of k (both positive and negative); find the value of k given the graphs. Experiment with cases and illustrate an explanation of the effects on the graph using technology. CCSS F.IF.7e Graph exponential and logarithmic functions, showing intercepts and end behavior, and trigonometric functions, showing period, midline, and amplitude.	CCSS SMP 5 On your calculator, graph $f(x) = 2^x$, $f(x) = 2^x + 2$, $f(x) = 2^x - 2$. Find the y-intercept, domain, and range of each function. 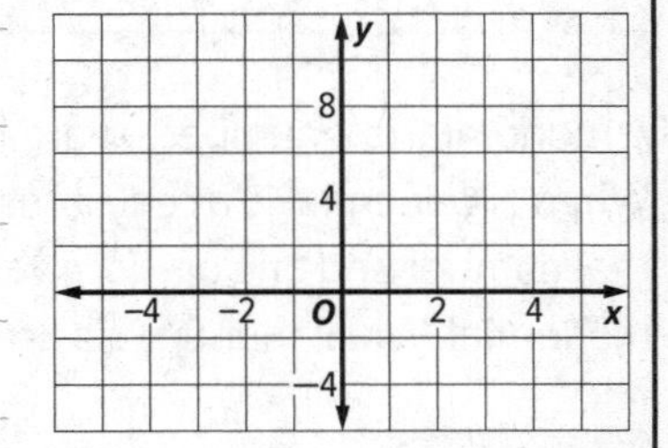
Lesson 7.6: Modeling: Exponential Functions	
CCSS N.Q.2 Define appropriate quantities for the purpose of descriptive modeling. CCSS A.SSE.1a Interpret parts of an expression, such as terms, factors, and coefficients. CCSS A.SSE.1b Interpret complicated expressions by viewing one or more of their parts as a single entity. CCSS F.BF.1b Combine standard function types using arithmetic operations. **Also addresses:** A.CED.1, F.BF.1a	CCSS SMP 4 A mouse population doubles every 3 months. If there were 5 mice in the original population, how many mice would there be after 6 years? Write and solve an equation to model the situation. Explain your results.
Lesson 7.7: Geometric Sequences as Exponential Functions	
CCSS F.BF.2 Write arithmetic and geometric sequences both recursively and with an explicit formula, use them to model situations, and translate between the two forms. CCSS F.LE.1c Recognize situations in which a quantity grows or decays by a constant percentage rate per unit interval relative to another. **Also addresses:** F.LE.2, F.LE.5, F.BF.1a	CCSS SMP 7 You decide to save money by depositing one penny and doubling your deposit every day. Use the explicit formula to write and solve an equation to determine how much money you will need to deposit on day 20.
Lesson 7.8: Recursive Formulas	
CCSS F.IF.3 Recognize that sequences are functions, sometimes defined recursively, whose domain is a subset of the integers. **Also addresses:** F.BF.2, A.SSE.2	CCSS SMP 7 A coat has been reduced in price with the following prices: \$184, \$147.20, \$117.76. Write a recursive formula for the price of the discounted coat and show the price of the coat after 2 more markdowns.

7.1 Multiplication Properties of Exponents

Content: F.IF.8b, A.SSE.3c
Practices: 2, 3, 4, 6, 7
Use with Lesson 7–1

Objectives

- Develop the multiplication properties of exponents.
- Multiply monomials to simplify expressions and solve problems.

A monomial is a number, a variable, or a product of a number and one or more variables. When a monomial is a real number, it is a constant. When a monomial is the product of a real number and one or more variables, the real number is called the coefficient of the monomial. When monomials are multiplied together, the result is another monomial

EXAMPLE 1 Product of Powers CCSS A.SSE.3c

a. USE STRUCTURE Name the parts of the term $5x^4$. CCSS SMP 7

5 ____________ x ____________ 4 ____________ x^4 ____________

b. CALCULATE ACCURATELY Use a calculator to evaluate each of the following expressions. CCSS SMP 6

$2^3 \cdot 2^3$ ____________ $2^2 \cdot 2^4$ ____________ $2^1 \cdot 2^5$ ____________ 2^6 ____________

c. FIND A PATTERN What pattern, if any, do you see in your answers to **part a**? CCSS SMP 7

____

d. FIND A PATTERN Rewrite the factors in each expression in expanded form, and use the Associative and Commutative Properties of Multiplication to find the product as a power in expanded form. Then write the product as a monomial in simplest form. CCSS SMP 7

$m^4 \cdot m^5$ ______________________________

$n^3 \cdot n^6$ ______________________________

$r^2 \cdot r^3 \cdot r^3$ ______________________________

$p^3q^2 \cdot p^2q^4$ ______________________________

$s^5t \cdot st^3$ ______________________________

$5n^4 \cdot 7n^2$ ______________________________

$8r^5s \cdot q^3rs^5 \cdot 9qs^3$ ______________________________

____

e. MAKE A CONJECTURE Compare your answers to those of other students. Then make a conjecture about the product of powers that have the same base. Consider how coefficients of monomials are handled when monomials are multiplied. CCSS SMP 3

____

____

____

f. USE STRUCTURE Use your conjecture to simplify the following monomial expressions without using expanded form. CCSS SMP 7

$w^8 \cdot w^{10} =$ ________ $x^7 \cdot x^2 =$ ________ $y^6z^5 \cdot yz^7 =$ ________

g. USE STRUCTURE Use your conjecture to simplify the following monomial expressions with coefficients without using expanded form. CCSS SMP 7

$5x^3 \cdot 9x^5 =$ ________ $11y \cdot x^4y^5 =$ ________ $3m^4np^2 \cdot \frac{1}{3}mn^5p^4$ ________

A simplified monomial expression is an equivalent expression in which each variable base appears only once, all coefficients are in simplest form, and there are no powers of powers.

EXAMPLE 2 Power of a Product CCSS A.SSE.3c

a. FIND A PATTERN Rewrite $(3x^5)^4$ in expanded form using $(3x^5)$ as a repeated factor. CCSS SMP 7

b. USE STRUCTURE Use the Commutative and Associative Properties of Multiplication to find the product of the coefficients and the product of the powers. Then write $(3x^5)^4$ as a monomial in simplest form. CCSS SMP 7

product of coefficients: ________ product of powers: ________ $(3x^5)^4 =$ ________

c. MAKE A CONJECTURE Compare your answers to those of other students. Then make a conjecture about the power of a product. CCSS SMP 3

d. MAKE A CONJECTURE Consider your conjecture from **part c**. Write an algebraic rule for raising a power to a power that starts with $(m \cdot n^x)^y$ CCSS SMP 3

e. CRITIQUE REASONING Nikki believes $(2j^7k^8)^5$ simplifies to $10j^{35}k^{40}$. Do you agree? Explain. CCSS SMP 3

f. REASON QUANTITATIVELY Explain why $(-y^4z^4)^7 = -y^{28}z^{28}$ but $(-y^7z^7)^4 \neq -y^{28}z^{28}$. CCSS SMP 2

g. CRITIQUE REASONING Hector believes that $[(2x^3)^5]^2$ is equivalent to $(2x)^{30}$. Do you agree? Explain. CCSS SMP 3

EXAMPLE 3 Multiplying Powers to Solve Problems 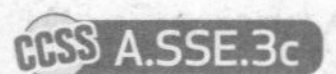

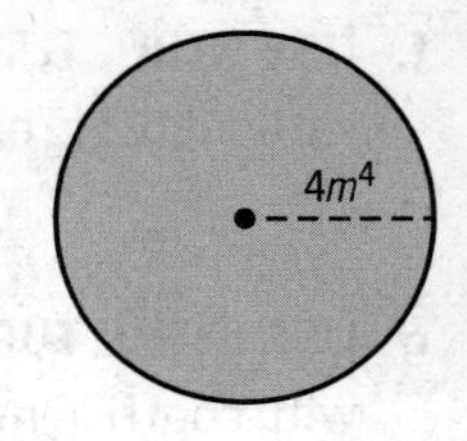

a. CALCULATE ACCURATELY The formula for the area of a circle is $A = \pi r^2$. Find the area, in simplest form, of the circle to the right. CCSS SMP 6

b. REASON QUANTITATIVELY A gardener is attempting to grow world-record tomatoes. A certain product indicates that it will double the size of your tomato for each ounce used. Write an expression that represents the factor by which the tomato increases if the gardener uses x ounces of the product. CCSS SMP 2

c. REASON QUANTITATIVELY The gardener in **part b** is thrilled with the results. She decides to apply the same x ounces of the product to her newest batch of tomatoes a multiple number of times y. What expression represents the factor by which each tomato will increase in size? Before the application of the product, one of her tomatoes has a volume of 4 in^3. What will be the volume of the tomato if she applies 2 ounces of the product 5 different times? CCSS SMP 2

PRACTICE

1. Sal and Jade are writing equivalent forms of g^5gh^6. CCSS A.SSE.3c

   a. CRITIQUE REASONING Jade believes $g^5gh^6 = (g \cdot g \cdot g \cdot g \cdot g) \cdot (g) \cdot (h \cdot h \cdot h \cdot h \cdot h \cdot h)$. Sal believes $g^5gh^6 = (g \cdot g \cdot g \cdot g \cdot g) \cdot (gh \cdot gh \cdot gh \cdot gh \cdot gh \cdot gh)$. Who is correct? Explain. CCSS SMP3

   b. USE STRUCTURE What is the simplified form of g^5gh^6? CCSS SMP 7

2. **CRITIQUE REASONING** Adrian believes $(p^3)(p^5)$ is positive for all nonzero values of p. Maya believes Adrian cannot make that claim because the product of 3 and 5 is an odd number. Do you agree with Maya or Adrian? Justify your answer. CCSS A.SSE.3c, SMP 3

3. **USE STRUCTURE** Determine whether each pair of expressions are equivalent. CCSS A.SSE.3c, SMP 7

   a. $(2^7)(2^7)(2^7)(2^7)$ and 8^7 ____________
   b. $(-j^9)(-j^9)(-j^9)$ and $-j^{27}$ ____________
   c. $(5p^3q)(4p^5q^9)$ and $20p^8q^{10}$ ____________
   d. $(6w^5)^2$ and $36w^{25}$ ____________
   e. $(x^{10})^3$ and $(x^3)^{10}$ ____________
   f. $[(2n^2)^3]^2$ and $(4n^4)^3$ ____________

4. a. **USE STRUCTURE** What must be true about p and q for $(m^p)(m^q) = (m^p)^q$? Give an example. CCSS A.SSE.3c, SMP 7

b. USE STRUCTURE Assuming m, p, and q are all positive integers, can $(m^p)(m^q) > (m^p)^q$ ever be true? CCSS A.SSE.3c, SMP 7

5. CRITIQUE REASONING The volume of a rectangular prism is the product of its length, width, and height. Chris believes the volume of the box shown to the right is $60x^{12}$ cubic units. What error, if any, is Chris making? CCSS A.SSE.3c, SMP 3

$5x^2$

$3x^2$

$4x^3$

6. Morgan must choose between two investment opportunities. Option 1 is expected to generate 6.2% interest compounded annually. Option 2 is compounded monthly at a rate of 0.51%. CCSS F.IF.8b

a. USE A MODEL Use $A = P(1 + r)^{nt}$ to write equations for the value of both options for P dollars invested at a rate of r, compounded n times a year for t years. CCSS SMP 4

b. REASON QUANTITATIVELY Which investment is the better option for Morgan? Justify your answer by substitution. CCSS SMP 2

7. USE STRUCTURE We know that both multiplication and addition are commutative and associative. We also know that multiplication distributes over addition. CCSS A.SSE.3c, SMP 7

a. What would it mean for the operation of raising one number to an exponent to be commutative? Decide and explain whether or not this operation is commutative.

b. What would it mean for the operation of raising one number to an exponent to be associative? Decide and explain whether or not this operation is associative.

c. What would it mean for the process of raising one exponent to another to distribute over addition? Decide and explain whether or not this is true.

d. What would it mean for the process of raising one exponent to another to distribute over multiplication? Decide and explain whether or not this is true.

7.2 Division Properties of Exponents

Objectives

- Develop the Division Properties of Exponents.
- Divide monomials to simplify expressions and solve problems.

Content: F.IF.8b, A.SSE.3c
Practices: 1, 2, 3, 4, 5, 7, 8
Use with Lesson 7-2

EXAMPLE 1 Developing the Quotient of Powers Property

a. USE TOOLS Use a calculator to evaluate each of the following expressions. CCSS SMP 5

$3^{10} \div 3^6$ ________ $3^{11} \div 3^7$ ________ $3^{12} \div 3^8$ ________ 3^4 ________

b. FIND A PATTERN What pattern, if any, do you see in your answers to **part a**? CCSS SMP 7

____

c. FIND A PATTERN What pattern, if any, do you see in the exponents in **part a**? Write another expression that would follow this pattern and check to see if it gives the same answer. CCSS SMP 7

____

d. USE STRUCTURE Rewrite each numerator and denominator in expanded form, and reduce the fraction. Then write the quotient as a monomial in simplest form. CCSS SMP 7

$\frac{m^6}{m^2}$ $\frac{n^{10}}{n^5}$

e. MAKE A CONJECTURE Compare your answers to those of other students. Then make a conjecture about the quotient of powers that have the same base. CCSS SMP 3

____

____

f. REASON QUANTITIVELY Use your conjecture to justify the Zero Exponent Property, for any nonzero number, a, $a^0 = 1$. CCSS SMP 2

____

____

g. CONSTRUCT ARGUMENTS Explain why the Zero Exponent Property does not apply when the base is 0. CCSS SMP 3

____

____

____

EXAMPLE 2 Developing the Negative Exponent Property

a. USE TOOLS Use a calculator to evaluate each of the following expressions. CCSS SMP 5

2^{-1} __________ 10^{-1} __________ 2^{-2} __________ 10^{-3} __________

b. FIND A PATTERN Rewrite your answers from **part a** as fractions, expressing the denominator as a power of 2 or 10 where possible. What pattern, if any, do you see in your answers to **part a**? CCSS SMP 7

c. USE STRUCTURE Use the Power of a Power Property to rewrite $\frac{1}{p^{-3}}$ with only positive exponents. CCSS SMP 7

$$\frac{1}{p^{-3}}$$

d. MAKE A CONJECTURE Compare your answers to those of other students. Then make a conjecture about rewriting powers with negative exponents. CCSS SMP 3

e. CRITIQUE REASONING Shawne believes that $y^6 \geq y^{-6}$ for all nonzero values of y. Do you agree? Explain. CCSS SMP 3

Definition: An exponential expression is **simplified** when it contains only positive powers, there are no powers of powers, all fractions are in simplest form, and each base appears exactly once.

EXAMPLE 3 Using Division Properties of Exponents to Solve Problems CCSS F.IF.8b, A.SSE.3c

a. REASON QUANTITATIVELY A research team studying plants found that the number of evergreen plants in a park can be modeled by the expression $p(0.98^n)$, where n is the number of months and p is the number of evergreens at the start of the study. By what percent does the population change each month? Is the number of plants increasing or decreasing? Explain how you know. CCSS SMP 2

b. INTERPRET PROBLEMS The expression $1 - \left(\frac{p(0.98)^{12}}{p(0.98)^{3}}\right)$ can be simplified to find the percent change in the population of plants between the end of month 3 and the end of month 12. Explain the role of the number 1 in the expression. Does the size of the starting evergreen population affect the answer? CCSS SMP 1

c. CRITIQUE REASONING Nadia used the expression in **part b** and believes the evergreen population decreased by approximately 7.76%. Do you agree? Justify your answer. CCSS SMP 3

PRACTICE

1. **CRITIQUE REASONING** Zoe says that $\left(\frac{p^7}{p^3}\right)^5$ is positive for all nonzero values of p. Becca disagrees because all of the exponents are odd. Who is correct? Explain. CCSS A.SSE.3c, SMP 3

2. **USE STRUCTURE** Are $4n^5$ and $4n^{-5}$ reciprocals? Explain. CCSS A.SSE.3c, SMP 7

3. David evaluated -5^0 on his calculator and believes his answer is a counterexample that disproves the Zero Exponent Property. CCSS A.SSE.3c

   a. **USE TOOLS** What answer did the calculator give David when he evaluated -5^0? CCSS SMP 5

   b. **EVALUATE REASONABLENESS** Is this a counterexample that contradicts the Zero Exponent Property? Explain. CCSS SMP 8

4. a. **USE STRUCTURE** Colleen and Tyler are using different steps to simplify $\left(\frac{x^2}{x^9}\right)^3$. Their work is shown in the box. Who is correct? Explain. CCSS A.SSE.3c, SMP 7

Colleen	Tyler
$\left(\frac{x^2}{x^9}\right)^3 = \frac{x^6}{x^{27}}$ $= \frac{1}{x^{21}}$	$\left(\frac{x^2}{x^9}\right)^3 = \left(\frac{1}{x^7}\right)^3$ $= x^{-21}$

   b. **USE STRUCTURE** Who has their answer in simplest form? Explain. CCSS A.SSE.3c, SMP 7

5. An investment is expected to increase in value by 4% every year. CCSS F.IF.8b, A.SSE.3c

a. USE A MODEL Write an expression that represents the value of the investment after t years if the initial value was n dollars. CCSS SMP 4

b. REASON QUANTITATIVELY By what percent does the value of the investment change between the end of year 2 and the end of year 8? Round your answer to the nearest tenth of a percent, and show your work. CCSS SMP 2

6. A poor investment is expected to decrease in value by 5% every year. CCSS F.IF.8b, A.SSE.3c

a. REASON QUANTITATIVELY If the initial value of the investment was $100, what does it mean for it to decrease in value by 5% in the first year? CCSS SMP 2

b. CRITIQUE REASONING Erik claims that rather than multiplying by 0.05 and subtracting, we can simply multiply the investment amount by 0.95. Is he correct? Explain. CCSS SMP 3

c. USE A MODEL Write an expression that represents the value of the investment after t years if the initial value was n dollars. CCSS SMP 4

d. REASON QUANTITATIVELY By what percent does the value of the investment change between the end of year 2 and the end of year 10? Round your answer to the nearest tenth of a percent, and show your work. CCSS SMP 2

7. CRITIQUE REASONING Jill claims that we never need to use the Division Property of Exponents. She says that exponents in the denominator can be written as negative exponents, and then the Multiplication Property of Exponents can be used instead. Is Jill correct? If so, give an example. If not, explain why. CCSS A.SSE.3c, SMP 3

7.3 Rational Exponents

Objectives

- Use the properties of exponents to rewrite expressions using radicals.
- Create exponential equations to model relationships between quantities.
- Graph exponential equations with rational exponents on a coordinate plane.

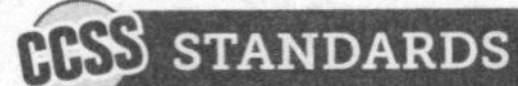

Content: N.RN.1, N.RN.2, A.CED.1, A.CED.2, F.IF.8b
Practices: 2, 3, 4, 5, 6, 7
Use with Lesson 7-3

The properties of exponents can be used to simplify and evaluate expressions. Suppose *m* and *n* are positive integers and *b* is a real number.

$b^m b^n = b^{m+n}$

$(b^m)^n = b^{mn}$

$\frac{b^m}{b^n} = b^{m-n}$ when $b \neq 0$

$b^{\frac{1}{n}} = \sqrt[n]{b}$ when $b \geq 0$ and $n > 1$ or $b < 0$ and n odd

A **simplified exponential expression** is an expression in which each variable base appears only once, all coefficients are in simplest form, and there are no powers of powers.

EXAMPLE 1 Rewriting Expressions

a. REASON QUANTITATIVELY Use the Power of a Power Property to explain why $7^{\frac{1}{5}} = \sqrt[5]{7}$. CCSS N.RN.1, N.RN.2, SMP 2

b. CRITIQUE REASONING Geoff believes the Power of a Power Property can also be used to explain why $h^{0.6} = \sqrt[5]{h^3}$. Do you agree? Explain. CCSS N.RN.1, N.RN.2, SMP 3

c. CRITIQUE REASONING Fran believes $b^{\frac{1}{n}} = \sqrt[n]{b}$ is only true when $b \geq 0$ and *n* is even because when $b < 0$, $b^{\frac{1}{n}} = \sqrt[n]{\frac{1}{|b|}}$. Do you agree? Explain. CCSS N.RN.1, N.RN.2, SMP 3

d. USE STRUCTURE Show how rational exponents and the properties of powers can be used to simplify $\sqrt[3]{27m^{20}n^{12}}$ CCSS N.RN.2, SMP 7

e. USE STRUCTURE Explain how rational exponents can be used to rewrite the expression $\sqrt[3]{\sqrt[4]{w}}$, $w > 0$, as an expression with a single radical. CCSS N.RN.1, N.RN.2, SMP 7

f. USE STRUCTURE Explain how rational exponents can be used to rewrite the expression $\sqrt[6]{y}(\sqrt[4]{y})$, $y > 0$, as an expression with a single radical.
CCSS N.RN.1, N.RN.2, SMP 7

g. MAKE A CONJECTURE Does order matter when evaluating expressions with rational exponents? When considering $m^{\left(\frac{a}{b}\right)}$ does it matter whether we first raise m to a power and then take the b root of that result, or take the b root of m and raise that result to the a power? Explain. CCSS N.RN.1, N.RN.2, SMP 3

An expression with a variable in the exponent is called an **exponential expression**. An equation with a variable in the exponent is called and **exponential equation**.

EXAMPLE 2 Exponential Equations

a. CRITIQUE REASONING Jason believes he can solve the equation $0.25^x = 32^{(2x-3)}$ by rewriting 0.25 and 32 as powers of 8 and applying the properties of powers. Do you agree? Explain. CCSS N.RN.2, A.CED.1, SMP 3

b. CALCULATE ACCURATELY The length, in meters, of a snake is modeled by the equation $m = n(4^{0.25t})$ where, m is the in snake's length in meters when the snake's age is t months, and n is the snake's length when it hatches. Use this formula to find the age of a 1.2-meter snake that was 0.15 meters long when it hatched. Show your work.
CCSS N.RN.2, A.CED.1, SMP 6

EXAMPLE 3 Exponential Equations

a. USE A MODEL A geologist is studying a 128-gram sample of an isotope with a half-life of 2 hours. Ricardo writes the following equation where y models the mass of the isotope remaining after x hours: $y = 128(2^{-0.5x})$. Explain why the exponent is negative. Find the mass of the isotope 4 hours after it was 128 grams. CCSS N.RN.2, A.CED.2, SMP 4

b. USE A MODEL Write a second equation that could be used to model the situation in **part a**. This equation should be equivalent to Ricardo's, but not use negative exponents. Use your equation to find the mass of the isotope 4 hours after it was 128 grams. CCSS SMP 3

c. USE TOOLS Use your function to complete the table. Then graph the equation on the coordinate plane. CCSS A.CED.1, A.CED.2, SMP 5

x	y
0	
2	
6	
10	
16	

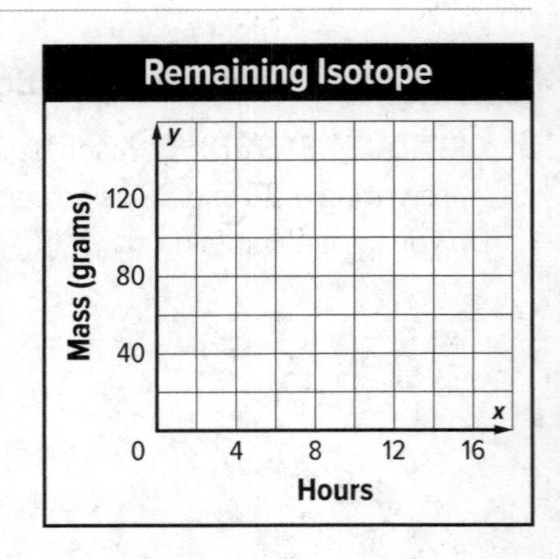

d. USE A MODEL Use your graph from **part c** to estimate when there will be approximately 16 grams of the isotope remaining. CCSS SMP 4

e. CALCULATE ACCURATELY Use Ricardo's equation from **part a** to find when there will be exactly 32 grams of the isotope remaining. Show your work. CCSS N.RN.2, A.CED.1, SMP 6

PRACTICE

1. **CRITIQUE REASONING** Sara and Makayla are evaluating $27^{\frac{2}{3}}$. Sara takes the cubed root of 729, while Makayla squares 3. Whose method is correct? Explain. CCSS N.RN.1, N.RN.2, SMP 3

2. **USE STRUCTURE** Use rational exponents to find an equivalent radical expression in simplest form. CCSS N.RN.2, SMP 7

   a. $\sqrt{w^{13}x^5z^8} =$ __________

   b. $\sqrt[4]{16m^{32}} =$ __________

   c. $(\sqrt{x})(\sqrt[3]{x}) =$ __________

   d. $\sqrt[3]{\sqrt{b}} =$ __________

3. **USE STRUCTURE** Simplify $\frac{4^{40}}{16^{20}}$. Justify your answer. CCSS N.RN.2, SMP 7

4. a. **CRITIQUE REASONING** Jada and Vic are using different bases when they rewrite $8^{(2+x)} = 16^{(2.5-0.5x)}$ to solve for x. Their rewritten equations are shown in the box to the right. Are both students' equations equivalent to $8^{(2+x)} = 16^{(2.5-0.5x)}$? Justify your answer. CCSS N.RN.1, N.RN.2, SMP 3

Jada	Vic
$2^{(8+4x)} = 2^{(10-2x)}$.	$4^{(3+1.5x)} = 4^{(5-x)}$.

b. **CALCULATE ACCURATELY** Solve $8^{(2+x)} = 16^{(2.5-0.5x)}$ for x. Show your work.
CCSS N.RN.2, SMP 6

5. **a.** **CRITIQUE REASONING** Javier believes the graph to the right shows the equation $y = x^{\frac{1}{2}}$. Do you agree? Explain. CCSS N.RN.1, N.RN.2, SMP 3

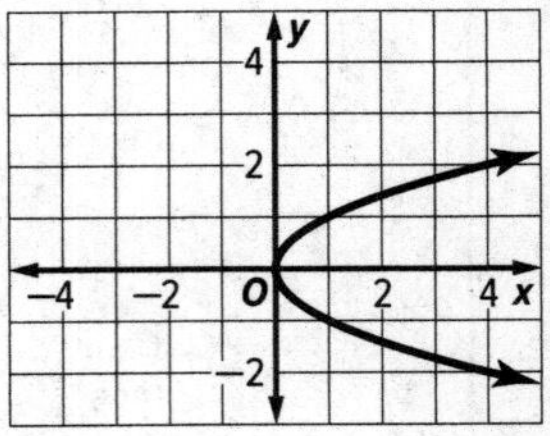

b. **CRITIQUE REASONING** Lara believes the graph to the right shows the equation $y = \left(\frac{1}{2}\right)^x$. Do you agree? Explain. CCSS N.RN.1, N.RN.2, SMP 3

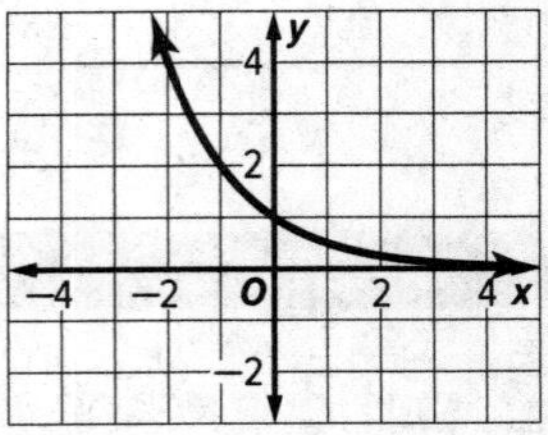

c. **CRITIQUE REASONING** Brad claims that $\sqrt{(|x|^2)} = x$. Rosa says that this is wrong and that $\sqrt{(|x|^2)} = |x|$. Who is correct? CCSS N.RN.1, N.RN.2, SMP 3

6. Without advertising, a Web site had 96 total visits. Today, the owners of the site are starting a new promotion which is expected to double the total number of visits to their Web site every 5 days.

a. **USE A MODEL** Write an equation that relates the total number of visits, *v*, to the number of days the promotion has been running, *d*. CCSS N.RN.2, A.CED.2, SMP 4

b. **CALCULATE ACCURATELY** Use your equation from **part a** to find how many days the promotion should be run in order to increase the traffic to the Web site to 12,288 total visits. Show your work. CCSS N.RN.2, A.CED.1. A.CED.2, SMP 6

c. USE STRUCTURE According to your model, what is the average daily percent of increase? Round your answer to the nearest whole percent and justify your answer. CCSS N.RN.2, A.CED.2, F.IF.8b, SMP 7

d. USE A MODEL After 50 days, the website owners discontinued the promotion. Once the promotion was stopped, the total number of visits continued to increase but at a rate of increase of 12% every 5 days. Write an equation that relates v, the total number of visits, to s, the number of days since stopping the promotion. CCSS N.RN.2, A.CED.2, SMP 4

e. CALCULATE ACCURATELY According to your model in **part d**, what is the average daily percent of increase after ending the promotion? Round your answer to the nearest tenth of a percent. CCSS N.RN.2, A.CED.2, F.IF.8b, SMP 6

7. USE STRUCTURE With the use of technology, graph each of the following functions. CCSS N.RN.1, N.RN.2, SMP 7

a. $y = x^{\frac{1}{3}}$

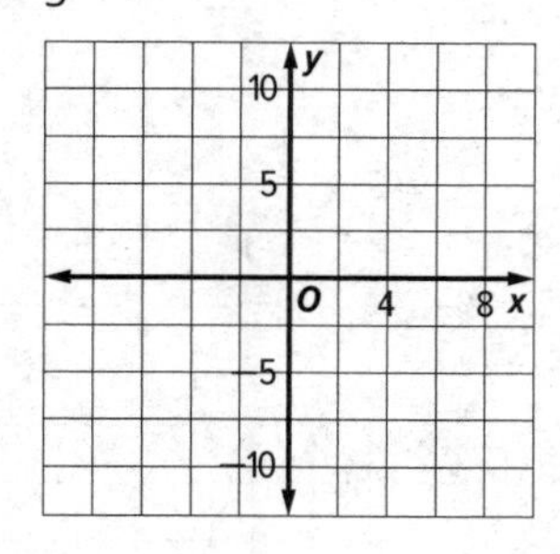

b. $y = x^{\frac{5}{3}}$

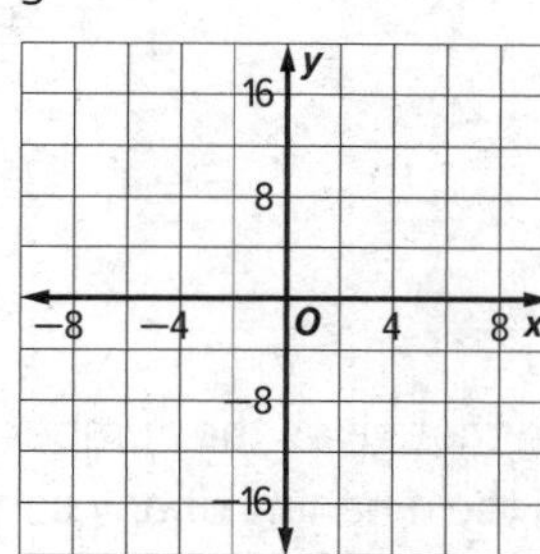

c. $y = x^{\frac{7}{3}}$

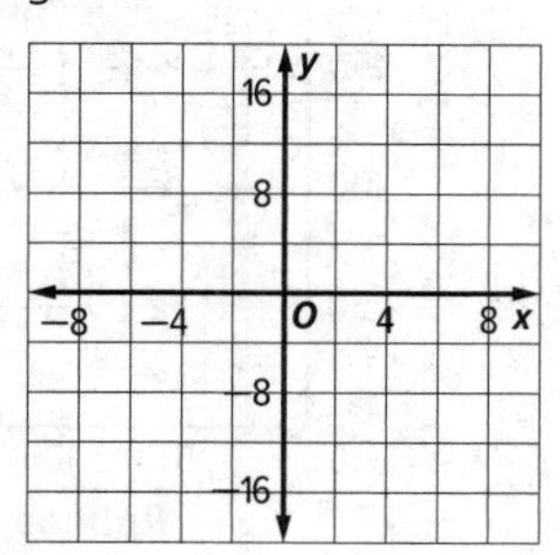

d. $y = x^{\frac{9}{3}}$

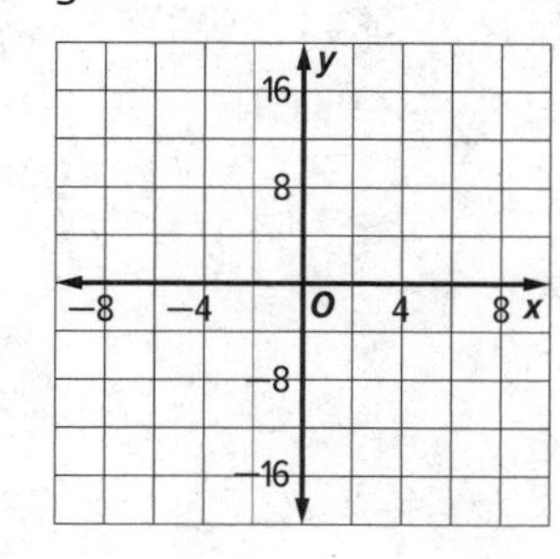

e. Describe the patterns that you see in the graphs as the exponents change.

f. The function $y = x^{\frac{3}{3}}$ was skipped in the pattern for **parts a–d**. Why? What is unique about this graph compared to the others?

7.4 Exponential Functions

Objectives

- Create exponential functions to model relationships between quantities.
- Graph exponential equations on a coordinate plane.

Content: F.IF.7e, F.IF.8b, F.LE.1a, F.LE.2, F.LE.5, A.REI.11
Practices: 3, 4, 7
Use with Lesson 7-5, 7-6

Functions in the form $y = a(1 + r)^x$ model exponential change. The constant a represents an initial value, and the constant r represents the rate of change. Positive values of r model growth, and negative values of r model decay.

EXAMPLE 1 Exponential Decay

EXPLORE 256 players entered a chess tournament. After each round of matches, half of the players are eliminated.

a. USE STRUCTURE Explain why an exponential function and not a linear function is the best model of the functional relationship between the numbers of rounds completed and players remaining in the tournament. CCSS F.LE.5, SMP 7

b. USE A MODEL Write an equation that gives the number of players $f(x)$ remaining after x rounds. Explain the meaning of any constant terms. Graph the equation and discuss the meaning of any intercepts. CCSS F.IF.7e, F.IF.8b, SMP 4

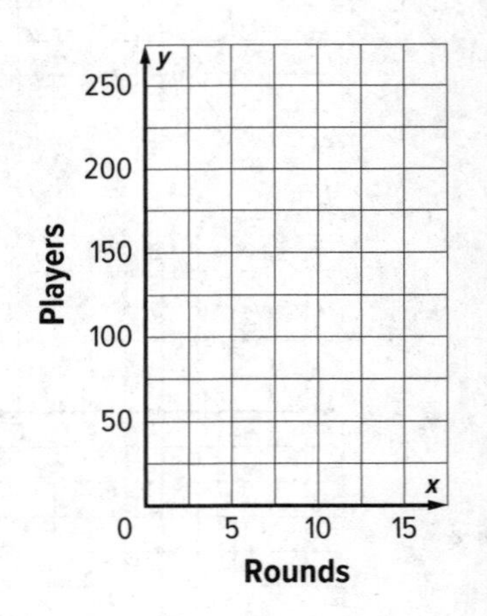

c. USE A MODEL The general model has the constant r in it. What is r in the case of the current model? Discuss its meaning and the meaning of its sign. CCSS F.LE.5, SMP 4

d. CRITIQUE REASONING Andre believes the domain of $f(x)$ is the set of nonnegative integers. Use the graph and equation in **part b** to critique Andre's conclusion. Do you agree? Explain. CCSS F.LE.5, SMP 3

EXAMPLE 2 Modeling a Table of Data

Julia bought a used car. The table to the right shows the expected value y in dollars of the car after she owns it for x years.

x	y
0	4500
1	3600
2	2880
3	2304

a. USE STRUCTURE Explain why an exponential equation should be used to model the relationship between the number of years she owns the car and its expected value and then write the equation that fits the data. CCSS F.LE.2, SMP 7

b. USE A MODEL Explain the meaning of the coefficient of the equation written in **part a**. What is the percent rate of change in the expected value of Julia's car? Does this equation represent exponential growth or exponential decay? CCSS F.IF.8b, SMP 4

c. CRITIQUE REASONING Austin says Julie's car is expected to decrease in value by 1.28% every month. Do you agree? Justify your answer. CCSS F.IF.8b, SMP 3

d. USE A MODEL Graph your equation from **part b** on the coordinate plane. Identify the x- and y-intercepts and describe what they represent about Julia's car. CCSS F.IF.7e, SMP 4

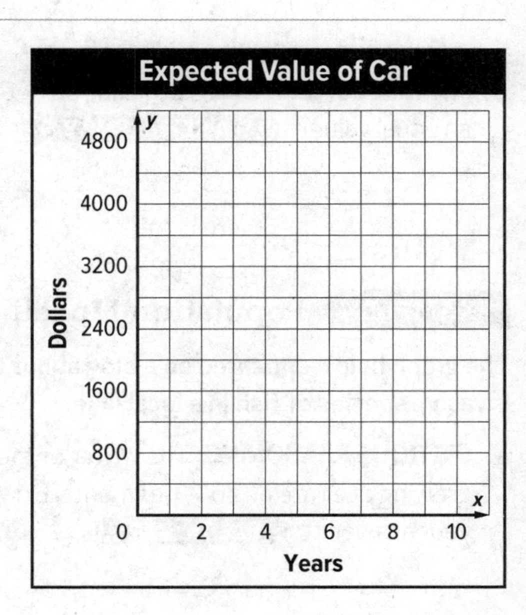

e. USE STRUCTURE Leon paid $3600 for a used car on the same day Julia bought her car. The value of his car is expected to decrease annually by the same percentage as Julia's car. How would a graph of the expected value of Leon's car compare to the graph from **part d**? Explain. CCSS F.IF.8b, SMP 7

EXAMPLE 3 Problem Solving

Samir and Marie each invested money in different accounts. The value of Samir's investment after x years is modeled by the equation $y = 500(1.065)^x$. The value of Marie's investment after x years is represented by the function graphed to the right.

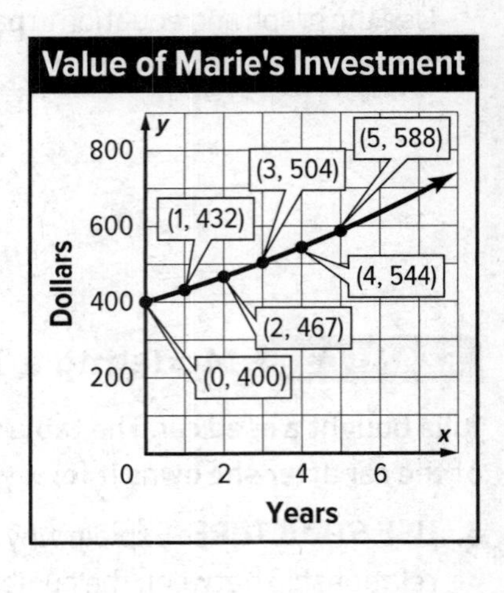

a. **USE STRUCTURE** Did Samir or Marie make the larger initial investment? How do you know? CCSS F.IF.7e, F.LE.5, SMP 7

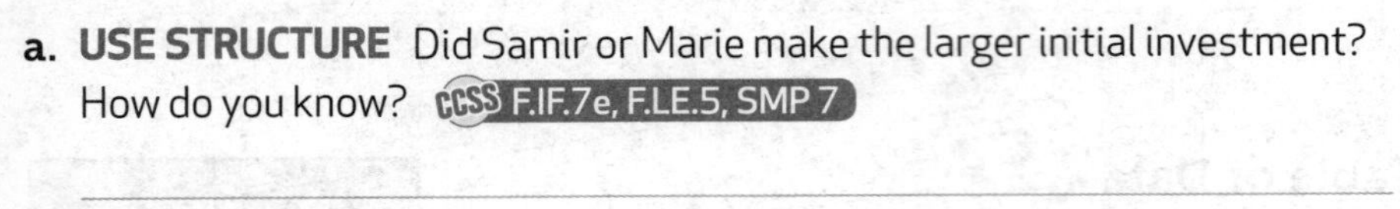

b. **USE A MODEL** Whose investment had the greater average annual increase in value over the first five years? Justify your answer. CCSS F.LE.5, SMP 4

c. **USE A MODEL** Whose investment pays the higher annual rate of interest? How do you know? CCSS F.IF.8b, SMP 4

d. **MAKE A CONJECTURE** How can you determine if and when the two investments have an equal value? CCSS A.REI.11, SMP 3

EXAMPLE 4 Population Modeling

The graph below appeared on a blog about the growth of the population of an invasive species of fish in a local lake.

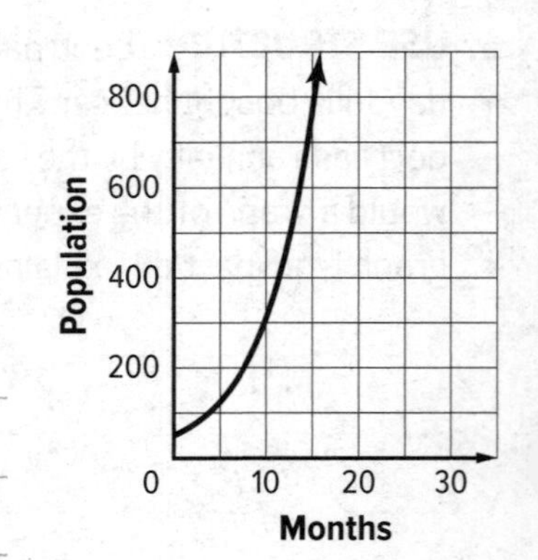

a. **CRITIQUE REASONING** The writer of the blog indicated that the population is growing at a rate of 20% per month. Do you agree? Use the graph to justify your answer. CCSS F.LE.1a, SMP 3

b. FIND A PATTERN The table shows the data that are used to create the graph. Complete the third column by finding the ratio of the number of fish in the row for the current month n to the number of fish in the row for the previous month $n - 1$. How is the change over an interval of 1 month found? What do you notice? CCSS F.LE.1a, SMP 7

Time Since Fish First Counted (months)	Number of Fish	Ratio $\#Fish_n/\#Fish_{n-1}$
0	50	NA
1	60	$\frac{60}{50}=1.20$
2	73	
3	86	
4	104	
5	126	
6	150	
7	180	
8	217	
9	260	
10	308	

c. MAKE A CONJECTURE Based on **parts a** and **b**, make a conjecture about how exponential functions change over equal intervals. How are the changes in the y values related to these changes? CCSS F.LE.1a, SMP 3

d. USE STRUCTURE Find the equation of the exponential curve on the graph, and describe how you found it. CCSS F.LE.2, SMP 7

PRACTICE

1. There is a leak in a container that holds a certain nontoxic gas. Each hour, it loses 10% of its volume.

a. USE A MODEL Write an equation that models the amount of gas left in the container after x hours, assuming there were 300 cubic centimeters in the container before the leak. Then use your equation to determine the amount of gas left in the container after 11 hours. Round your answer to the nearest tenth. CCSS F.LE.2, SMP 4

b. USE A MODEL Harry believes a graph of this function should be a scatter plot instead of a continuous curve. Do you agree? Explain how this relates to the domain of the function. CCSS F.LE.1a, SMP 4

c. **CONSTRUCT ARGUMENTS** Tricia believes the domain of this function is infinite because as x get large, $h(x)$ approaches but never equals 0. Explain why Tricia's reasoning may not fit the context of the function. CCSS F.LE.5, SMP 3

2. The scatter plot to the right shows the projected student enrollment at East Hills High School over the next ten years

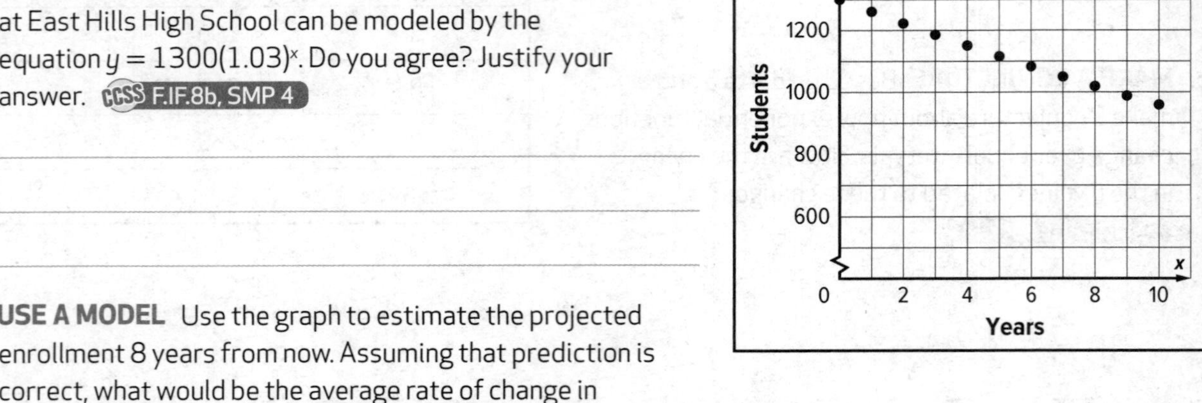

a. **USE A MODEL** Ben believes the projected enrollment at East Hills High School can be modeled by the equation $y = 1300(1.03)^x$. Do you agree? Justify your answer. CCSS F.IF.8b, SMP 4

b. **USE A MODEL** Use the graph to estimate the projected enrollment 8 years from now. Assuming that prediction is correct, what would be the average rate of change in enrollment at the school? CCSS F.IF.7e, SMP 7

3. **USE A MODEL** Graph the equation $y = 100(2^{\frac{x}{3}})$. Then describe a real-world situation that can be modeled by the equation. CCSS F.IF.7e, F.LE.5, SMP 4

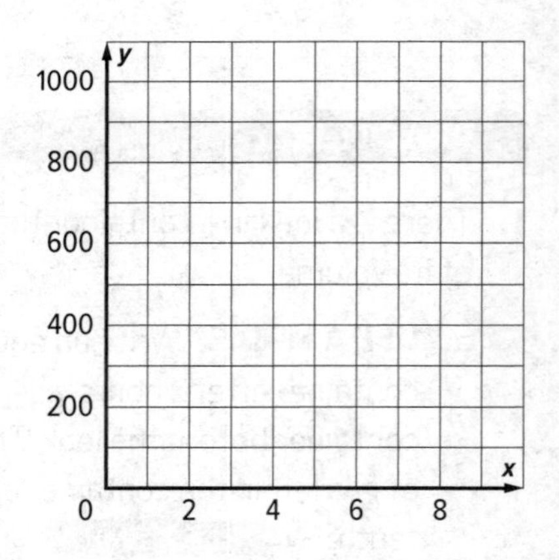

4. **USE STRUCTURE** Graham invested money to save for a car. After x years, the value of Graham's investment can be modeled by the equation $y = 2400(0.95)^x$. How much did Graham originally invest? Is the value of his investment increasing or decreasing? How do you know? Use technology to find when the investment will be worth half of its starting value. CCSS F.IF.8b, SMP 7

5. A wildlife researcher is studying the population of deer in a forest.

Years of Study	0	1	2	3
Population	128	160	200	250

a. USE A MODEL The table shows the estimated number of deer in the forest over a 3-year period. Write an exponential function which fits this data and can be used to predict the deer population in future years. CCSS F.LE.2, SMP 4

b. CALCULATE ACCURATELY What was the average rate of change in population during those three years? CCSS F.LE.5, SMP 6

c. MAKE A CONJECTURE If the population growth follows the model from **part a**, do you expect the deer population to continue to increase by the value you came up with in **part b**? Explain? CCSS F.IF.6, SMP 3

d. USE STRUCTURE Use the values in the table to show how you know the function is exponential, not linear. CCSS F.LE.1a, SMP 7

e. USE A MODEL The same year the researcher began studying the deer population, she also began analyzing their food supply. She found that when t is the number of years she has been studying the deer and their food supply, the function $F(t) = 750(0.98)^t$ models the number of deer that the forest can sustain. Graph $F(t)$ and your function from **part a** on the coordinate plane. CCSS F.IF.7e, A.REI.11, SMP 4

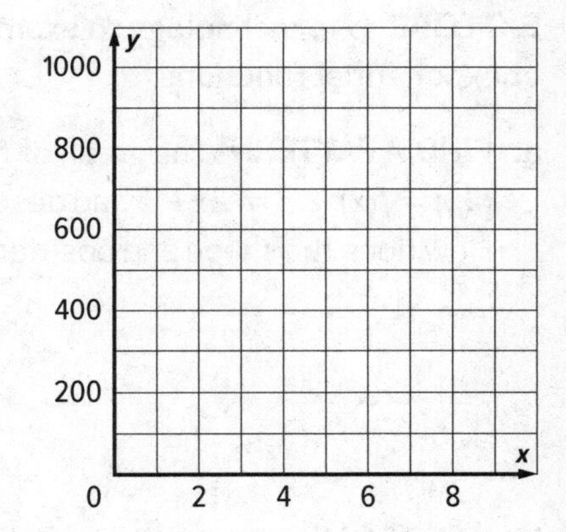

f. MAKE A CONJECTURE Estimate the point of intersection of the two curves. Interpret this point in the context of the problem. Then make a conjecture about the validity of the model after this point. CCSS A.REI.11, SMP 3

g. USE A MODEL Using your function from **part a**, by what percent does the deer population change between years 2 and 7. Round your answer to the nearest percent, and show your work. CCSS F.LE.2, SMP 4

7.5 Transforming Exponential Functions

Objectives

- Identify and analyze the effect of transformations on the graph of the exponential function $y = b^x$.

Content: F.BF.3, F.IF.7e
Practices: 2, 5, 6
Use with Explore 7-5

An **exponential function** is a function of the form $y = ab^x$, where $a \neq 0$, $b > 0$, and $b \neq 1$. In this section, you will be exploring the graphs of exponential functions as the equations are transformed by

a. Adding a constant k to the function: $y = ab^x + k$

b. Adding a constant k to the exponent: $y = ab^{x+k}$

c. Multiplying a constant k to the function: $y = k\,ab^x$

d. Multiplying a constant k to the exponent: $y = ab^{kx}$

EXAMPLE 1 Transformations of Exponential Functions: Adding a constant to the function and adding a constant to the exponent

EXPLORE Use technology to examine the behavior of the graphs of exponential functions.

a. FIND A PATTERN The graph of $f(x) = 2^x$ is shown at the right. Graph $g(x) = f(x) + 3 = 2^x + 3$ and describe how the outputs of the function change. How does the shape and position of the graph change? CCSS F.BF.3, F.IF.7e, SMP 6

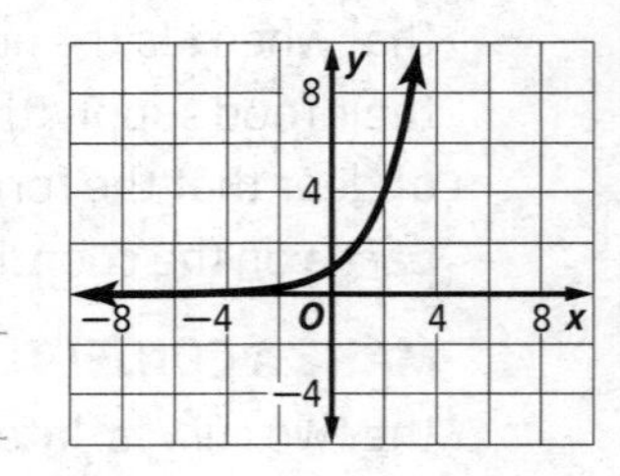

b. USE TOOLS Use technology to graph $f(x)$ and $h(x) = 2^x - 3$. Describe how the graph of $h(x)$ is related to the graphs of $f(x)$ and $g(x)$. CCSS F.BF.3, SMP 5

c. FIND A PATTERN The graph of $f(x) = 3^x$ is shown at the right. Graph $g(x) = f(x + 2) = 3^{x+2}$ and describe how the outputs of the function change. How does the shape and position of the graph change? CCSS F.BF.3, SMP 6

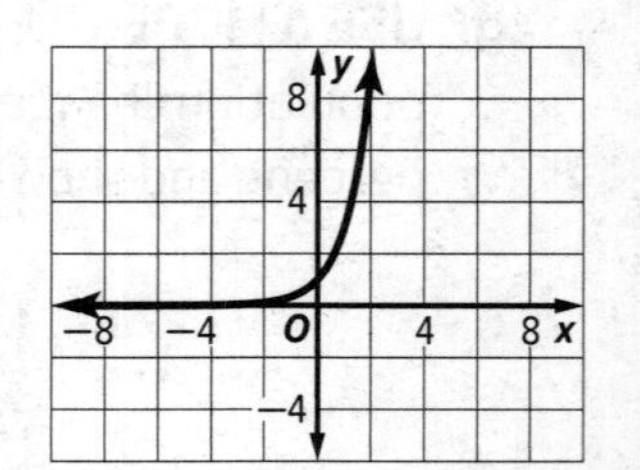

d. USE TOOLS Use technology to graph $f(x)$ and $h(x) = 3^{x-2}$. Describe how the graph of $h(x)$ is related to the graphs of $f(x)$ and $g(x)$. CCSS F.BF.3, SMP 5

e. USE REASONING Consider your work in **parts a–d**. Develop a general rule for the transformations $f(x) + k$ and $f(x + k)$. CCSS F.BF.3, SMP 2

EXAMPLE 2 Transformations of Exponential Functions: Multiplying a constant to the function and multiplying a constant to the exponent.

a. USE TOOLS The graph of $f(x) = 2^x$ is shown at the right. Use technology to graph $g(x) = 5 \cdot 2^x$ and reproduce the graph on the grid. Describe how the graph of $g(x)$ is related to the graph of $f(x) = 2^x$. Discuss the shape, intercept and asymptote. CCSS F.BF.3, SMP 5

y
8
4
−8 −4 O 4 8 x
−4

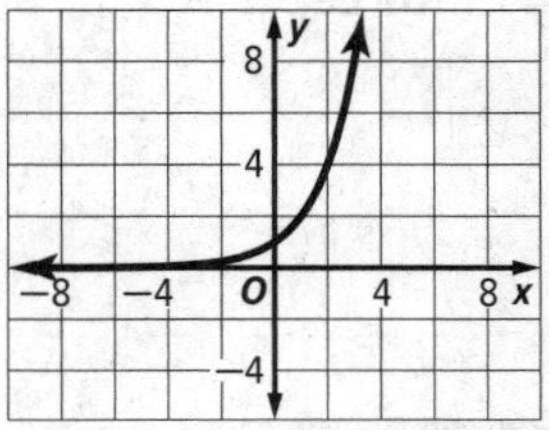

b. USE TOOLS The graph of $f(x) = 2^x$ is shown at the right. Use technology to graph $g(x) = 2^{3x}$ and reproduce the graph on the grid. Describe how the graph of $g(x)$ is related to the graph of $f(x) = 2^x$. Discuss the shape, intercept and asymptote. CCSS F.BF.3, SMP 5

y
8
4
−8 −4 O 4 8 x
−4

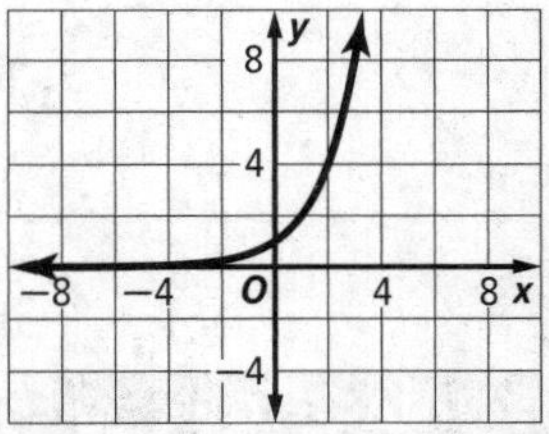

c. USE REASONING Consider your work in **parts a–b**. Develop a general rule for the transformations $kf(x)$ and $f(kx)$ for $k > 1$. What would you expect to happen for $0 < k < 1$? CCSS F.BF.3, SMP 2

EXAMPLE 3 Reflections of Exponential Functions

a. USE TOOLS Sketch the graphs on the coordinate grid and answer the question below. CCSS F.BF.3, SMP 5

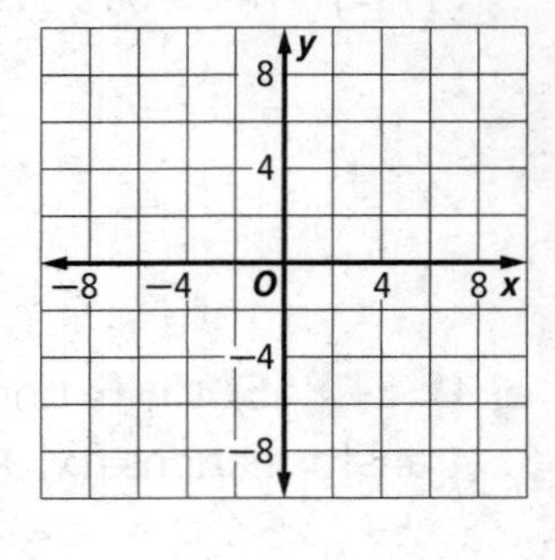

$f(x) = 3^x$ $\quad$ $g(x) = -3^x$

Describe the similarities and differences among the graphs, including the intercepts and the end behaviors.

b. USE TOOLS Sketch the graphs on the coordinate grid and answer the question below. CCSS F.BF.3, SMP 5

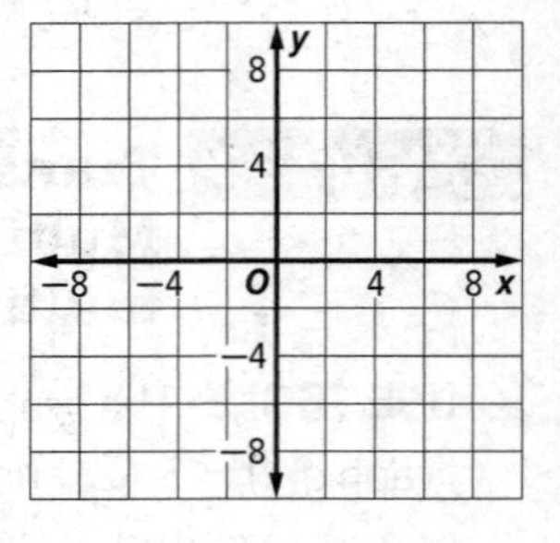

$f(x) = 3^x$ $\quad$ $g(x) = 3^{-x}$

Describe the similarities and differences among the graphs, including the intercepts and the end behaviors.

EXAMPLE 4 Multiple Transformations of Exponential Equations

COMMUNICATE PRECISELY Explain how to use transformations to graph the equations $y = 2^{x+1} - 3$ and $y = 6 - 3^{x-2}$. CCSS F.BF.3, SMP 3

PRACTICE

1. **REASON ABSTRACTLY** Each graph shows a transformation of the function $y = 2^x$. Write an equation that describes each function and explain your reasoning. CCSS F.BF.3, SMP 2

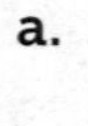

a.

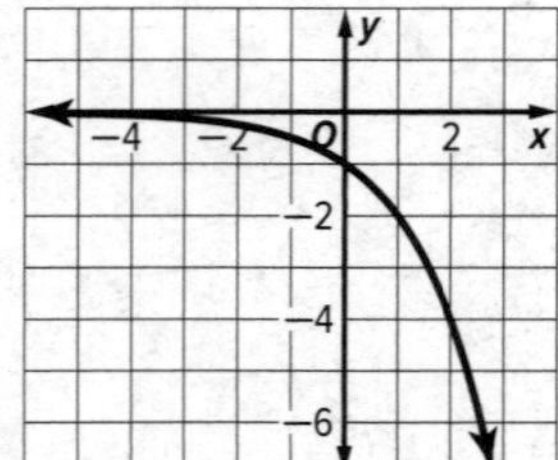

b.

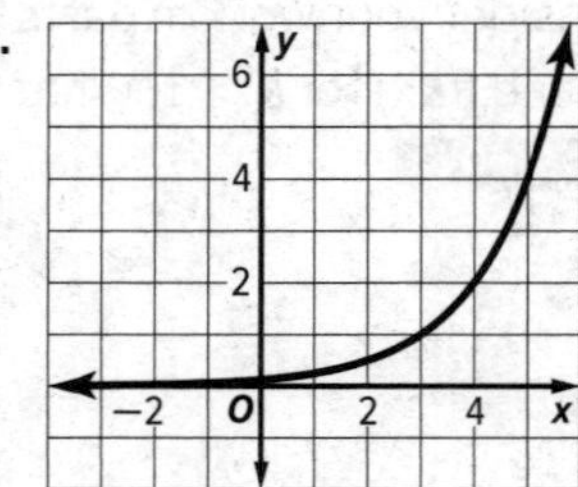

c.

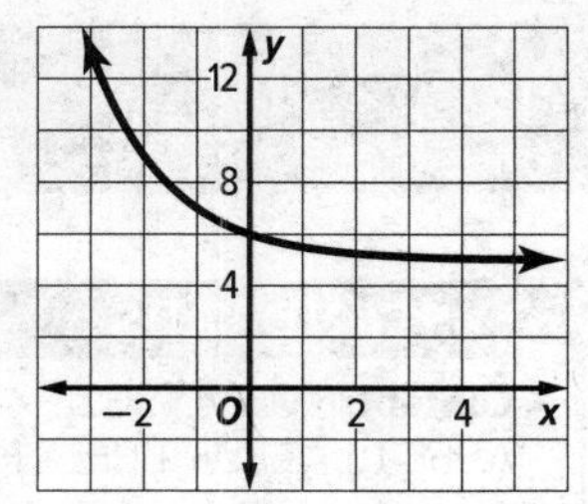

d.

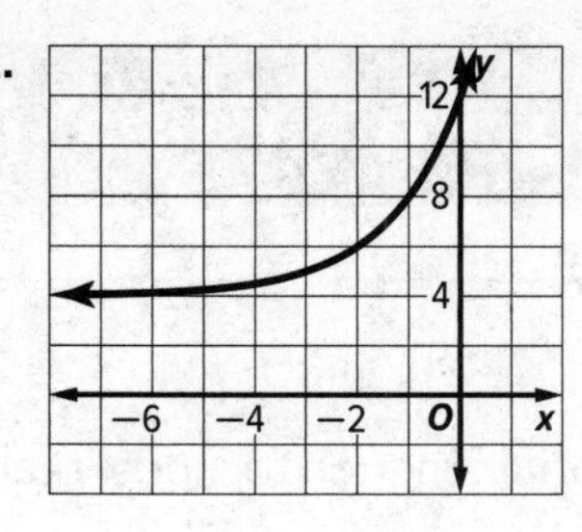

2. **REASON ABSTRACTLY** What would happen to the shape of the graph of an exponential function if the function is multiplied by a number between 0 and -1? What would happen to its shape if the exponent is multiplied by a number between 0 and -1? CCSS F.BF.3, SMP 2

3. **REASON ABSTRACTLY** Each graph below has a parent function provided. Sketch the required transformed function without using graphing technology. Explain the transformations applied. CCSS F.BF.3, F.IF.7e, SMP 2

Parent: $y = 3^x$
Transformed: $y = 5 \cdot 3^{x+2} - 4$

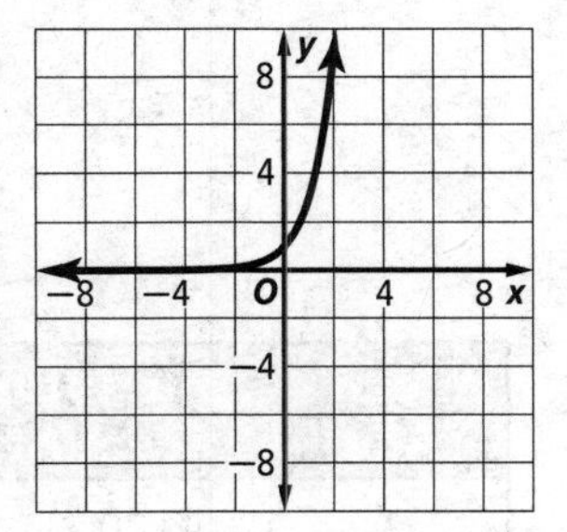

Parent: $y = 2^x$
Transformed: $y = -3 \cdot 2^{-x} + 1$

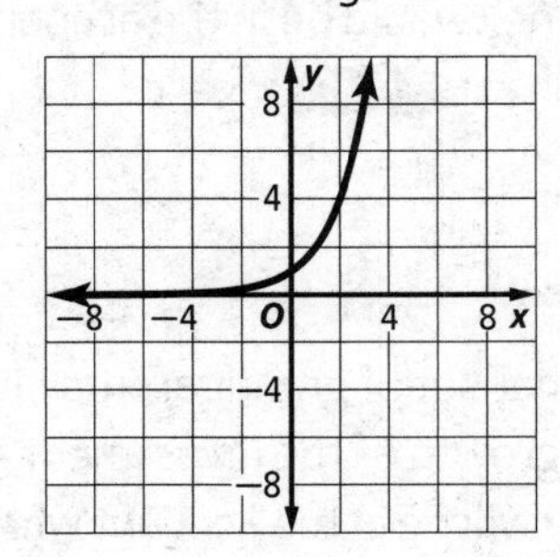

4. **REASON ABSTRACTLY** Jennifer claims that the graph of $y = 2(2^x)$ is a graph that rises more rapidly than its parent function $y = 2^x$. James claims that it is actually the parent graph shifted to the left 2 units. Who is correct? Explain your reasoning. CCSS F.BF.3, SMP 2

7.6 Modeling: Exponential Functions

Objectives

- Define an exponential function to model real-world growth and decay.
- Use and interpret exponential models correctly in a real-world context.

Content: N.Q.2, A.SSE.1a, A.SSE.1b, A.CED.1, F.BF.1a, F.BF.1b
Practices: 1, 2, 4, 5, 6, 7
Use with Lesson 7-6

EXAMPLE 1 Modeling with Exponential Functions

Exponential functions can be used to model the growth of investments over time. In this example you will compare different investments using exponential equations.
CCSS N.Q.2, A.SSE.1a, A.SSE.1b, A.CED.1, F.BF.1a, F.BF.1b

EXPLORE When Aileen was born, her grandparents established college funds for her.

a. USE STRUCTURE One set of grandparents invested \$10,000 in an account. If the account earns an average of 12% each year compounded annually, how much will be in the account after 18 years? Justify your solution. CCSS SMP 7

b. USE STRUCTURE If interest was compounded monthly, what would the new equation look like? How much money would be in the account after 18 years? Explain any changes to your equation. CCSS SMP 7

c. USE TOOLS Aileen's other set of grandparents invested \$8000 in a different account. They track the growth of the investment each year. Based on the initial values, how much money will be in the account when Aileen turns 18? Describe the method you used to make your prediction. CCSS SMP 5

Aileen's Age	Amount in Account
1	\$9120
2	\$10,396.80
3	\$11,852.35
4	\$13,511.68

d. USE STRUCTURE What is the annual interest for each of the three account option. In other words, what percent of the initial amount is earned after one year? Which account offers the best annual rate? CCSS SMP 7

e. PLAN A SOLUTION When Aileen is 10, her parents win a contest and receive a $5000 prize. They decide to invest this money into the college fund with the higher rate of return. How much more money will be in this particular account when Aileen turns 18 as a result of this investment? Explain your reasoning. CCSS SMP 1

f. USE STRUCTURE When Aileen was born, the average cost of attending a state university was $23,000 per year. Assume that this college cost increases at an average rate of 3% each year. Write an equation to describe cost as a function of time, and use your equation to complete the table. CCSS SMP 7

Year in college	Aileen's Age	Annual Costs
1	18	
2	19	
3	20	
4	21	
Total Costs of 4 years:		

g. CALCULATE ACCURATELY Based on these models, when Aileen turns 18, will there be enough money to pay for four years of college? CCSS SMP 6

PRACTICE

1. REASON QUANTITATIVELY Mr. and Mrs. Ramirez buy a house for $180,000.
CCSS N.Q.2, A.CED.1, F.BF.1a, SMP 2

a. Mr. and Mrs. Ramirez made a 20% down payment and financed the balance with a 15-year mortgage having monthly payments of $1065. What is the total amount that will be paid at the end of 15 years?

b. In their community, the value of houses has increased at a rate of 6% each year. Write an equation that represents the future value of their house as a function of time if the value increases at the same rate. Use this equation to predict the value of their home in 15 years.

c. If Mr. and Mrs. Ramirez sell their house in 15 years, what is the return on their investment?

2. Mr. Johnson's class is conducting an experiment to understand how hot liquids cool. A beaker of water is heated to boiling, and the temperature of the water is measured every 5 minutes. The measurements are recorded in the table below. CCSS N.Q.2, A.SSE.1, F.BF.1a, F.BF.1b

Time (min)	0	5	10	15	20	25	30
Temperature (°F)	212	186.6	165.8	148.8	134.9	123.5	114.2

Time (min)	35	40	45	50	55	60
Temperature (°F)	106.5	100.3	95.1	90.9	87.5	84.7

a. **PLAN A SOLUTION** List different methods that you could use to analyze the cooling trend represented by the data. CCSS SMP 1

b. Use the method of your choice. Is this trend more likely to be linear or exponential?

The rate at which an object cools is related to the temperature of the surrounding environment. At the time of the experiment, Mr. Johnson's classroom temperature was 72° F. The approximate temperature of the water at time t in minutes in Mr. Johnson's classroom is predicted by the function $T(t) = 72 + (212 - 72)2.72^{-0.4t}$ where $-0.4°$ per minute is defined as the rate of cooling.

c. **REASON QUANTITATIVELY** Rewrite this function so that the coefficient of t in the exponent is 1. CCSS SMP 2

d. REASON QUANTITATIVELY Over time, what is the minimum temperature that you expect the water to reach? Explain your reasoning. CCSS SMP 2

e. USE STRUCTURE Look at the numbers in the formula above. Do you recognize any of them from the information provided for the problem? Make a conjecture about how this formula was derived, and then write an equation that models the cooling of a cup of water that is 180° when it is placed in a refrigerator at 40° if the rate of cooling is −0.52° per minute. CCSS SMP 7

3. **a. REASON ABSTRACTLY** How are growth and decay related? Compare and contrast exponential functions for growth and decay. CCSS A.CED.1, SMP 2

b. USE A MODEL Shaniece is buying a new car for $20,000. The car loses 20% of its value every year. Describe an exponential function that can be used to model the value of her car over time. Use the function to determine the value of her car after 4 years. CCSS A.CED.1, SMP 4

c. INTERPRET PROBLEMS Josiah is an urban planner. He has developed two mathematical models that describe the trends in population over time for two different counties. The model for county A is $f(x) = 200{,}000\,(1.03)^x$. The model for county B is $g(x) = 500{,}000\,(0.96)^x$. If x is time in years, describe what is happening in regards to population in each county? Will the counties ever have the same population? How do you know? CCSS A.SSE.1a, SMP 1

d. USE A MODEL Monetary inflation causes the value of money to decrease. Ms. Watson wins $50,000 in the lottery. She puts her money in her mattress and takes it out 20 years later. Assuming inflation averages 2% per year over this period, how much are Ms. Watson's winnings worth when she takes it out of the mattress? CCSS F.BF.1a, SMP 4

7.7 Geometric Sequences as Exponential Functions

Objectives

- Describe and create geometric sequences.

STANDARDS

Content: F.BF.2, F.LE.1c, F.LE.2, F.LE.5, F.BF.1a
Practices: 1, 2, 3, 4, 7
Use with Lesson 7–7

A **geometric sequence** is a numerical pattern in which the first term is nonzero and each term after the first is found by multiplying the previous term by a nonzero constant r called the **common ratio**. Examples are 2, 6, 18, 54, 162 ... (a geometric sequence with a first term 2 and a common ratio of 3) and 200, 100, 50, 25, 12.5, 6.25 ... (a geometric sequence with a first term 200 and a common ratio of 0.5).

EXAMPLE 1 The Relationship Between Exponential Functions and Geometric Sequences

EXPLORE Consider a $100 investment that earns 6% interest compounded annually.

a. USE STRUCTURE Write the exponential function that models the situation and use that model to write the amount the investment would be worth after each of the first six years (the first year being year 0, the day the investment was made). CCSS F.BF.2, F.BF.1a, SMP 7

b. REASON QUANTITATIVELY Given the provided description of a geometric sequence, is the sequence of numbers produced in **part a** a geometric sequence? How do you know? What are its characteristics? How do the characteristics of the sequence relate to the characteristics of an exponential function? CCSS F.LE.5, SMP 2

c. USE STRUCTURE The following sequence is geometric: 1200, 960, 768, 614.4, 491.52 .... What is the first term and common ratio? What is the corresponding exponential equation? Use the exponential equation to find the 10th term in the sequence. Explain. CCSS F.LE.5, SMP 7

EXAMPLE 2 Comparing Geometric Sequences CCSS F.BF.2, F.LE.1c, F.LE.2, F.LE.5, F.BF.1a

The table shows the populations of two colonies of bacteria at different points in time.

Time	Colony A	Colony B
Initial	2000	1000
After 1 hour	2200	1200
After 2 hours	2420	1440
After 3 hours	2662	1728

a. REASON QUANTITATIVELY Assume that the bacteria populations continue to grow at approximately the same rates. Determine whether the populations grow as geometric, algebraic, or in some other type of sequence. CCSS SMP 2

b. MAKE A CONJECTURE The initial population for Colony B is less than that for Colony A. Will Colony B always be the smaller colony? Justify your reasoning. CCSS SMP 3

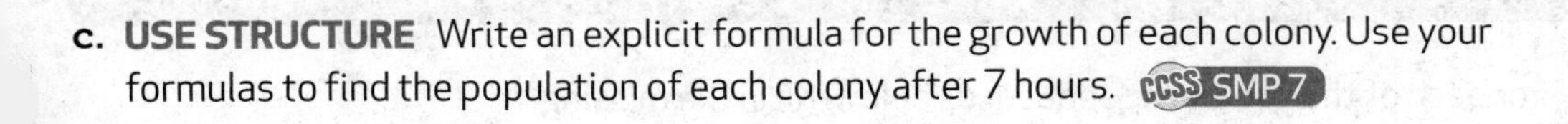

c. USE STRUCTURE Write an explicit formula for the growth of each colony. Use your formulas to find the population of each colony after 7 hours. CCSS SMP 7

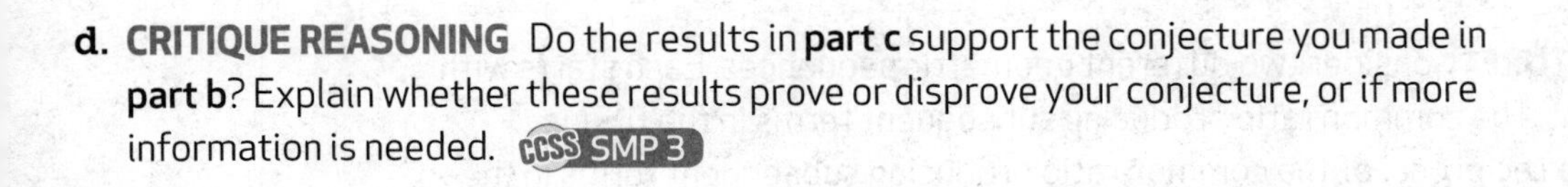

d. CRITIQUE REASONING Do the results in **part c** support the conjecture you made in **part b**? Explain whether these results prove or disprove your conjecture, or if more information is needed. CCSS SMP 3

e. USE A MODEL Use the graphs of exponential functions to verify the conjecture you made in **part b**. Justify your procedure. CCSS SMP 4

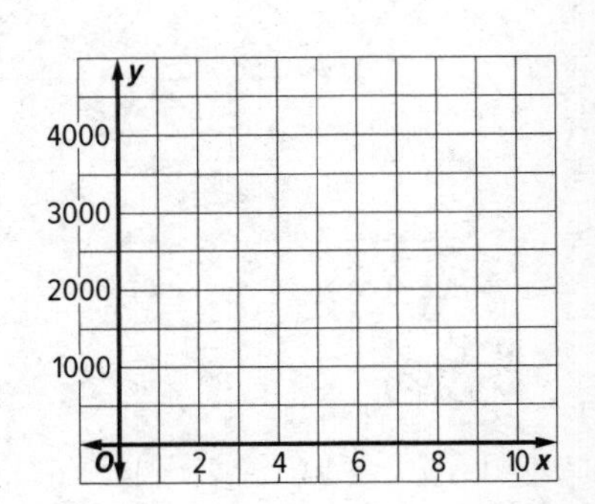

EXAMPLE 3 Define a Geometric Sequence CCSS F.BF.2, F.LE.1c, F.LE.2, F.LE.5, F.BF.1a

You can create a geometric sequence and its exponential equation if given non-consecutive terms of a sequence. Suppose you know that a certain geometric sequence has its third term (a_3) = −24 and its eighth term (a_8) = 768.

a. CONSTRUCT ARGUMENTS Determine the common ratio between any two terms in the sequence. Explain your process. CCSS SMP 3

b. CONSTRUCT ARGUMENTS Given the common ratio and the third term, find the initial term. Write the exponential equation that defines the geometric sequence. Explain. CCSS SMP 3

PRACTICE

1. **USE STRUCTURE** For each of the geometric sequences below, fill in the missing terms, write the corresponding exponential equation and use the exponential equation to determine the 10th term in the sequence. CCSS F.LE.2, SMP 7

   a. 0.5 6 ____ ____ ____ $f(x) =$ ____ 10th term ____

   b. ____ 10 ____ 40 ____ $g(x) =$ ____ 10th term ____

   c. ____ −4 ____ ____ 1.6875 $h(x) =$ ____ 10th term ____

2. **MAKE A CONJECTURE** Consider two different geometric sequences. Each starts with the same constant. The common ratio producing subsequent terms in the first is positive and is the reciprocal of the common ratio producing subsequent terms in the second. How would the graphs of the two sequences compare? Think about intercepts, asymptotes, and symmetry. Then graph an example of the situation. CCSS F.LE.1c, SMP 3

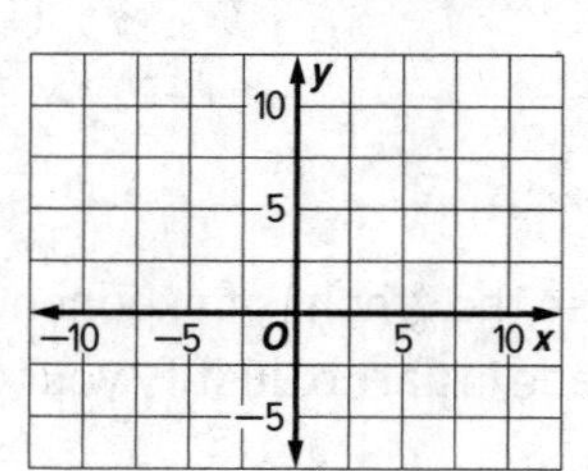

3. You are planning to work part time. Your potential employer offers two different methods of payment. They are shown in the table. CCSS F.BF.1a, F.BF.2, F.LE.5

Month	Method 1 Payment	Method 2 Payment
1	\$100.00	\$0.01
2	\$108.00	\$0.02
3	\$116.00	\$0.04
4	\$124.00	\$0.08

a. **INTERPRET PROBLEMS** Describe the two different methods of payment being offered. CCSS SMP 1

b. **USE STRUCTURE** What kind of mathematical equations can you use to model each situation? How do you know? Write each equation. CCSS SMP 7

c. **REASON QUANTITATIVELY** You are planning to work at this job for two years. Your manager promises you that she will continue raising your salary the way it is described in the table, as long as you work for her. Which payment plan do you choose and why? CCSS SMP 2

4. **CONSTRUCT ARGUMENTS** The terms of a geometric sequence are defined by the equation $a_n = 512(0.5)^x$. A second sequence contains the terms $b_3 = 7168$ and $b_7 = 28$. CCSS F.BF.2, SMP 3

a. Determine which sequence has the greater common ratio.

b. What is the initial term of each sequence? Explain your reasoning.

5. **CRITIQUE REASONING** The sum of the interior angles of a triangle is 180°. The interior angles of a pentagon add to 540°. Is the relationship between the number of sides in a polygon and the sum of interior angles a geometric sequence? Use the sum of the measures of the interior angles of a square to justify your answer. CCSS F.BF.1a, SMP 3

7.8 Recursive Formulas

Objectives

- Recognize sequences as functions that are sometimes recursively defined.
- Write and convert between recursive and explicit formulas for arithmetic and geometric sequences.

Content: F.IF.3, F.BF.2, A.SSE.2
Practices: 1, 2, 3, 4, 5, 6, 7, 8
Use with Lesson 7-8

A **sequence** of real numbers, such as 4, 7.5, 11, 14.5, 18, ..., can be thought of as a function that maps the first number or 1 to $a_1 = 4$, 2 to $a_2 = 7.5$, and so on.

EXAMPLE 1 Express a Sequence Algebraically CCSS F.IF.3

In some types of plants, each new branch skips the first possible branch point, then starts to branch at the second branch point as shown.

a. USE A MODEL Find and describe a relationship between the number of branches at each horizontal level and the numbers of branches at the two levels below it. CCSS SMP 4

____

____

Level 7
6
5
4
3
2
1

b. FIND A PATTERN Write this relationship as a formula, using a_{n+1}, a_n, and a_{n-1} to stand for the number of branches at levels $n + 1$, n, and $n - 1$. CCSS SMP 7

____

c. EVALUATE REASONABLENESS Check this formula by drawing the next level and comparing the number of branches drawn to the formula's predicted value. CCSS SMP 8

____

Sequences can be expressed as functions in two ways: using an **explicit formula,** such as $a_n = 0.5 + 3.5n$, or using a **recursive formula,** such as $a_n = a_{n-1} + 3.5$, with $a_1 = 4$ specified as the starting value. Sometimes it is easier to work with the recursive formula because the function is defined. For instance, you could find a_n by computing a_2 from a_1, a_3 from a_2, a_4 from a_3, and so on.
The recursive formula for an arithmetic sequence with common difference d is $a_n = a_{n-1} + d$.
The recursive formula for a geometric sequence with common ratio r is $a_n = r \cdot a_{n-1}$.

EXAMPLE 2 Use and Write Recursive Formulas CCSS F.BF.2, F.IF.3

Use a recursive formula to find values in an arithmetic or geometric sequence, and conversely use the initial values of a sequence to write its recursive formula.

a. REASON QUANTITATIVELY A geometric sequence is defined by $a_1 = -3$ and $a_n = 2a_{n-1}$. Write the next three terms in the sequence, showing your calculations. CCSS SMP 2

$a_2 =$ ____________ $a_3 =$ ____________ $a_4 =$ ____________

b. FIND A PATTERN The first three terms of a sequence are -2, 3, and 8. Identify the sequence as arithmetic or geometric. Then, write a recursive formula. Explain your reasoning. CCSS SMP 7

c. DESCRIBE A METHOD How many values do you need to decide if a nonconstant sequence is arithmetic or geometric? Explain. CCSS SMP 8

The explicit formula for an arithmetic sequence with common difference d is $a_n = a_1 + (n - 1)d$. The explicit formula for a geometric sequence with common ratio r is $a_n = r^{n-1} \cdot a_1$.

EXAMPLE 3 Convert Between Recursive and Explicit Formulas CCSS F.BF.2, F.IF.3, A.SSE.2

FIND A PATTERN Convert between recursive and explicit formulas for arithmetic and geometric sequences. CCSS SMP 7

a. Write recursive and explicit formulas for the arithmetic sequence 5, 1, -3, ...,. Explain your reasoning.

b. Write an explicit formula for the sequence $a_n = -3a_{n-1}$ and $a_1 = -3$. Explain your reasoning.

c. COMMUNICATE PRECISELY Write the explicit formulas for general arithmetic and geometric sequences in terms of n rather than $n - 1$. Using these versions of the formulas, describe the type of function involved with each type of sequence. CCSS SMP 6

EXAMPLE 4 Analyze an Investment Decision CCSS F.BF.2, F.IF.3, A.SSE.2

USE A MODEL Rachelle has $1500 to invest and is looking at two options. The first is a municipal bond that offers 0.4% interest calculated on the original investment each month. The second is a savings account that offers 0.35% interest per month, calculated on the current value of the savings account and includes all previous interest earned.

a. Write a recursive formula for each investment. Simplify each expression. CCSS SMP 4

b. FIND A PATTERN Convert each recursive formula in **part a** to an explicit formula. 

c. CALCULATE ACCURATELY Using $n = 61$, recommend an investment for Rachelle, based on a period of five years. CCSS SMP 6

d. PLAN A SOLUTION The graph compares the values of the two investments from five years on. Would you change your recommendation for a longer period of time? Explain. CCSS SMP 1

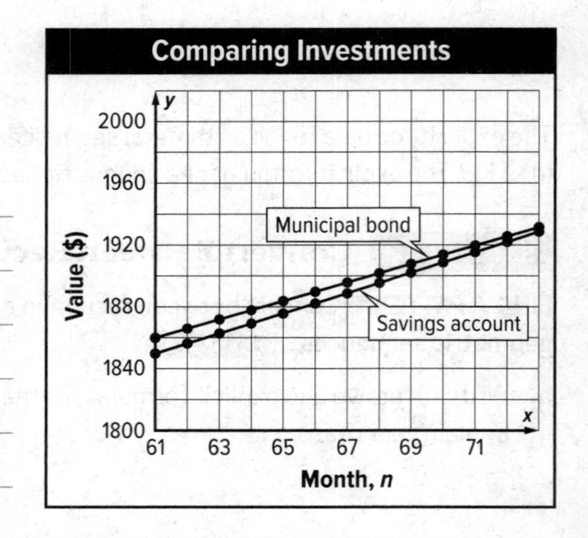

PRACTICE

1. **FIND A PATTERN** Write recursive and explicit formulas for each arithmetic or geometric sequence, given its first three values. Show your work. CCSS F.BF.2, F.IF.3, A.SSE.2, SMP 7

   a. 3, 6, 12

   b. −3, 6, 15

2. **FIND A PATTERN** Write a recursive formula for the sequence defined by each explicit formula. Show your work. CCSS F.BF.2, F.IF.3, A.SSE.2, SMP 7

   a. $a_n = 12 - 7n$

   b. $- a_n = -2(-3)^n$

3. USE A MODEL In 2013, a county had a population of 1.3 million people. The largest factory in the area produced 1700 million widgets per year. The county produces the most widgets per person in the country. The county's population is projected to grow at 1.2% per year and the number of widgets is expected to grow by 10 million per year. CCSS F.IF.3, F.BF.2, SMP 4

a. INTERPRET PROBLEMS Develop explicit formulas for the population and daily annual production, in millions, as functions of the number of years n after 2013. CCSS SMP 1

b. USE TOOLS The graph shows annual widget production per person for the first ten years after 2013 for this county and the next highest widget-producing county which stays at a constant 1200 widgets per person. Use a graphing calculator to extend the graph and find the year when the county will no longer be the leader. Explain your results. CCSS SMP 5

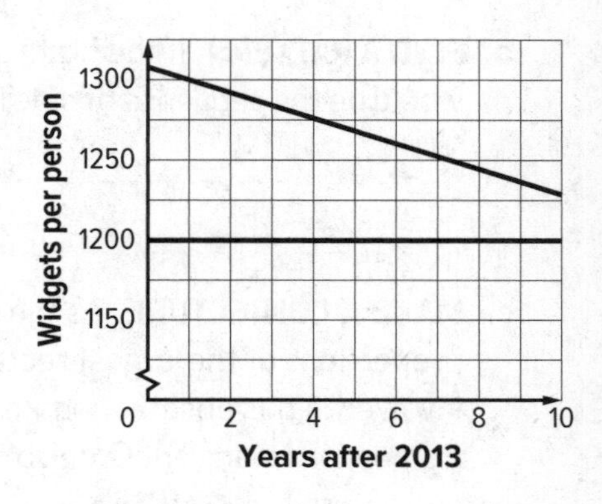

4. Carl Friedrich Gauss, a German mathematician of the 1700s, was asked as a young boy for the sum of the integers from 1 to 100 and unhesitatingly replied with the correct answer. How did he do this? CCSS F.BF.2, A.SSE.2

a. FIND A PATTERN Identify the type of the sequence 1, 2, 3, ... 100, and explore a way to find its sum based on grouping pairs of numbers from each end of the sequence. CCSS SMP 7

b. MAKE A CONJECTURE Find an explicit formula for the sum S of n terms of an arithmetic sequence whose first term is a_1 and whose nth, or last, term is a_n. CCSS SMP 3

5. a. USE TOOLS The first ten numbers in the Fibonacci sequence can be defined by $a_{n+1} = a_n + a_{n-1}$, and each ratio $\frac{a_n}{a_{n-1}}$ can be computed using a spreadsheet (see column C). Which spreadsheet formulas could have been used to calculate the entries in cells B3 and C2? CCSS F.IF.3, SMP 5

Fibonacci sequence

◇	A	B	C
1	1	1	
2	2	1	1
3	3	2	2
4	4	3	1.5
5	5	5	1.666667
6	6	8	1.6
7	7	13	1.625
8	8	21	1.615385
9	9	34	1.619048
10	10	55	1.617647

Sheet 1 Sheet 2 Sheet 3

b. INTERPRET PROBLEMS Compute the ratio $\frac{a_n}{a_{n-1}}$ up to $n = 50$. What do you observe, and how does this relate to **part b**? CCSS SMP 1

6. a. INTERPRET PROBLEMS The diagram shows the first three stages of a construction involving squares with sides of integer lengths. It is based on the Fibonacci Sequence: 1,1,2,3,5,8 ....., as the area of each square is a term of the sequence. Continue the construction for another four stages, and state any properties you observe. CCSS F.IF.3, SMP 1

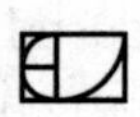

b. FIND A PATTERN If the first square has a side length of 1 centimeter, then what would be the sequence of side lengths in this construction? Explain your reasoning. 

c. MAKE A CONJECTURE As the construction continues, what can you say about the proportions of the largest rectangle at each stage of the construction? (The value involved in this observation, known as the golden ratio, was regarded as an "ideal" proportion by ancient Greek philosophers, and occurs often in art, architecture, and nature.) CCSS SMP 3

7. a. COMMUNICATE PRECISELY Explain in terms of common ratios and differences why the sequence $a_1 = 1, a_n = 3a_{n-1} + 2$ is neither arithmetic nor geometric. CCSS F.IF.3, F.BF.2, SMP 6

b. USE STRUCTURE Write formulas for a_n in terms of a_{n-2} and of a_{n-3}. Do not simplify expressions such as $3(2) + 2$. CCSS F.IF.3, A.SSE.2, SMP 7

c. USE STRUCTURE Write an explicit formula for this sequence in terms of a sum of powers of 3. CCSS F.IF.3, A.SSE.2, SMP 7

d. FIND A PATTERN Write a shorter explicit formula for this sequence, using the fact that $k^n - 1 = (k - 1)(k^{n-1} + k^{n-2} + \ldots + k + 1)$. CCSS F.IF.3, A.SSE.2, SMP 7

8. **REASON QUANTITATIVELY** Ramon has been tracing his family tree with his parents. He claims that he has over 250 great- great- great- great- great- great-grandparents. Is this possible? Write both an explicit and recursive formula for this situation.
CCSS F.BF.2, SMP 2

9. **INTERPRET PROBLEMS** Janelle runs 3 miles each Saturday. She plans to run a marathon. If she increases her distance by 10% each week, how many weeks will it take to prepare for the 26.2-mile race? Write both an explicit and recursive formula for this situation. CCSS F.BF.2, SMP 2

10. **USE TOOLS** There is a famous puzzle called the "Tower of Hanoi." There are three pegs, and a certain number of disks of varying sizes can be set on each peg. The puzzle starts with the disks in a stack on the leftmost peg with the heaviest disk on the bottom and the disks getting smaller as they are stacked. The goal is to move the disks from the leftmost peg to the rightmost peg while obeying three rules. First, only one disk can be moved at a time. Second, only the topmost disk on any peg can be moved. Third, at no time can a larger disk be placed on a small disk. CCSS F.BF.2, SMP 5

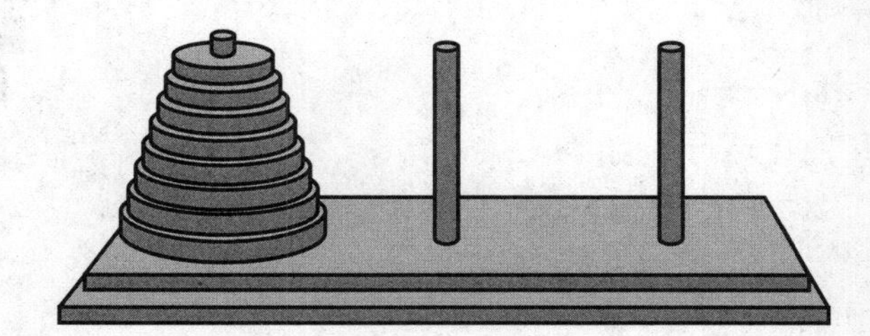

a. If a_n is the number of moves it takes to solve a puzzle consisting of n disks, discuss why the recursive formula $a_n = a_{n-1} + 1 + a_{n-1}$ makes sense.

b. Simplify the recursive formula. What is a_1? Why?

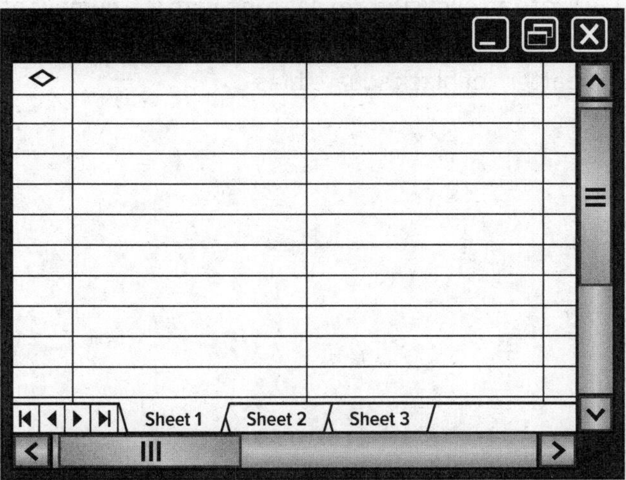

c. Use a spreadsheet to find the number of moves it takes for a game of 8 disks.

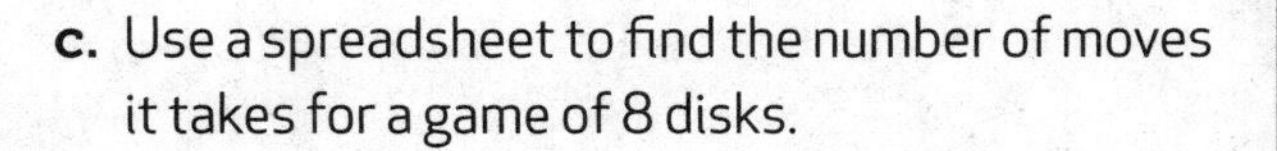

d. Use the information from the spreadsheet to write an explicit formula for a_n.

Performance Task

Making a Design

Provide a clear solution to the problem. Be sure to show all of your work, include all relevant drawings, and justify your answers.

Rosa is making a design with small circular white tiles and small square black tiles. The diagram shows the first four stages of the design.

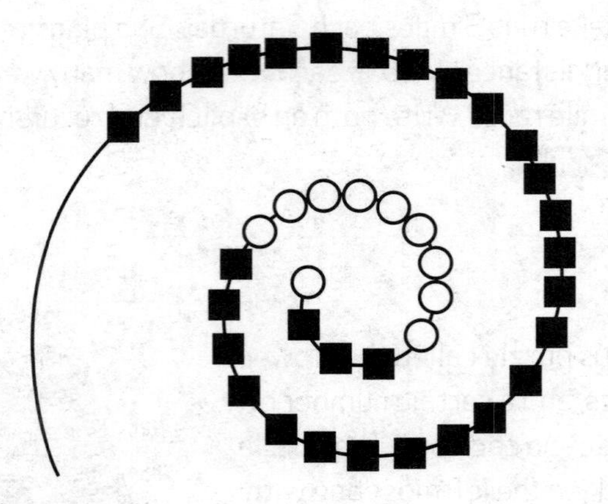

Part A

Write an explicit formula to indicate the number of white circular tiles placed along the spiral each time white circular tiles are needed. Then write a recursive formula for the pattern of white circular tiles.

Write an explicit formula to indicate the number of black square tiles placed along the spiral each time black square tiles are needed. Then write a recursive formula for the pattern of black square tiles.

Part B

Rosa has decided to extend the design shown in **Part A** using only complete sets of tiles. (A complete set of tiles is a set of tiles of one type whose number is determined by the patterns suggested by the diagram.) Here are examples of complete sets.

second complete set of circular tiles

second complete set of square tiles 

Rosa has budgeted $250.00 to spend on tiles.

- Each circular tile costs $0.02.
- Each square tile costs $0.03.

What is the greatest number of complete sets of circular tiles and square tiles that she can buy and not exceed her budget?

How many tiles will she buy in all?

Part C

Rosa finds that, on average, it takes her 10 seconds to attach one tile to the spiral. She can work on the design for 2.5 hours each day. At that rate, how many days will it take Rosa to complete the design from **Part B**?

Performance Task

Planning to Buy a Boat

Provide a clear solution to the problem. Be sure to show all of your work, include all relevant drawings, and justify your answers.

Four friends are thinking about buying a boat.

- They are considering a boat that will cost between $7000 and $8000.
- Each friend will make their initial deposit at the same time and leave the money in their account for 3 years.
- Fred, Juanita, and Rosanna are thinking of the savings plans shown in this table. Robert has not yet decided on the amount of his initial deposit.

	Initial Deposit	Annual Interest Rate	Compounding Period
Fred	$1200.00	4%	annually
Juanita	$2000.00	2%	quarterly
Rosanna	$1500.00	3%	semiannually
Robert	?	2%	annually

Part A

What should Robert's initial deposit be if the boat costs $7000? $8000?

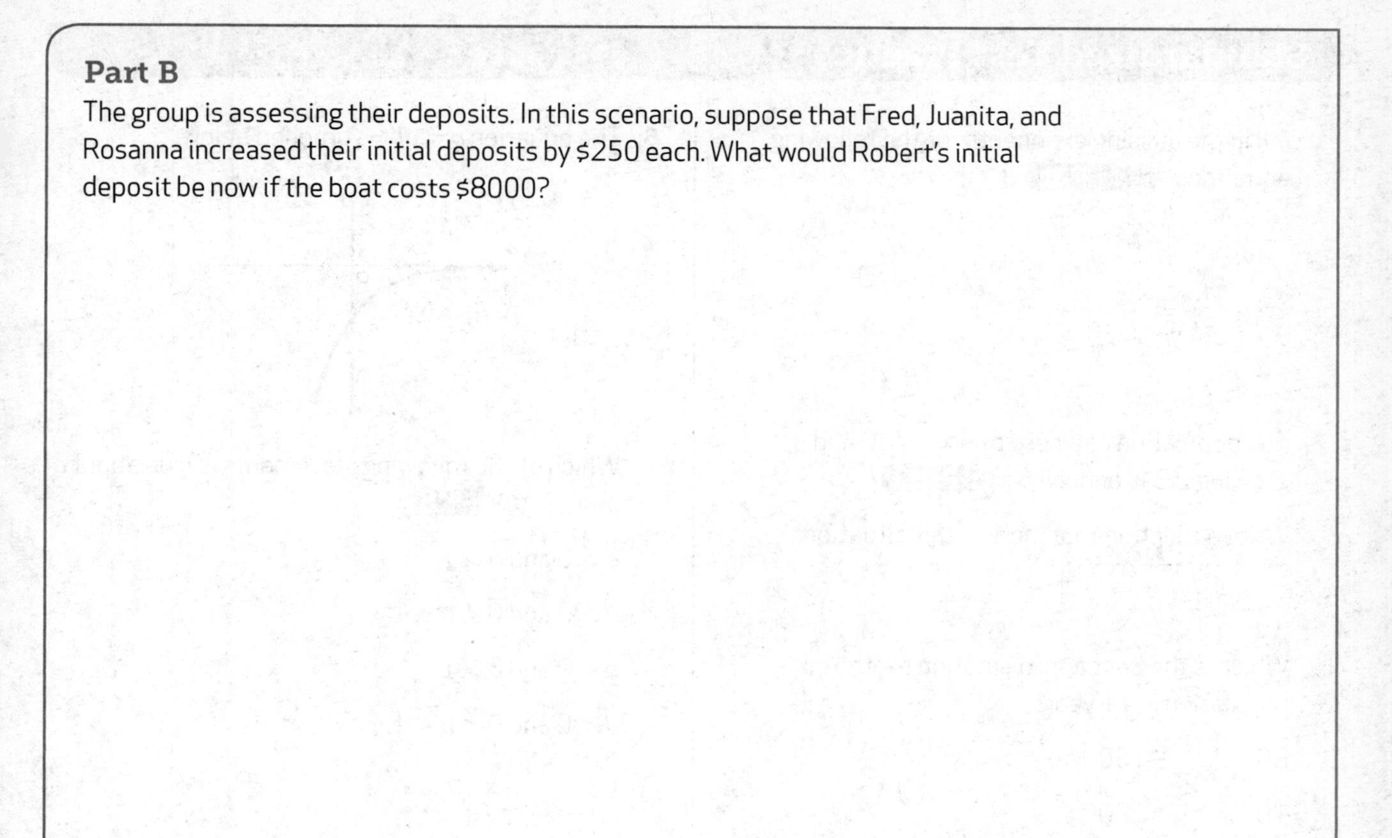

Part B

The group is assessing their deposits. In this scenario, suppose that Fred, Juanita, and Rosanna increased their initial deposits by $250 each. What would Robert's initial deposit be now if the boat costs $8000?

Part C

The group is considering an alternative. Suppose that the boat they want to buy costs $8000.

In this scenario:

- all four friends will make the same initial deposit
- all four friends put their money into one account that pays 2.5% interest compounded annually
- they leave the money in the account for 3 years

Write and solve an equation to determine each initial deposit. Compare this result with the four initial deposits in the table and in **Part A**.

Standardized Test Practice

1. Fill in the missing exponents for the following equations. CCSS F.IF.8b

$\frac{(a^4 b^{\square})^3}{a^5 b} = a^{\square} b^5$

$x^{-2}(y^2 z)^3 = \frac{x^4 y^{\square} z^{\square}}{x^{\square} y^{-1}}$

2. The population of a herd of elk is 780 and growing 2.3% annually. CCSS F.LE.2

Write a function that models this situation.

☐

Which is the best approximation for the elk population in 11 years?

800 1600

1000 2000

Approximately how many years will it take for the elk population to double?

20 40

30 50

3. Simplify each expression below. CCSS N.RN.2

$27^{\frac{2}{3}} - 16^{\frac{3}{4}} = \square$

$\left(\frac{4}{25}\right)^{-\frac{3}{2}} = \square$

$\left(\frac{1}{4}\right)^{-2} \frac{(xy^2)^{-2}}{(36x^6)^{\frac{1}{2}}} = \square$

4. A geometric sequence is given by the recursive formula $a_1 = 1000, a_n = \frac{3a_{n-1}}{5}$. What are the third and fourth terms of the sequence? CCSS F.IF.3

600 and 360

360 and 216

600 and 216

216 and 129.6

5. The equation $y = ab^x$ is graphed below.

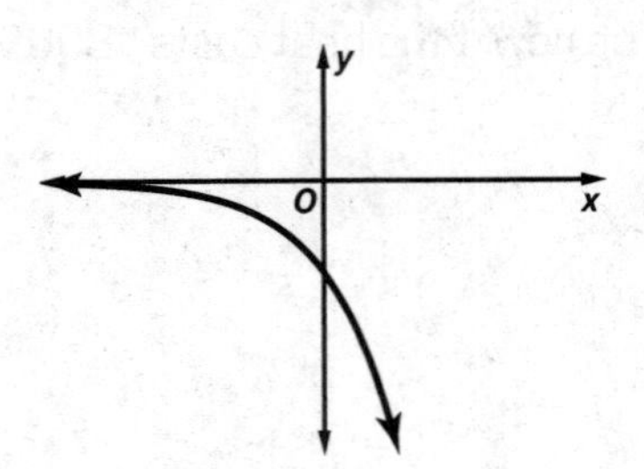

Which of the following statements is true about a and b? CCSS F.IF.1

$a > 0$ and $b > 1$

$a > 0$ and $0 < b < 1$

$a < 0$ and $b > 1$

$a < 0$ and $0 < b < 1$

6. A ball is dropped from a height of 500 feet. The table shows the heights of the first three bounces. CCSS F.BF.2

Bounce	1	2	3
Height (ft)	400	320	256

The geometric sequence that represents the heights of the ball is

☐

The recursive formula for this sequence is

☐

The height of the ball, rounded to the nearest tenth of a foot, on the 10th bounce is ☐ feet.

7. What function has a graph that is shifted 2 units left and 5 units down from the graph of $y = 4^x$? CCSS F.BF.3

☐

8. The function $f(x)$ is an exponential function in the form $f(x) = ab^x$, where $f(x)$ is the number of milligrams of a substance x hours after the beginning of an experiment. The table shows two ordered pairs for $f(x)$.

x	2	5
$f(x)$	5	135

a. Write an equation for $f(x)$. Show your work. 

b. What do the values of a and b mean in the context of this situation? 

c. How many milligrams of the substance are there after 7 hours? Explain two ways to determine the amount. CCSS A.CED.1

9. What function represents the sequence 5, 6, 7.2, 8.64, ...? Justify your answer. CCSS F.IF.3

10. The 2010 model of a certain type of car travels 32 miles on a gallon of gas. The manufacturer increased that car's gas mileage by 6% each model year. On a 1000 mile trip, how many fewer gallons of gas, to the nearest tenth of a gallon, does a 2015 model use than a 2011 model? Show your work. CCSS N.Q.2, N.Q.3

11. Graph $f(x) = 2^x$ and $g(x) = 3(2)^{x+1}$. Describe the graph of $g(x)$ as a transformation of $f(x)$. CCSS F.BF.3

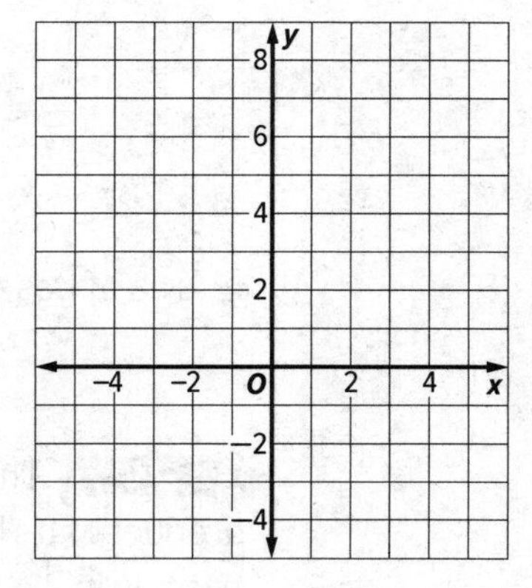

8 Polynomials

CHAPTER FOCUS Learn about some of the Common Core State Standards that you will explore in this chapter. Answer the preview questions. As you complete each lesson, return to these pages to check your work.

What You Will Learn	Preview Question
Lesson 8.1: Adding and Subtracting Polynomials	
CCSS A.APR.1 Understand that polynomials form a system analogous to the integers, namely, they are closed under the operations of addition, subtraction, and multiplication; add, subtract, and multiply polynomials.	CCSS SMP 3 Describe and correct the error in finding the difference of the polynomials. $(2x^2 + 6x - 12) - (3x^2 - 10x + 4)$ $= 2x^2 + 6x - 12 - 3x^2 - 10x + 4$ $= -x^2 - 4x - 8$ CCSS SMP 7 A triangle has sides of length $4x + 7$, $3x^2 + 5x + 1$, and $x^2 - 2$. What is the perimeter of the triangle?
Lesson 8.2: Multiplying a Polynomial by a Monomial	
CCSS A.APR.1 Understand that polynomials form a system analogous to the integers, namely, they are closed under the operations of addition, subtraction, and multiplication; add, subtract, and multiply polynomials.	CCSS SMP 5 The formula for the volume of a cylinder is *area of the base* times *the height*. Calculate the volume of the cylinder below. 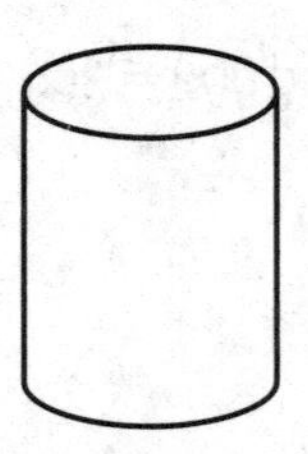 Height = $3y^3$ cm Area of base = $4y^2 + 7y - 10$ cm^2 CCSS SMP 7 The area of a triangle is equal to $\frac{1}{2}$ the product of its height and its base. The height of a triangle is $2x^2$. The base is $x^2 - x + 2$. What is the area of the triangle?

What You Will Learn	Preview Question
Lesson 8.3: Multiplying Polynomials	
CCSS A.APR.1 Understand that polynomials form a system analogous to the integers, namely, they are closed under the operations of addition, subtraction, and multiplication; add, subtract, and multiply polynomials.	CCSS SMP 7 Find the area of the shaded region. 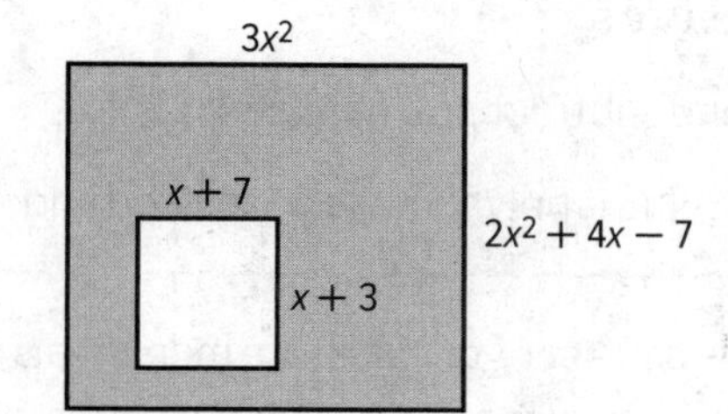 CCSS SMP 7 What is the product of $x - 2$ and $x^2 + 2x + 4$?
Lesson 8.4: Special Products	
CCSS A.APR.1 Understand that polynomials form a system analogous to the integers, namely, they are closed under the operations of addition, subtraction, and multiplication; add, subtract, and multiply polynomials. CCSS A.SSE.2 Use the structure of an expression to identify ways to rewrite it.	CCSS SMP 3 Carlos says that when he multiplied two binomials, the product was also a binomial. How is this possible? In your answer give an example. CCSS SMP 7 Maida was asked to find the product of $(x + 3)(x - 2)(x - 3)(x + 2)$. Her work is shown. $(x + 3)(x - 2)(x - 3)(x + 2)$ $= (x^2 + x - 6)(x^2 - x - 6)$ $= x^2(x^2 - x - 6) + x(x^2 - x - 6) - 6(x^2 - x - 6)$ $= (x^4 - x^3 - 6x^2) + (x^3 - x^2 - 6x)$ $+ (-6x^2 + 6x + 36)$ $= x^4 - 13x^2 + 36$ Jorge says that he can find this product without using paper and pencil. Show how he might do this.

8.1 Adding and Subtracting Polynomials

Objectives

- Add and subtract polynomials.
- Determine if polynomials are closed under addition and subtraction.

Content: A.APR.1
Practices: 1, 2, 3, 6, 7
Use with Lesson 8-1

Polynomials consist of one or more monomials that are combined by addition and subtraction. Each monomial is a number, a variable, or a product of a number and one or more variables. Sometimes the monomials in a polynomial are referred to as terms, and the order in which these terms appear in the polynomial does not matter. To add or subtract polynomials, rearrange the terms to match up monomials with the same exponent. These are called like terms. Then add or subtract the coefficients of each like term.

EXAMPLE 1 Adding Polynomials

EXPLORE Find the sum of the polynomials.

a. USE STRUCTURE Compute the following polynomial sums. CCSS SMP 7

$(x^3 - 2x^2 + x + 4) + (2x^3 + 2x^2 - 3x + 1)$

$(-3x^4 + 5 + 2x - x^2) + (x - 3x^3 + x^4 - 3)$

b. COMMUNICATE PRECISELY Using the polynomials $x^2 - 4x + 3$ and $-6x^2 + x - 1$, determine if the Commutative Property applies to addition of polynomials. CCSS SMP 6

c. COMMUNICATE PRECISELY Using the polynomials $3x^2 - 2$, $4x^2 + 2x$, and $5 - 7x$, determine if the Associative Property applies to addition of polynomials. CCSS SMP 6

d. REASON ABSTRACTLY How does the behavior of polynomials compare to the behavior of integers under addition? CCSS SMP 2

The process of subtracting polynomials is similar to adding polynomials. Recall that subtracting an integer is equivalent to adding its additive inverse. The same is true when subtracting polynomials. To find the additive inverse of a polynomial, change the sign of each term.

EXAMPLE 2 Subtracting Polynomials

a. USE STRUCTURE Compute the following polynomial differences. CCSS SMP 7

$(4x^3 - 2x^2 + 5x - 6) - (2x^3 + x^2 - 7x + 2)$

$(8x - 3x^2 + x^4) - (-2x^4 + x - 5x^3 - 3x^2)$

b. COMMUNICATE PRECISELY Using the polynomials $12x^2 - 3x + 5$ and $x^3 - 2x^2 + 7$, determine if the Commutative Property applies to subtraction of polynomials. CCSS SMP 6

c. COMMUNICATE PRECISELY Using the polynomials $3x^2 - 2$, $4x^2 + 2x$, and $5 - 7x$, determine if the Associative Property applies to subtraction of polynomials. CCSS SMP 6

d. REASON ABSTRACTLY How does the behavior of polynomials compare to the behavior of integers under subtraction? CCSS SMP 2

Notice from the first two examples that addition and subtraction of polynomials is similar to addition and subtraction of integers. Considered next is if closure on the set of polynomials under addition and subtraction is similar to that of the integers.

A set of numbers is considered closed under an operation if when the operation is performed on any numbers in the set, the result belongs in the same set. The set of integers is closed under addition and subtraction because the result of adding or subtracting any integers is also an integer. The set of real numbers is also closed under addition and subtraction: the result of adding or subtracting real numbers is a real number.

EXAMPLE 3 Closure

a. **COMMUNICATE PRECISELY** Make conjectures regarding the closure of the set of polynomials under addition and under subtraction. CCSS SMP 6

b. **CONSTRUCT ARGUMENTS** Give reasoning for the answers to **part a**. CCSS SMP 3

PRACTICE

1. **USE STRUCTURE** Compute the following sums or differences. Show your work. CCSS SMP 7

   a. $(-6x^3 - x^2 + 4 - 3x) + (2 - 5x^3 + 8x)$

   b. $(7x^5 - 3x^2 + 2x^3 - 6x + 4) - (-x^2 - 9x^3 + 2x^4 + 4x - 2)$

   c. $(4.2x^2 - 3.7x + 15.32) + (5.1x - 1.65x^2 - 1.42)$

   d. $(17.63x^3 - 24.19x + 3.28x^2 + 9.21) - (-4.73x^2 + 3.12x^3 + 19.23x - 21.62)$

   e. $(4x^3 + 2x^2 - x + 1) - (4x^3 - 2x^2 + x + 1)$

2. **USE STRUCTURE** A company delivers their product in cubic boxes that have volume x^3. When the company begins to manufacture a second product, manufacturing designs a new shipping box that is 3 inches longer in one dimension and 1 inch shorter in another dimension. The volume of the new box is $x^3 + 2x^2 - 3x$. CCSS SMP 7

   a. What is the total volume of 4 of each kind of box?

   b. Find an expression that shows the difference in volume between the two boxes.

3. **CRITIQUE REASONING** Jonathan says that polynomials are not closed under addition and gives this counterexample, which he says is not a polynomial: $(x^2 - 2x) + (-x^2 + 2x) = 0$. Explain the error in Jonathan's reasoning. CCSS SMP 3

4. **INTERPRET PROBLEMS** The volume of a sphere with radius x is $\frac{4}{3}\pi x^3$ units³ and the volume of a cube with side length x is x^3 units³. CCSS SMP 1

   a. Find the total combined volume of the sphere and cube.

   b. How much more volume does the sphere contain?

5. **REASON ABSTRACTLY** Compute the following differences. CCSS SMP 2

   a. $(5x^2 - 3x + 7) - (18x^2 - 2x - 3)$

   b. $(18x^2 - 2x - 3) - (5x^2 - 3x + 7)$

   c. What do you notice about the second difference compared to the first? How does this relate to the structure of the integers?

6. **REASON ABSTRACTLY** In the set of integers, every integer has an additive inverse. In other words, for every integer n, there is another integer $-n$ such that $n + (-n) = 0$. Is this true in the set of polynomials? Does every polynomial have an additive inverse? Demonstrate with an example. CCSS SMP 2

8.2 Multiplying a Polynomial by a Monomial

Objectives

- Multiply a polynomial by a monomial to form another polynomial.
- Solve problems involving polynomial-monomial products.

Content: A.APR.1
Practices: 1, 3, 5, 6, 7, 8
Use with Lesson 8–2

You can use the Distributive Property to multiply a polynomial expression by a monomial expression, forming another polynomial expression. Often it is useful to find the maximum or minimum of an expression. In the context of this lesson, these values occur at the vertex of the graph.

EXAMPLE 1 Model Revenue Using a Polynomial

EXPLORE A company developed a GPS navigation app for smartphones and they are trying to decide the cost. A sales model suggests that if the app costs $0.5m$ dollars, they will sell $25{,}000 - 1000m$ units. How much revenue can they expect?

a. INTERPRET PROBLEMS Revenue is the product of the price and the number of sales. Develop an expression for the revenue, based on their model. CCSS SMP 1

b. USE TOOLS The graph shows revenue as a function of m. Using technology, determine the sale price to maximize revenue. Explain your method. CCSS SMP 5

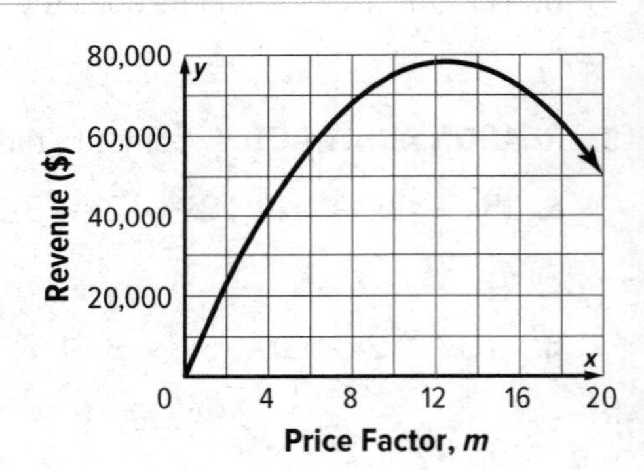

c. CRITIQUE REASONING A sales manager thinks that the price found in **part b** is too low. She argues that the more money they charge, the more revenue they will make. Do you agree or disagree? Explain. Assume that the model is reasonable. CCSS SMP 3

d. USE TOOLS As the price continues to climb, the graph seems to indicate that revenue will eventually become $0.00. When does this occur, and why would this happen? CCSS SMP 5

EXAMPLE 2 Expand Expressions Involving Products of Monomials and Polynomials

USE STRUCTURE Expand each expression. CCSS SMP 7

a. $3x(4x^3 - 2x + 1)$

b. $7p^2(2p + 5) + 5(3p^2 - 3)$

c. $3a^7(2 - 5a^2) - a^4(5a^5 + 3a^3)$

d. $2k(3 - k^2 + 5k) + 7k(3k^2) - 5k^2(1 - 3k)$

e. DESCRIBE A METHOD Describe the steps you used in simplifying **parts a–d**. Justify each step with a property. CCSS SMP 8

EXAMPLE 3 Areas and Volumes Involving Products of Monomials and Polynomials

USE STRUCTURE Solve each equation. CCSS SMP 7

a. An area is enclosed by the fence shown at right, represented by the solid line segments. Write and simplify an expression for the area of the enclosure.

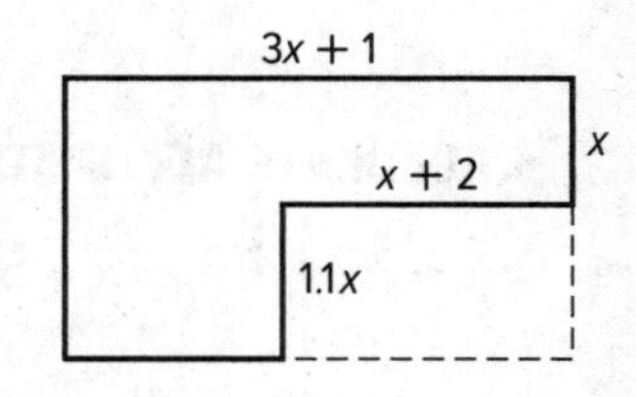

b. The volume of a cylinder with radius r and height h is $V = \pi r^2 h$. A cylinder has radius r and height $5 - 2r^2$. Write and simplify an expression for the volume of the cylinder, then use appropriate technology to find the radius that maximizes the volume.

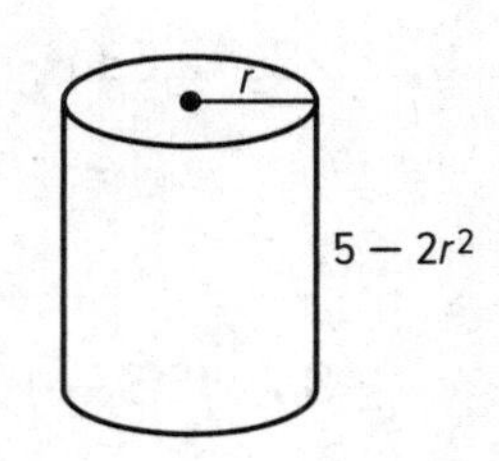

c. The surface area of a cylinder with radius r and height h is $2\pi rh + 2\pi r^2$. Find the expression for the surface area of the cylinder in **part b**, and find the radius that maximizes the surface area.

EXAMPLE 4 Prove Product Properties of Monomials and Polynomials

Use what you know about the products of monomials to prove product properties of monomials and polynomials.

a. CONSTRUCT ARGUMENTS Prove that the product of a monomial and a polynomial is a polynomial. CCSS SMP 3

b. MAKE A CONJECTURE Do you think the product of two polynomials will be a polynomial as well? Write your answer as a statement about closure. CCSS SMP 3

PRACTICE

1. USE STRUCTURE Expand each product. Show your work. CCSS A.APR.1, SMP 7

a. $5x^2(1 - x + 3x^2)$

b. $-7z^5(z - 5) + 2z(6z^2 - z)$

c. $-3b^4(-b^2 + 2b) - 12b^2(2b^3 - 3b^4)$

d. $2h(3 - h^3 + 7h) - 2h^2(4 + h)$

2. CALCULATE ACCURATELY Expand each product. CCSS A.APR.1, SMP 6

a. $-2.74x^2(3.21x - 5.76) + 8.27x(1.03x^3 + 9.82)$

b. $5.28y^3(2.63y^2 - 4.72y + 7.60)$

c. $-9.7x^4(5.3 - 4.7x^3) - 8.9x^2(-4.1 - 6.2x^5)$

d. $-x^3(8 - 4x^3) - 2x^2(2x^4 - 4x)$

3. The base lengths of this trapezoid are given by polynomial expressions in terms of h, the trapezoid's height. CCSS A.APR.1

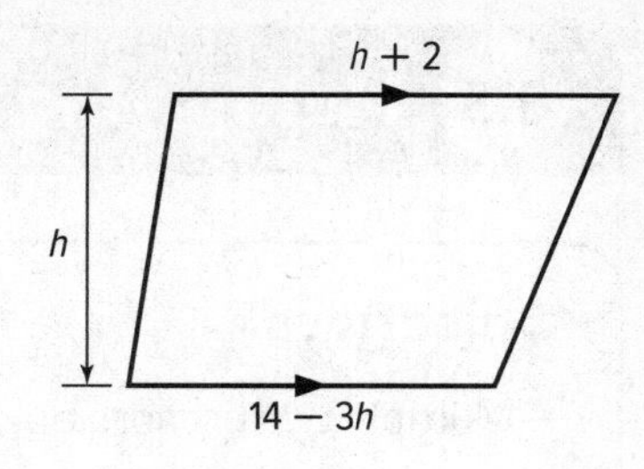

a. USE STRUCTURE Write and simplify an expression for the area of the trapezoid. CCSS SMP 7

b. USE TOOLS Use appropriate technology to determine the maximum area of the trapezoid and the height that gives this area. CCSS SMP 5

4. CRITIQUE REASONING Jodie was simplifying the expression $12y^2(3y - 2y^2) - 3y^3(4 - 2y)$. Identify and correct any errors she made. CCSS A.APR.1, SMP 3

$$12y^2(3y - 2y^2) - 3y^3(4 - 2y)$$
$$36y^3 - 24y^2 - 12y^3 - 6y^4$$
$$24y^3 - 24y^2 - 6y^4$$

5. INTERPRET PROBLEMS The diagram shows the dimensions of a right rectangular prism. Write and simplify an expression for the volume of the prism. Use technology to find the value of h that gives the maximum volume. What is the maximum volume? CCSS A.APR.1, SMP 1

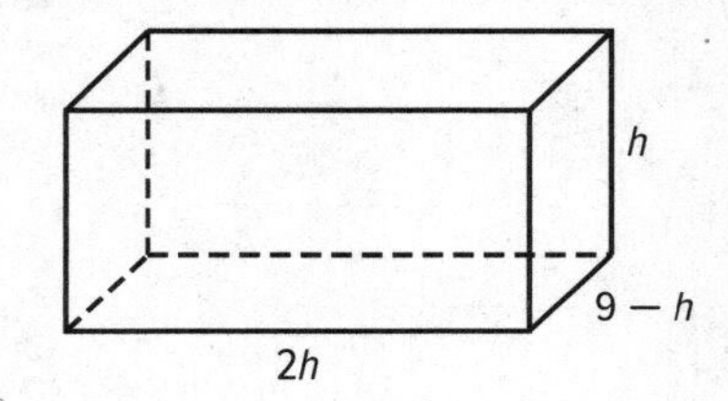

6. Through market research, a company finds that it can expect to sell $45 - 5x$ products if each is priced at $1.25x$ dollars. CCSS A.APR.1

a. INTERPRET PROBLEMS Develop an expression for the expected revenue. CCSS SMP 1

b. USE TOOLS Use technology to determine the price, to the nearest cent, at which the company should they sell its product to maximize revenue.  CCSS SMP 5

7. USE STRUCTURE An area is enclosed by the fence shown at the right, represented by the solid line segments. Write and simplify an expression for the area of the enclosure. CCSS A.APR.1, SMP 7

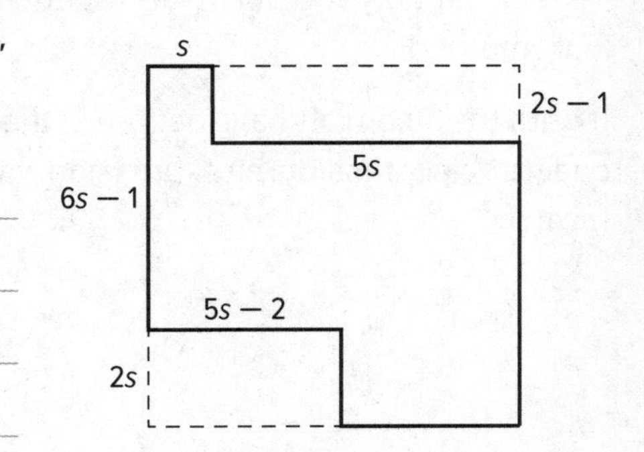

8.3 Multiplying Polynomials

Objectives

- Multiply two or more polynomials to form another polynomial.
- Solve problems involving products of polynomials.

CCSS STANDARDS

Content: A.APR.1
Practices: 1, 2, 3, 4, 5, 7, 8
Use with Lesson 8-3

You can use the Distributive Property of Equality to multiply two or more polynomial expressions and form another polynomial expression.

EXAMPLE 1 Find the Product of Two Binomials CCSS A.APR.1

EXPLORE Use algebra tiles to form the product of two binomial expressions.

	x	x	x	-1	-1	-1	-1
x	x^2	x^2	x^2	$-x$	$-x$	$-x$	$-x$
x	x^2	x^2	x^2	$-x$	$-x$	$-x$	$-x$
1	x	x	x	-1	-1	-1	-1
1	x	x	x	-1	-1	-1	-1
1	x	x	x	-1	-1	-1	-1

a. USE TOOLS Interpret this arrangement of algebra tiles as a product of binomials, and explain how it works in terms of the Distributive Property. CCSS SMP 5

b. INTERPRET PROBLEMS Write and simplify expressions for each quadrant of the product diagram, beginning with $(3x)(2x)$. Then, write the complete binomial product that the diagram represents. CCSS SMP 1

c. USE STRUCTURE Express the product $(ax + b)(cx + d)$ as a polynomial with three terms. CCSS SMP 7

You can multiply two binomials together using the FOIL method:

If both the binomials have degree 1, the result is a **quadratic expression:** an expression with the general form $ax^2 + bx + c$.

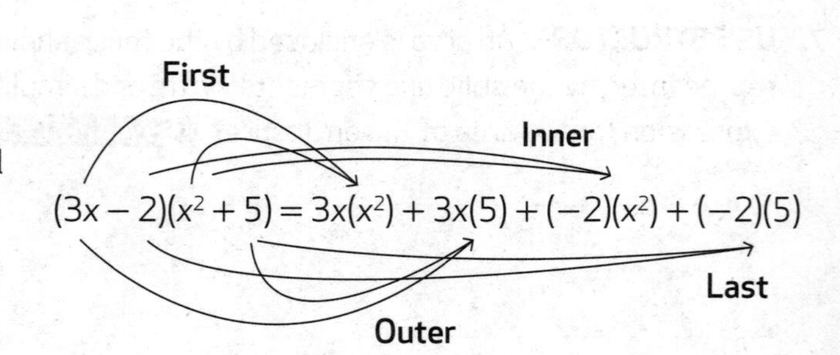

EXAMPLE 2 Solve a Problem Involving the Product of Binomials CCSS A.APR.1

USE A MODEL Miguel is planning to frame one of his paintings. He is going to place a white matting over top of his painting and then place the picture and the matting inside of a frame. The matting will be twice as wide as the frame. CCSS SMP 4

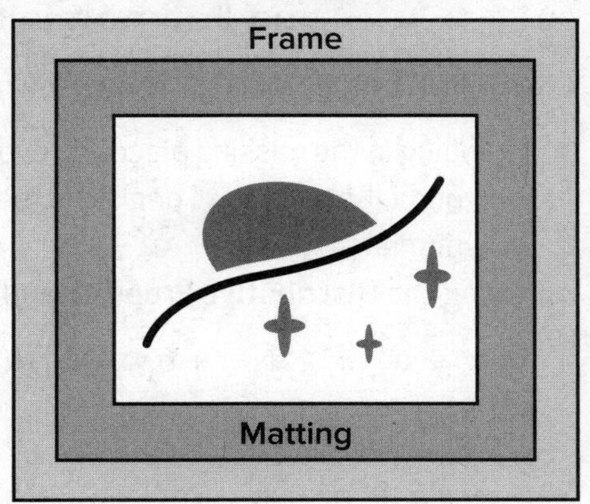

a. REASON QUANTITATIVELY Miguel's painting has dimensions 55 centimeters by 35 centimeters. Given that the width of the frame is x centimeters, write simplified expressions for the overall dimensions of the framed painting and for the dimensions of the painting plus the matting. CCSS SMP 2

b. INTERPRET PROBLEMS If the frame is 4 centimeters, find the area of matting by developing a quadratic expression. CCSS SMP 1

Two polynomials can be multiplied together using either the vertical or the horizontal method.

EXAMPLE 3 Find Products of Polynomials CCSS A.APR.1

USE STRUCTURE Find each product using the method given. CCSS SMP 7

a. $(3x^2 - 2x - 5)(4x - 3)$, using the vertical method

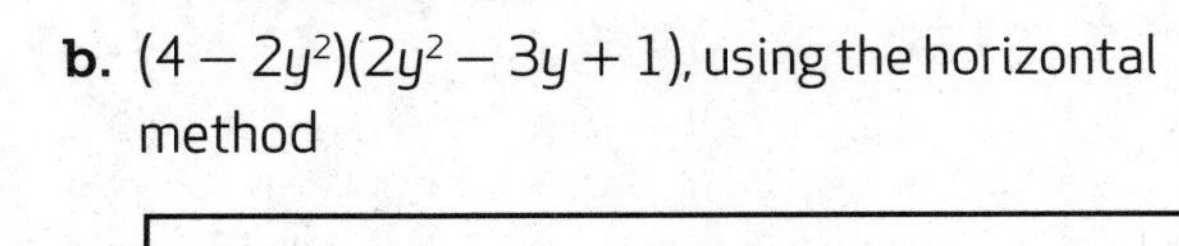

b. $(4 - 2y^2)(2y^2 - 3y + 1)$, using the horizontal method

c. DESCRIBE A METHOD What are some of the advantages and disadvantages of each method? Which method do you prefer? Justify your answer. CCSS SMP 8

EXAMPLE 4 Prove Properties of the Product of Monomials and Polynomials CCSS A.APR.1

CONSTRUCT ARGUMENTS Prove key facts about the product of two polynomials. CCSS SMP 3

By filling in the missing pieces, prove that the product of two polynomials is a polynomial, and that its degree is the sum of the degrees of the two original polynomials.

Using the Distributive Property and simplifying, the product is

$(a_nx^n + a_{n-1}x^{n-1} + \ldots + a_1x + a_0)\,(b_mx^m + b_{m-1}x^{m-1} + \ldots + b_1x + b_0)$

$= a_nx^n(\qquad\qquad\qquad) + a_{n-1}x^{n-1}(\qquad\qquad\qquad) + \ldots$

$+ a_1x\,(b_mx^m + b_{m-1}x^{m-1} + \ldots + b_1x + b_0) + \quad(\qquad\qquad\qquad)$

$= a_nx^n(b_mx^m) + a_nx^n(\qquad) + \ldots + \quad(\quad) + \quad(\quad) + \ldots$

$+ a_{n-1}x^{n-1}(\qquad) + \quad(b_{m-1}x^{m-1}) + \ldots + \quad(\quad) + \quad(b_0) + \ldots$

$+ a_1x(\qquad) + a_1x(\qquad) + \ldots + \quad(\quad) + \quad(\quad) + \ldots$

$+ \quad(\qquad) + \quad(\qquad) + \ldots + \quad(\quad) + \quad(\quad)$

$= a_nb_mx^{n+m} + (a_nb_{m-1} + a_{n-1}\,b_m)\,x^{n+m-1} + \ldots + (a_1b_0 + a_0b_1)\,x + a_0b_0$

This is a polynomial, with degree

PRACTICE

1. **CALCULATE ACCURATELY** Write each expression as a simplified polynomial. CCSS A.APR.1, SMP 6, SMP 7

a. $(3c - 2)(4c^2 - c^3 + 3)$

b. $(5x - y)(3x^2 - 2xy) + (2x + y)(y^2 - 4x^2)$

c. $-2x(3 - x^2)(2x + 4)$

d. $(z - 1)(2 - z)(z + 1)$

2. The dimensions of the composite figure are given in terms of h, the triangle's height, in the diagram. CCSS A.APR.1

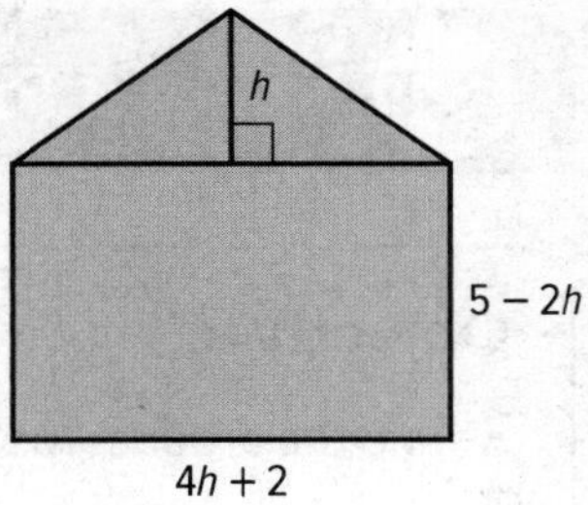

h

$5 - 2h$

$4h + 2$

a. USE STRUCTURE Write and simplify a quadratic expression for the area of the figure. CCSS SMP 7

b. USE TOOLS Use appropriate technology to determine the maximum area of the figure, and the triangle height that gives this area. CCSS SMP 5

3. The relationship between monthly profit P, monthly sales n, and unit price p is $P = n(p - U) - F$, where U is the unit cost per sale and F is a fixed cost that does not depend on the number of sales. For an online business advice service, the unit cost is \$30 per hour-long session and the monthly fixed cost is \$3000. CCSS A.APR.1

a. USE A MODEL Given a model for monthly sales of $n = 5000 - 40p$ for a given price p per session, write and simplify a quadratic expression for P in terms of p. CCSS SMP 1, SMP 4

b. USE TOOLS What is the best price to charge for the service, and what is the monthly profit at this price? CCSS SMP 5

4. a. USE STRUCTURE Use the law of exponents to simplify $x^{4p+1}(x^{1-2p})^{2p+3}$. CCSS A.APR.1, SMP 7

b. REASON QUANTITATIVELY Find any integer values of p that make this expression equal to 1 for all values of x. CCSS A.APR.1, SMP 2

5. USE STRUCTURE Find and simplify an expression for the volume of the rectangular prism pictured. CCSS A.APR.1, SMP 7

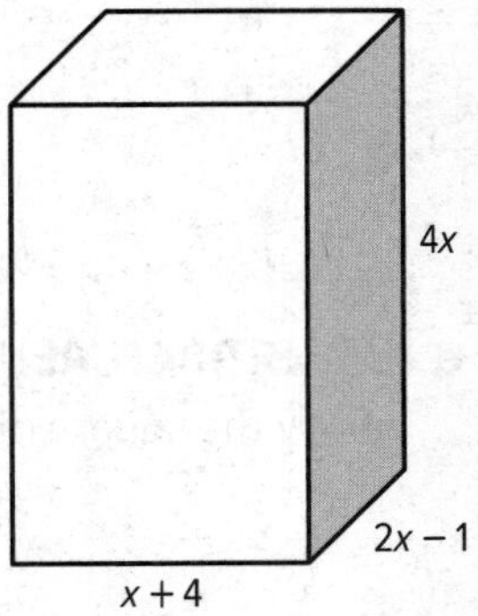

$4x$

$2x - 1$

$x + 4$

8.4 Special Products

Objectives

- Find the product for the square of a binomial.
- Find the product of the sum and difference of two terms.

Content: A.APR.1, A.SSE.2
Practices: 1, 2, 4, 7, 8
Use with Lesson 8–4

By polynomial multiplication the patterns for the square of a difference and the square of a sum can be found. The square of the sum $a + b$ is $(a + b)^2 = a^2 + 2ab + b^2$ and the square of the difference $a - b$ is $(a - b)^2 = a^2 - 2ab + b^2$.

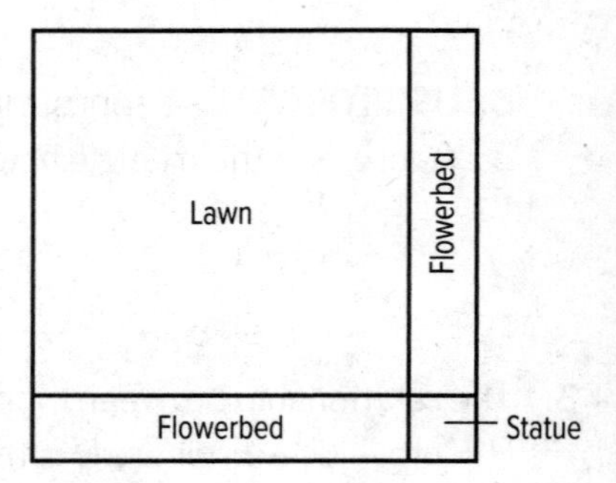

EXAMPLE 1 Squares of Sums and Differences

CCSS A.APR.1, A.SSE.2

EXPLORE Grace is planning a garden that will have a square grass lawn, two rectangular flowerbeds of equal width, and a square concrete base for a statue.

a. USE A MODEL If the length of the entire layout is a units on each side and the width of each flowerbed is b units, find the dimensions of the lawn, flowerbeds, and square concrete base. Label the lengths on a separate diagram. CCSS SMP 4

b. CHECK REASONABLENESS Use the fact that the area of the lawn is the area of entire layout minus the area of the two flowerbeds and the square concrete base to verify the square of the difference identity $(a - b)^2 = a^2 - 2ab + b^2$. CCSS SMP 8

c. USE A MODEL What area does the quantity ab represent? Use this to explain in terms of areas why the area of the lawn is $a^2 - 2ab + b^2$. CCSS SMP 4

d. USE STRUCTURE Use the square of a difference pattern $(a - b)^2 = a^2 - 2ab + b^2$ to verify the square of a sum pattern by replacing b with $-b$. CCSS SMP 7

EXAMPLE 2 Use the Squares of Sums and Differences to Find Areas CCSS A.APR.1, A.SSE.2

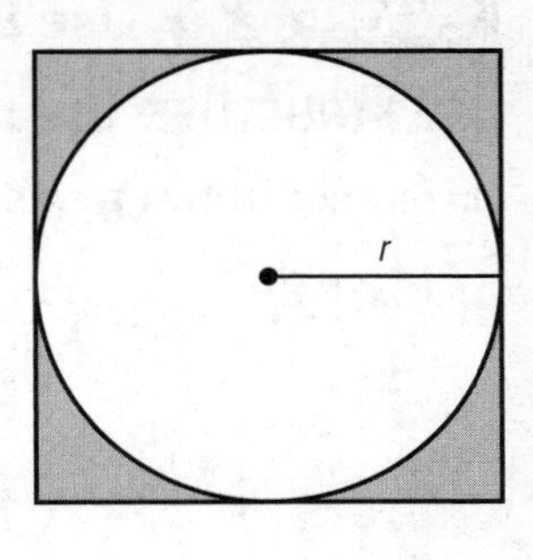

A circle is inscribed in a square as pictured. The radius of the circle is r.

a. USE STRUCTURE Write an expression for the area inside the square not inside the circle (the shaded region) in terms of the radius r. CCSS SMP 7

b. INTERPRET PROBLEMS Suppose the radius is increased by 3 units. Use the square of a sum pattern to help find and expand an expression for the new shaded area. CCSS SMP 1

c. INTERPRET PROBLEMS Suppose the radius is decreased by 2 units. Use the square-of-a-difference pattern to help find and expand an expression for the new shaded area. 

d. DESCRIBE A METHOD Using the same process, generalize your result from **part b** by finding and expanding an expression for the new shaded area if the radius is increased by a units. CCSS SMP 8

The pattern for the product of a sum and a difference of two terms is $(a + b)(a - b) = a^2 - b^2$.

EXAMPLE 3 Area of a Transformed Square CCSS A.APR.1, A.SSE.2

A square has side length a units.

a. INTERPRET PROBLEMS If the length of the square is decreased by b units and the width is increased by b units, find the dimensions and area of the rectangle created by these changes. Use the pattern for a product of a sum and difference of two terms to simplify your answer. Draw and label a diagram to illustrate the changes. CCSS SMP 1

b. USE A MODEL Can you determine from your diagram if the area of the rectangle is greater than or less than the area of the square? Explain algebraically why the area of a rectangle with dimensions $(a + b)$ units by $(a - b)$ units must have less area than the square with side length a units. CCSS SMP 4

EXAMPLE 4 Use Special Products to Rewrite Expressions CCSS A.APR.1, A.SSE.2

USE STRUCTURE **Use squares of sums and differences and the product of the sum and difference of two terms to expand the following expressions.** CCSS SMP 7

a. $(3x + 2)^2$

b. $(2x^3 - 5x)^2$

c. $(x - 4)(x + 4)$

d. $(2x + 3b)(2x - 3b)$

PRACTICE

1. **USE STRUCTURE** For each expression state which pattern applies (square of a difference, square of a sum, product of a sum and difference of two terms) then use the pattern to expand the expression. CCSS A.APR.1, A.SSE.2, SMP 7

   a. $(x - 5)^2$

   b. $(y + 2)^2$

   c. $(2z - 3)^2$

   d. $(-a - 2)^2$

   e. $(2b - 3)(2b + 3)$

   f. $(4n - 2m)(4n + 2m)$

2. Consider the product $(a + b)(a - b)(a + b)(a - b)$.

   a. **REASON QUANTITATIVELY** Show the steps required to determine the product. CCSS A.SSE.2, SMP 2

   b. Evaluate the original expression for $a = 5$ and $b = 2$.

   c. Evaluate the simplified expression you wrote in **part a** for $a = 5$ and $b = 2$. What does the result indicate about the expression you wrote in **part a**?

3. **USE STRUCTURE** Tanisha is investigating growth patterns for the area of a square. She begins with a square of side length s and looks at the effects of enlarging the side length by one unit at a time. CCSS A.APR.1, A.SSE.2, SMP 7

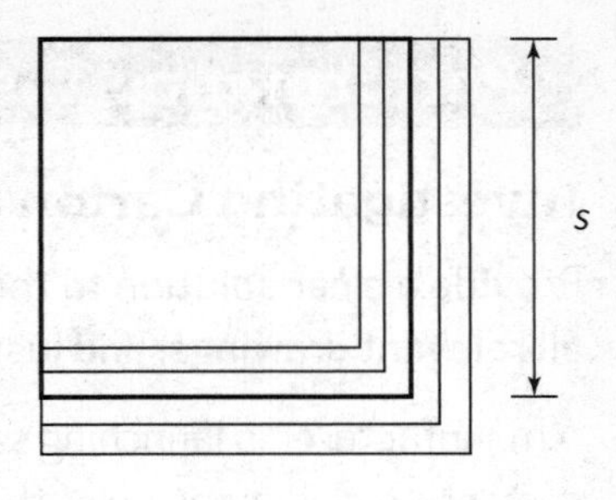

a. How much more area does the square with side length $s + 1$ have compared to the square with side length s?

b. How much more area does the square with side length $s + 2$ have compared to the square with side length $s + 1$?

c. How much more area does the square with side length $s + 3$ have compared to the square with side length $s + 2$?

d. Based on **parts a–c**, how much more area will a square with side length $a + 1$ have compared to the square with side length a? Test your answer by finding the difference in areas algebraically.

4. **USE STRUCTURE** Use the patterns for the squares of sums and differences to find the following products. CCSS A.APR.1, A.SSE.2, SMP 7

a. $(x + 3)^3$

b. $(x - 3)^3$

c. $(x - 2)^3$

Performance Task

Investigating Carton Sizes

Provide a clear solution to the problem. Be sure to show all of your work, include all relevant drawings, and justify your answers.

A manufacturer is launching sales of a new product for which they need shipping cartons. Each dimension of the chosen carton must be a whole number of inches. No dimension is to be less than 2 inches. The dimensions of the product are shown.

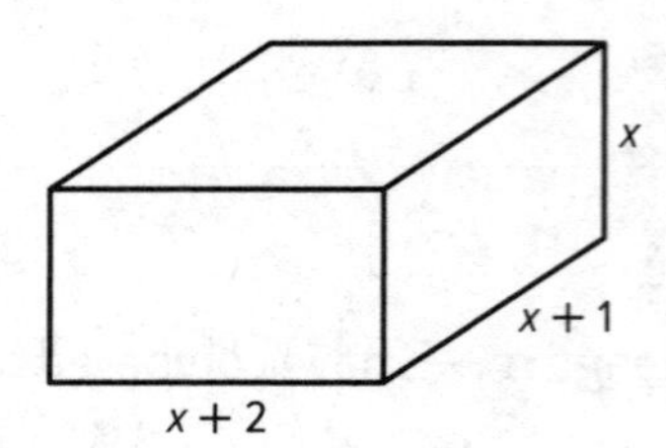

The shipping department is considering the three cartons shown.

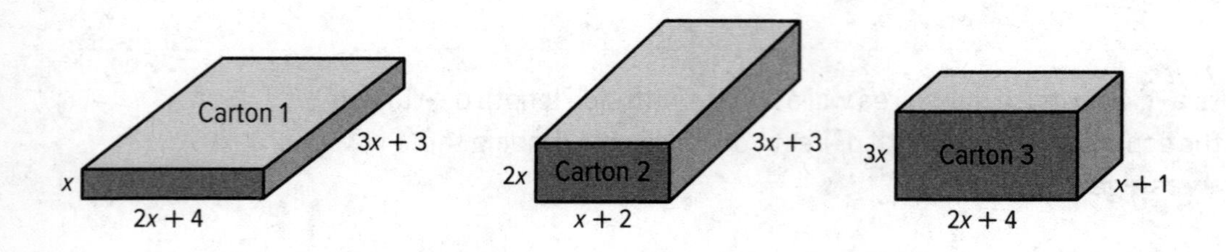

Part A

Confirm with drawings and calculations that 6 boxes of the new product will fit in each of the shipping cartons.

Part B

- Write polynomials to represent the surface areas of the three shipping cartons. If $x \geq 2$, which carton has the least surface area?
- If the flaps for each carton are equal in area to the top and bottom of each carton, which carton will require the least amount of cardboard? Is it the same as the carton with the least surface area? Explain.

Part C

The manufacturer wants to put advertising on the top and sides of the shipping carton, but not on the flaps or on the bottom of the carton.

- Printing costs \$0.50 per square foot.
- The cost per carton for printing must not exceed \$10.
- They decide to use the carton with the least surface area.

Determine the greatest possible dimensions for the carton.

Performance Task

Planning a Garden Walkway

Provide a clear solution to the problem. Be sure to show all of your work, include all relevant drawings, and justify your answers.

The Jameson family is planning to put a new garden in their backyard.

- The four shaded rectangular regions will be congruent flower and herb gardens.
- The white space inside the large rectangle will be walkways covered with small stones.
- They have not decided on the width x of the walkway, but they know that it will be 3 inches deep.

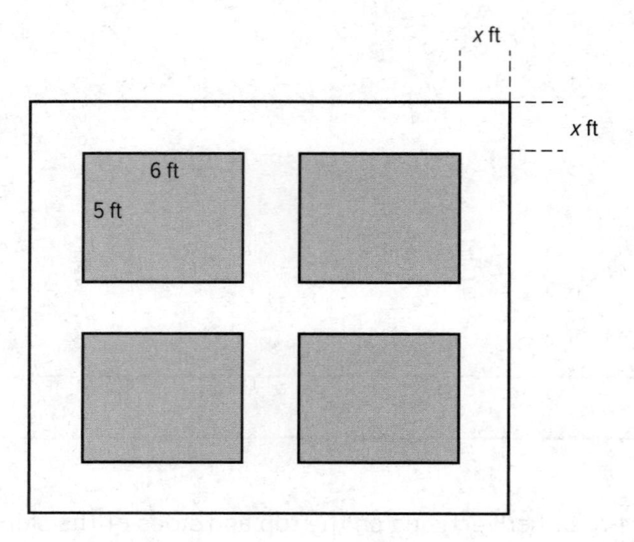

Part A

Write a polynomial in x to represent the total area of the walkway in square feet.
Write a polynomial in x to represent the volume of the stones in cubic feet.

Part B

Suppose that the Jamesons consider making the walkway 3 feet wide. This table shows quantities and costs for different ways to buy stone.

Supplier 1	Supplier 2
$75 per cubic yard	$3.77 per 0.5 cubic foot

How much will each supplier charge for the total amount of stone?

Part C

The Jamesons have decided to use the cheaper buying plan based on the results of **Part B**. They want to pay $265.00 or less for the stone. Determine the width of the walkway, to the nearest quarter of a foot, that will get them closest to their estimated cost.

Standardized Test Practice

1. The perimeter of the rectangle shown below is $6ab^2 + 4a - 4$. CCSS A.APR.1

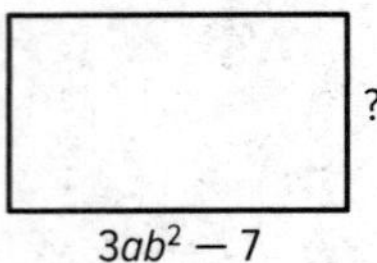

Which is an expression for its width? Circle the correct answer.

$2a - 9$	$4a + 10$
$2a + 5$	$4a - 18$

2. If $(x - y)^2 = 49$ and $xy = 18$, what is $x^2 + y^2$? CCSS A.SSE.2

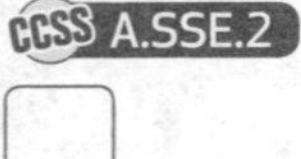

3. The surface area of a sphere is given by the formula $A = 4\pi r^2$. What is the surface area of the sphere shown below? Give your answer in terms of π. CCSS A.APR.1

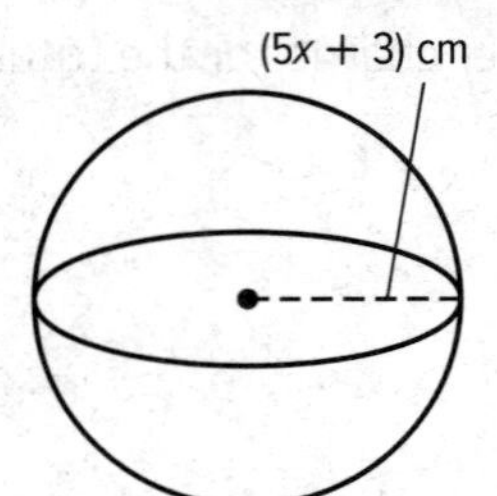

4. Simplify the following expressions. CCSS A.APR.1

$(6g^3 - 4g + g^2) + (9g^2 + g) =$

$(5x^2y + 3xy^2) - (x^2y^2 - 7xy^2) =$

$(-2x^2 + 8y^2 - 4x) + 5x(3y + 4x + 1) =$

$7a(4a + 5) + 3(5a^2 - 2a) =$

5. Multiply the following expressions. CCSS A.APR.1, A.SSE.2

$-7yz^5(3y^2 - 2yz) =$

$(5t + s)^2 =$

$(2a + b)(3a - 4b + 7) =$

$(8m + 3n)(8m - 3n) =$

6. Show your work to multiply $(x^4 + y^4)(x^2 + y^2)(x + y)(x - y)$. CCSS A.SSE.2

7. Two plastic containers in the shape of rectangular prisms have the dimensions shown in the diagram below. CCSS A.APR.1

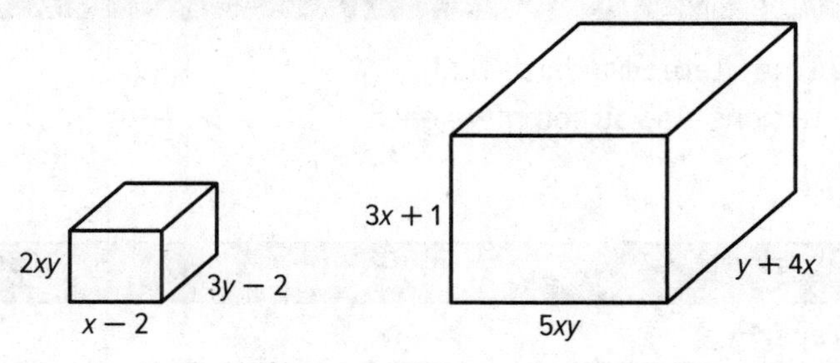

a. What is the volume of the small container? Show your work.

b. What is the volume of the large container? Show your work.

c. The small container is filled with water and emptied into the large container. How much water must be added to fill the large container? Show your work.

8. Write a polynomial that represents the area of the figure below. Show your work. CCSS A.APR.1

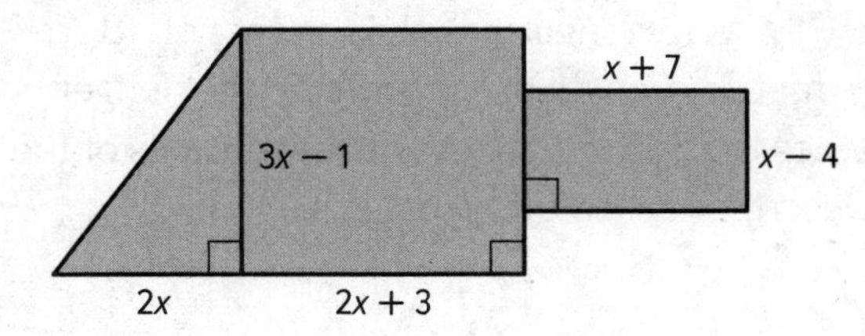

9. A square has side length x. A rectangle has sides of length $x + 1$ and $x - 1$. Which figure has the greater area? How much greater is the area if $x = 4$? How much greater is the area if $x = 7$? CCSS A.APR.1

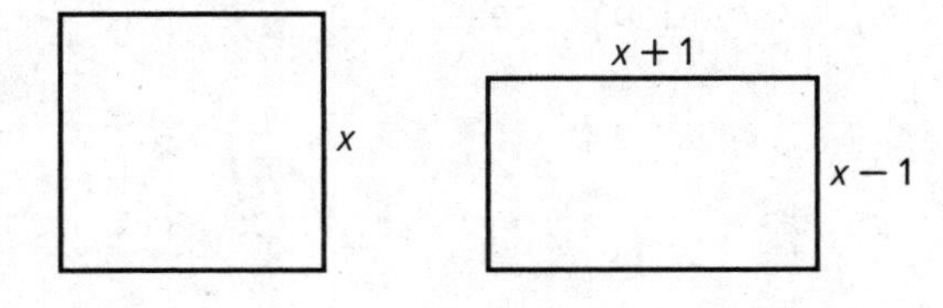

9 Quadratic Equations

CHAPTER FOCUS Learn about some of the Common Core State Standards that you will explore in this chapter. Answer the preview questions. As you complete each lesson, return to these pages to check your work.

What You Will Learn	Preview Question
Lesson 9.1: Solving by Factoring: The Distributive Property	
CCSS A.SSE.3a Factor a quadratic expression to reveal the zeros of the function it defines. **Also addresses:** A.REI.4b	CCSS SMP 7 Solve the equation $3x^2 + 6x - x - 2 = 0$ by factoring by grouping. Describe your solution process.
Lesson 9.2: Solving by Factoring: $x^2 + bx + c = 0$	
CCSS A.REI.4b Solve quadratic equations by inspection (e.g., for $x^2 = 49$), taking square roots, completing the square, the quadratic formula and factoring, as appropriate to the initial form of the equation. Recognize when the quadratic formula gives complex solutions and write them as $a \pm bi$ for real numbers a and b. **Also addresses:** A.SSE.3a	CCSS SMP 4 A rectangular piece of poster board is 10 inches longer than its width. If the area of the poster board is 144 in^2, find its length. Describe your solution process.
Lesson 9.3: Solving by Factoring: $ax^2 + bx + c = 0$	
CCSS A.REI.4b Solve quadratic equations by inspection (e.g., for $x^2 = 49$), taking square roots, completing the square, the quadratic formula and factoring, as appropriate to the initial form of the equation. Recognize when the quadratic formula gives complex solutions and write them as $a \pm bi$ for real numbers a and b. **Also addresses:** A.SSE.3a, A.CED.1, A.CED.2	CCSS SMP 4 A rectangular rug has an area of 48 square feet. The length of the rug is 10 feet longer than twice the width. What are the dimensions of the rug? Describe your solution process.
Lesson 9.4: Solving by Factoring: Differences of Squares	
CCSS A.REI.4b Solve quadratic equations by inspection (e.g., for $x^2 = 49$), taking square roots, completing the square, the quadratic formula and factoring, as appropriate to the initial form of the equation. Recognize when the quadratic formula gives complex solutions and write them as $a \pm bi$ for real numbers a and b. **Also addresses:** A.SSE.3a	CCSS SMP 7 Solve the equation $4x^2 - 2.25 = 0$. Describe your solution process.

What You Will Learn	Preview Question
Lesson 9.5: Solving by Factoring: Perfect Squares	
CCSS A.REI.4b Solve quadratic equations by inspection (e.g., for $x^2 = 49$), taking square roots, completing the square, the quadratic formula and factoring, as appropriate to the initial form of the equation. Recognize when the quadratic formula gives complex solutions and write them as $a \pm bi$ for real numbers a and b. **Also addresses:** A.SSE.3a	CCSS SMP 7 Solve the equation $9x^2 + 30x + 25 = 0$ by factoring. Describe your solution process.
Lesson 9.6: Solving by Graphing	
CCSS A.REI.4b Solve quadratic equations by inspection (e.g., for $x^2 = 49$), taking square roots, completing the square, the quadratic formula and factoring, as appropriate to the initial form of the equation. Recognize when the quadratic formula gives complex solutions and write them as $a \pm bi$ for real numbers a and b. CCSS F.IF.7a Graph linear and quadratic functions and show intercepts, maxima, and minima. **Also addresses:** F.IF.8a	CCSS SMP 1 Use the graph of the function $y = 2x^2 + 2x - 12$, shown below, to solve the related equation $2x^2 + 2x - 12 = 0$. Justify your solution. 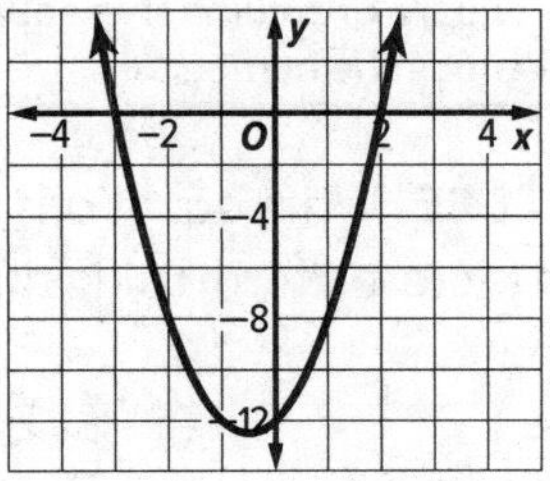
Lesson 9.7: Solving by Completing the Square	
CCSS A.REI.4a Solve quadratic equations in one variable. Use the method of completing the square to transform any quadratic equation in x into an equation of the form $(x - p)^2 = q$ that has the same solutions. Derive the quadratic formula from this form. **Also addresses:** F.IF.8a, A.SSE.3b	CCSS SMP 7 Solve the equation by completing the square $x^2 - 3x + \frac{5}{4} = 0$.
Lesson 9.8: Solving by Using the Quadratic Formula	
CCSS A.REI.4b Solve quadratic equations by inspection (e.g., for $x^2 = 49$), taking square roots, completing the square, the quadratic formula and factoring, as appropriate to the initial form of the equation. Recognize when the quadratic formula gives complex solutions and write them as $a \pm bi$ for real numbers a and b. **Also addresses:** A.CED.1	CCSS SMP 1 Janine throws a baseball from a height of 1 meter. The height of the baseball is modeled by the equation $y = -0.05x^2 + 1.25x + 1$, where x is the horizontal distance in meters. If the baseball hit the ground, how far from Janine did it land, to the nearest tenth of a meter?

9.1 Solving by Factoring: The Distributive Property

Objectives

- Apply the Distributive Property to factor quadratic equations.
- Solve equations by factoring quadratic expressions and using the Zero Product Property.

STANDARDS

Content: A.SSE.3a, A.REI.4b
Practices: 1, 2, 3, 7, 8
Use with Lesson 8-5

EXAMPLE 1 Factoring by Using the Distributive Property

The work in this section will help develop skills that can be applied when solving certain quadratic equations. These skills will build on your understanding of factoring, the distributive property and the properties of multiplying by zero.

EXPLORE Elena's teacher writes the following statements on the board:

The square of a number is equal to 4 times the number.	Twice the square of a number is equal to 8 times the number.

a. USE STRUCTURE Write equations to represent each statement. Then write equivalent equations in factored form using the Distributive Property. CCSS A.SSE.3a, SMP 7

b. CONSTRUCT ARGUMENTS What value do you need to multiply $(m - 4)$ by in order to produce an answer of zero? Will this work regardless of the value of $(m - 4)$? Explain. CCSS A.REI.4b, SMP 3

c. CONSTRUCT ARGUMENTS What will the expression $m(m - 4)$,equal if m is set equal to 4? Why is this the case? Is there any other value for m that will make this expression equal 0? Explain. CCSS A.REI.4b, SMP 3

d. DESCRIBE A METHOD Consider your responses to **parts b** and **c** and solve your equations from **part a** to find the numbers. Explain what you did or show your work. CCSS A.REI.4b, SMP 8

e. USE STRUCTURE The zeros of a function are those values that cause the function to equal zero. Use factoring and your understanding of the properties of multiplying by zero to find the zeros of the function $y = x^2 - 7x$. What are the zeros of the function $y = 2x^2 - 14x$? CCSS A.SSE.3a, A.REI.4b, SMP 7

f. MAKE A CONJECTURE Based on your answers from **parts a–c**, make a conjecture about the zeros of the function $y = ax^2 - abx$, where a and b are nonzero real numbers. Prove your conjecture. CCSS A.SSE.3a, A.REI.4b, SMP 3

g. USE STRUCTURE Write a statement similar to the statements Elena's teacher wrote, with the same solutions. Show that you are correct. CCSS A.SSE.3a, A.REI.4b, SMP 7

KEY CONCEPT Factoring by Grouping

A polynomial can be factored by grouping only if all of the following conditions exist.

- There are four or more terms
- Terms have ________________ that can be grouped together.
- There are two common factors that are identical or ________________ of each other.

EXAMPLE 2 Solve with Factoring by Grouping

Denise has increased the length of one side of her square garden to make it a rectangular plot. The area of the new plot, in square feet, is represented by the equation $x^2 - 6x + 11x = 66$. The side length of the original garden measures x feet.

x

a. USE STRUCTURE Set the equation equal to zero then solve to find the possible values of x. Describe how you solved the equation. CCSS A.SSE.3a, A.REI.4b, SMP 7

b. EVALUATE REASONABLENESS Give all possible side lengths of the original garden. Explain your reasoning. What are the dimensions of the new garden? 

c. USE STRUCTURE For a second garden, the area is represented by $x^2 - 5x + 15 = 3x$. Set the equation equal to zero and find all possible values of x. CCSS SMP 7

d. EVALUATE REASONABLENESS Give all possible side lengths of the new garden. Why is this different from the first garden? CCSS SMP 8

PRACTICE

1. **CRITIQUE REASONING** Kayla says the zeros of the equation $2m^2 - 12m = 0$ are 0 and -12. Is she correct? Explain your reasoning. CCSS A.SSE.3a, SMP 3

Solve each equation by factoring. CCSS A.SSE.3a, A.REI.4b, SMP 1

2. $4x^2 - 6x = 0$

3. $4x^2 = -1.2x$

4. $3x^2 + 24x - 1.5x - 12 = 0$

5. $2x^2 - 0.6x + 3x = 0.9$

6. **PLAN A SOLUTION** Mr. Billings writes the equation $x^2 - 3x = 0$ on the board. He then writes $x = w + 7$. Solve for w using the given information. Explain your solution. CCSS A.SSE.3a, A.REI.4b, SMP 1

7. **DESCRIBE A METHOD** In this section some problems have been solved by factoring using the distributive property and others by factoring by grouping. What are the similarities in regards to each technique? CCSS A.REI.4b, SMP 8

8. **REASON QUANTITATIVELY** Choose a value, q, so that $2x^2 + qx + 7x = 21$ can be solved using factoring by grouping and the Zero Product Property. Then solve the equation. CCSS A.SSE.3a, A.REI.4b, SMP 2

9. **CRITIQUE REASONING** Given the quadratic equation $3x^2 + 7x - 18x - 42 = 0$, Theresa says you need to factor by grouping using the binomials $(3x^2 - 18x)$ and $(7x - 42)$. Connor says you need to use the binomials $(3x^2 + 7x)$ and $(-18x - 42)$. Are either of them correct? Justify your answer. CCSS A.SSE.3a, A.REI.4b, SMP 3

10. **USE STRUCTURE** Juan's teacher gave the class the following problem to solve: $2x^2 - 10xy + 3x - 15y = 0$. Write the equation in factored form and solve for x. CCSS A.REI.4b, SMP 7

11. **CRITIQUE REASONING** Alexis and Calvin are solving $3m^2 = 12m$. Are either of them correct? Explain your reasoning. CCSS A.SSE.3a, A.REI.4b, SMP 3

Alexis	Calvin
$3m^2 = 12m$	$3m^2 = 12m$
$\frac{3m^2}{m} = \frac{12m}{m}$	$3m^2 - 12m = 0$
$3m = 12$	$3m(m - 4) = 0$
$m = 4$	$3m = 0$ or $m - 4 = 0$
	$m = 0$ or $m = 4$

9.2 Solving by Factoring: $x^2 + bx + c = 0$

Objectives

- Solve quadratic equations of the form $x^2 + bx + c = 0$ by factoring.

STANDARDS

Content: A.SSE.3a, A.REI.4b
Practices: 1, 3, 4, 7, 8
Use with Lesson 8-6

You can use algebra tiles to factor a quadratic expression of the form $x^2 + bx + c$. If a rectangular array can be formed by algebra tiles, then the length and width of the rectangle are factors of the quadratic expression that represents the area of the rectangular array.

EXAMPLE 1 Factor Quadratic Expressions (CCSS A.SSE.3a)

Factor the expression represented by the model shown.

x^2	x	x	x	x	x	x	x	x
x	1	1	1	1	1	1	1	1
x	1	1	1	1	1	1	1	1
x	1	1	1	1	1	1	1	1

a. USE A MODEL Write an expression, in the form $x^2 + bx + c$, that represents the area of the rectangular array shown. What are the values of b and c? (CCSS SMP 4)

____

b. USE STRUCTURE The 1 tiles also form a rectangular array. Describe the relationship between the columns and rows of 1 tiles and the values of b and c. (CCSS SMP 7)

____

____

c. USE A MODEL Write expressions for the length and width of the rectangular array. Use these to write an expression for the area of the rectangular array. (CCSS SMP 4)

____

d. USE STRUCTURE Compare the expression in **part a** that represents the area of the model to the expression in **part c**. What do you notice? (CCSS SMP 7)

____

e. USE STRUCTURE Use **part d** to help you solve the equation $x^2 + 11x + 24 = 0$. (CCSS SMP 7)

____

KEY CONCEPT

Use the information from the exploration to complete each sentence.

Words	To factor trinomials in the form $x^2 + bx + c$, find two integers, m and p, with a ________ of b and a ________ of c. Then write $x^2 + bx + c$ as $(x + m)(x + p)$.
Example	$x^2 + 6x + 8 = (x +$ ____$)(x +$ ____$)$, because $2 + 4 = 6$ and $2 \cdot 4 = 8$.

When quadratic expressions have the form $x^2 + bx + c$, the values of b and c can each be positive or negative.

EXAMPLE 2 Determine the Factors of Quadratic Expressions CCSS A.SSE.3a

Complete these steps to factor $x^2 - 11x + 24$.

a. **USE A MODEL** In this quadratic expression, $b = -11$ and $c = 24$. Complete the table to make a list of the factors of 24 and find their sum. CCSS SMP 4

Factors of 24	Sum of Factors

b. **USE STRUCTURE** Write $x^2 - 11x + 24$ in the form $(x + m)(x + p)$. Explain how you determined the values of m and p. CCSS SMP 7

c. **EVALUATE REASONABLENESS** Explain how you can check your answer. CCSS SMP 8

d. **USE STRUCTURE** Sort the quadratic expressions into appropriate categories. Write the expressions in factored form. CCSS SMP 7

$x^2 + x - 30$ $\quad x^2 - 5x - 36$ $\quad x^2 + 11x + 30$ $\quad x^2 - 13x + 36$

$x^2 + 13x + 36$ $\quad x^2 - 11x + 30$ $\quad x^2 + 5x - 36$ $\quad x^2 - x - 30$

b is positive and *c* is positive; factors are the same sign	*b* is negative, *c* is positive; factors are the same sign	*b* is negative and *c* is negative; factors have opposite signs	*b* is positive, *c* is negative; factors have opposite signs

e. **USE STRUCTURE** What can you conclude about the signs of the factors when c is positive? When c is negative? Why does this make sense? CCSS SMP 7

A **quadratic equation** can be written in the standard form $ax^2 + bx + c = 0$ where $a \neq 0$. Some equations of the form $x^2 + bx + c = 0$ can be solved by factoring and then using the Zero Product Property.

EXAMPLE 3 Solve Problems by Factoring CCSS A.REI.4b

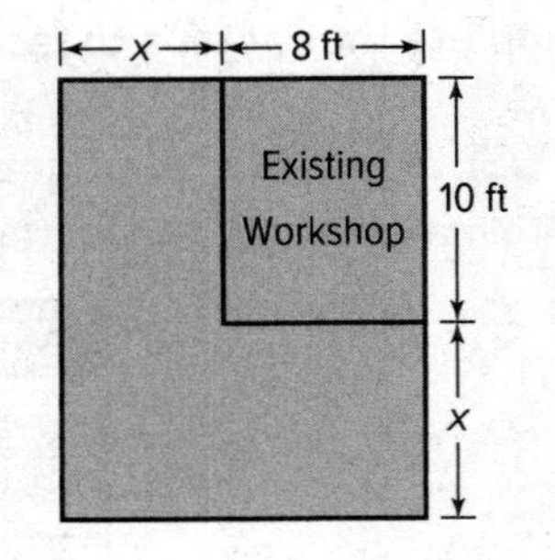

Alex is building a new workshop that is 1.5 times the area of his old workshop. He wants to increase the length and width of the old workshop by the same amount.

a. INTERPRET PROBLEMS Write an equation that represents the area of the new workshop. CCSS SMP 1

b. CONSTRUCT ARGUMENTS Write the equation in standard form. Explain each step. CCSS SMP 3

c. DESCRIBE A METHOD What property can you use to solve the problem? Solve for x. Which values of x make sense in this situation? CCSS SMP 8

d. INTERPRET PROBLEMS What dimensions should Alex use for the new workshop? Justify your answer. CCSS SMP 1

e. MAKE A CONJECTURE Suppose Alex wanted to build a new workshop that is two times the area of the old workshop. Would you be able to solve the problem by factoring? Explain. CCSS SMP 3

PRACTICE

1. **PLAN A SOLUTION** Write an expression for the perimeter of a rectangle that has an area $A = x^2 + 20x + 96$. Explain how you solved the problem. CCSS A.SSE.3a, SMP 1

2. **INTERPRET PROBLEMS** A triangle has an area of 64 square feet. If the height of the triangle is 8 feet more than its base, x, what are its height and base? CCSS A.REI.4b, SMP 1

3. **CONSTRUCT ARGUMENTS** A triangle's height is 10 feet more than its base. The area of the triangle is 100 square feet. Use factoring to find the dimensions. CCSS A.REI.4b, SMP 3

4. **CRITIQUE REASONING** Ira solves $x^2 - 2x = 63$ and finds that $x = 7$ or $x = -9$. Is Ira correct? If not, explain his error. CCSS A.REI.4b, SMP 3

5. **DESCRIBE A METHOD** Explain how to use the Zero Product Property to solve a quadratic equation. CCSS A.REI.4b, SMP 8

6. **MAKE A CONJECTURE** The equations in this lesson can also be written in standard form, $ax^2 + bx + c = 0$. What is the value of a in these equations? What is the coefficient of x in the factors $(x + m)$ and $(x + p)$? What do you think is the connection between the value of a and the coefficients of x? CCSS A.SSE.3a, SMP 3

7. **CRITIQUE REASONING** Jared claims that the quadratic equation $x^2 - 7x + 5 = 0$ has no solution because the left side does not factor. CCSS A.SSE.3a, SMP 3

   a. Is Jared correct about not being able to factor $x^2 - 7x + 5$? Explain your reasoning.

   b. Is Jared correct about the equation having no solution? Explain your reasoning.

   c. Use technology to graph the equation and find the solutions to this equation.

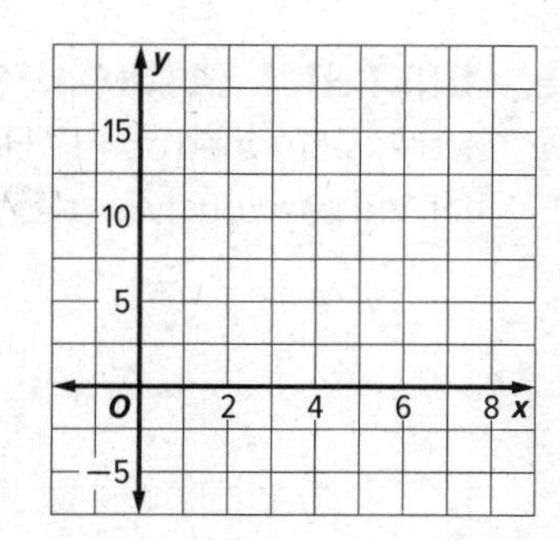

9.3 Solving by Factoring: $ax^2 + bx + c = 0$

Objectives

- Solve quadratic equations of the form $ax^2 + bx + c = 0$ by factoring.

CCSS STANDARDS

Content: A.SSE.3a, A.REI.4b, A.CED.1, A.CED.2

Practices: 1, 2, 3, 4, 6, 7, 8

Use with Lesson 8-7

EXAMPLE 1 Understand Factoring Quadratic Expressions CCSS A.SSE.3a

Factor the expression represented by the model.

a. USE A MODEL Write a quadratic expression for the model. How is this quadratic expression similar to expressions you have seen previously? How is it different? CCSS SMP 4

x^2	x^2	x	x	x	x	x
x	x	1	1	1	1	1
x	x	1	1	1	1	1
x	x	1	1	1	1	1

b. USE A MODEL What are the dimensions of the rectangle formed by the algebra tiles? How are the dimensions related to the quadratic expression? Justify your answer. CCSS SMP 4

c. USE STRUCTURE To factor a quadratic expression of the form $ax^2 + bx + c$, you need to find two integers, m and p, such that $mp = ac$ and $m + p = b$. What are possible values for m and p? Explain your choice. Can you use the values of m and p to write the factorization as $(x + m)(x + p)$ as in the previous lesson? Explain. CCSS SMP 7

d. DESCRIBE A METHOD Use the values you found for m and p in **part c** to rewrite the quadratic expression in the form $ax^2 + mx + px + c$. Can you factor this expression? If so, factor the expression and explain your work. CCSS SMP 8

e. CONSTRUCT ARGUMENTS Can you use the form $ax^2 + mx + px + c$ to factor an expression when $a = 1$? How does the strategy in this lesson relate to strategies learned previously? CCSS SMP 3

KEY CONCEPT Factoring $ax^2 + bx + c$

To factor trinomials in the form $ax^2 + bx + c$, find two integers, m and p, with a sum of ______________ and a product of ______________.

Then write $ax^2 + bx + c$ as $ax^2 + mx + px + c$ and factor by ______________.

EXAMPLE 2 Finding the zeros of a polynomial with a coefficient in front of the quadratic term other than one CCSS A.SSE.3a

Consider the equation $y = 6x^2 - 2x - 20$

a. PLAN A SOLUTION The zeros of the polynomial are those points where the graph crosses the x-axis. How can you find the zeros of the given polynomial? CCSS SMP 1

b. CALCULATE ACCURATELY Find the zeros of the equation given above. CCSS SMP 6

c. REASON ABSTRACTLY Will you always be able to use factoring to determine the zeros of a quadratic trinomial? Explain. CCSS SMP 3

d. USE A MODEL Explain how can you use algebra tiles to determine whether factoring can be used to find the zeros of a polynomial. CCSS SMP 4

EXAMPLE 3 Solve Equations by Factoring

Two parks have the same area, but different dimensions. The Jones Street Park has an area of 1200 square yards. The length of River Run Park is 20 yards more than four times its width w.

River Run Park

a. INTERPRET PROBLEMS Label the dimensions of River Run Park on the diagram. Then write an equation to represent the area of the park. Explain your reasoning. CCSS A.CED.1, SMP 1

b. USE STRUCTURE Find the dimensions of River Run Park. Explain how you solved the problem. CCSS A.REI.4b, SMP7

c. CRITIQUE REASONING Roberto suggests that the possible dimensions of Jones Street Park are $2w$ yards and $3w - 5$ yards, where w is the width of River Run Park. Is Roberto correct? Explain. CCSS A.REI.4b, SMP 3

EXAMPLE 4 Solve Equations by Factoring CCSS A.REI.4b

Jake is starting a small business where he washes cars. The equation $p = 8x^2 - 20x$ represents the profit he will make if he washes x cars. During the first five days of his business he washes one more car than the previous day. On the third day his profit is $100.

a. PLAN A SOLUTION Describe how you can determine how much profit Jake makes during the first five days. CCSS SMP 1

b. INTERPRET PROBLEMS How many cars does Jake wash for each of the first five days? Explain how you solved the problem. CCSS SMP 1

c. CALCULATE ACCURATELY How much profit did he make during each of the first five days? What was his total profit after the first five days? CCSS SMP 6

d. INTERPRET PROBLEMS Were there any results or solutions that you chose to disregard based on the problem? Explain. CCSS SMP 1

EXAMPLE 5 Factoring $ax^2 + bx + c$ when $a < 0$ CCSS A.SSE.3a

USE STRUCTURE Consider the equation $y = -2x^2 + 5x - 2$. CCSS SMP 7

a. What is the value of b in this equation? What is the value of ac?

b. Find the values of m and p. Explain your reasoning.

c. Factor the equation and find the zeros.

d. Eileen claims that she can solve $-2x^2 + 5x - 2 = 0$ without having to deal with the negative coefficient of x^2. Is she correct? If not, explain why. If so, how would you solve it?

e. Is the function $y = -2x^2 + 5x - 2$ the same function as $y = 2x^2 - 5x + 2$? Plot both functions on a coordinate plane, and comment on their similarities and differences.

y
2
O 1 2 3 x
−2

PRACTICE

1. **USE STRUCTURE** A square has an area of $4x^2 + 16xy + 16y^2$ square inches. The dimensions are binomials with positive integer coefficients. Find the perimeter of the square. Is the perimeter a multiple of $(x + 2y)$? Explain. CCSS A.CED.2, SMP 7

2. Both $2x^2 + qx - 12 = 6x$ and $3y^2 + 11y - q = 0$ can be solved by factoring. Follow the steps to find possible values for q that will make both equations true. CCSS A.REI.4b

   a. **COMMUNICATE PRECISELY** For the first equation, how can you use number relationships to predict values for q? CCSS SMP 6

   b. **COMMUNICATE PRECISELY** For the second equation, how can you use number relationships to predict values for q? CCSS SMP 6

   c. **REASON QUANTITATIVELY** What value of q makes both equations true? Rewrite both equations using the value of q. CCSS SMP 2

   d. **REASON QUANTITATIVELY** Find the values of x and y. CCSS SMP 2

3. **PLAN A SOLUTION** The rectangle at the right has a perimeter of $14x + 30$ centimeters and area of 225 square centimeters. Find the dimensions of the rectangle. Explain how you solved the problem. CCSS A.SSE.3a, A.REI.4b, A.CED.1, SMP 1

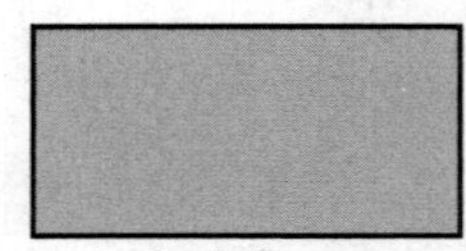

$6x + 15$ cm

4. The height of a projectile is given by $h = -16t^2 + vt + h_0$, where h is the height in feet, t is the time in seconds, v is the initial upward velocity in feet per second, and h_0 is the initial height in feet. A stunt man is propelled from ground level into the air at an initial velocity of 16 feet per second and lands on a platform that is 3 feet above the ground.

a. INTERPRET PROBLEMS Write an equation to determine how much time the stunt man is in the air. CCSS A.CED.1, SMP 1

b. INTERPRET PROBLEMS How many seconds is the stunt man in the air before landing on the platform? Explain. CCSS A.SSE.3a, A.REI.4b, SMP 1

c. MAKE A CONJECTURE How would the solution method and answer change if the stunt man is propelled from a platform that is more than 3 feet off the ground? CCSS A.REI.4b, SMP 3

5. INTERPRET PROBLEMS A model rocket is launched from a height of 5 feet with an initial velocity of 64 feet per second. Use the height of a projectile equation in **Exercise 4** to find when the rocket will reach a height of 53 feet. Explain. CCSS A.SSE.3a, A.REI.4b, SMP 7

6. CALCULATE ACCURATELY Find the zeros of each polynomial. CCSS A.SSE.3a, A.REI.4b, SMP 6

a. $y = -2x^2 + 2x + 4$

b. $y = -3x^2 + 11x - 10$

9.4 Solving by Factoring: Differences of Squares

Objectives

- Factor binomials that are the differences of squares.
- Use the differences of squares to solve real-world problems.

Content: A.SSE.3a, A.REI.4b
Practices: 1, 2, 3, 4, 7, 8
Use with Lesson 8-8

EXAMPLE 1 Model the Differences of Squares CCSS A.SSE.3a

Sometimes factoring can be more efficient if you are able to recognize and use a pattern. In this section you will learn to recognize when an expression can be categorized as a difference of squares and practice quickly factoring these types of expressions.

EXPLORE Use the model to factor binomials that are differences of squares.

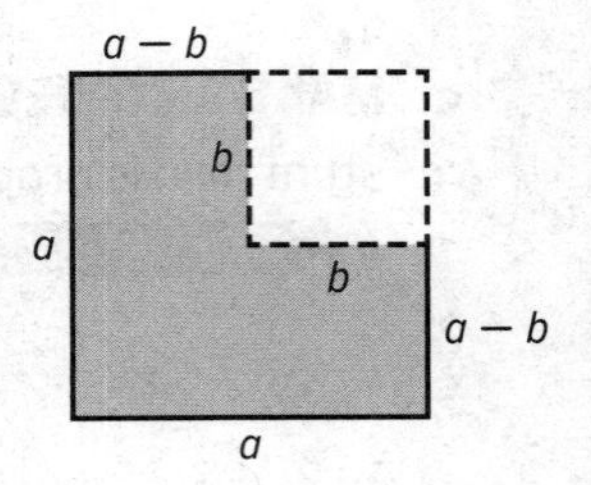

a. USE A MODEL Write an expression that represents the area of the shaded region in the model. Why might this expression be known as a difference of squares? CCSS SMP 4

b. USE A MODEL Use tracing paper to copy the shaded region of the figure. Cut the shaded region into two congruent parts and rearrange the parts to form a triangle. You may have to recopy the figure and experiment with different configurations. Sketch the shaded region below and indicate how to cut the figure into two congruent parts. Sketch the shaded parts in the form of a rectangle. Label all side lengths. CCSS SMP 4

c. USE STRUCTURE What are the dimensions of the rectangle? Express the area of the rectangle as the product of two binomial factors. Then multiply the factors to find the product. CCSS SMP 7

d. DESCRIBE A METHOD Choose values for a and b and substitute those values into the expressions developed in **parts a** and **b**. Show that they are equivalent. Explain how to use what you learned to factor the expression $x^2 - 25$? CCSS SMP 8

e. USE STRUCTURE Jake separated the shaded region into two rectangles. Write an expression to represent the sum of the areas of the shaded rectangles. Can you factor the expression into two binomial factors? Explain. CCSS SMP 7

KEY CONCEPT

Factor Differences of Squares

The difference of two squares, $a^2 - b^2$ can be factored as the product of the sum and difference of a and b.

$$a^2 - b^2 = (a + b)(a - b)$$

Sometimes to completely factor a polynomial, you need to apply several different factoring strategies.

EXAMPLE 2

Completely Factor a Polynomial CCSS A.REI.4b

Factor the expression $2m^3 + 2m^2 - 128m - 128$.

a. REASON QUANTITATIVELY What is the greatest common factor of all the terms in the expression? How can you use the Distributive Property to factor the expression? CCSS SMP 2

b. DESCRIBE A METHOD How can you factor the cubic polynomial into two binomials? Factor and explain your steps. CCSS SMP 8

c. CRITIQUE REASONING Emma says that each of the factors are prime factors, and so the expression cannot be factored further. Is she correct? If not, explain how to continue factoring. CCSS SMP 3

Factor the expression $x^8 - 81y^4z^{16}$.

d. USE STRUCTURE In order to use the difference of squares pattern, both terms on either side of the minus sign must be perfect squares. Find the square roots of the terms, call them a and b and factor. CCSS SMP 7

e. **DESCRIBE A METHOD** How can you factor the expression calculated in **part d** completely? Explain why you cannot use this method for both of the factors. Write the complete factored expression. CCSS SMP 8

EXAMPLE 3 Solve Problems with Differences of Squares CCSS A.SSE.3a

Mr. Alberto's backyard is a square separated into two parts, a grassy area and a square patio. Mrs. Baker's rectangular backyard has the same area as the grassy area of Mr. Alberto's backyard. Mrs. Baker's yard is 8 ft longer than Mr. Alberto's. What is the difference between the length and width of Mrs. Baker's backyard?

4*n* ft

4*n* ft

8 ft

8 ft

patio

a. **USE STRUCTURE** Write an expression to represent the grassy area of Mr. Alberto's backyard. Does this expression represent the differences of squares? If yes, state the values for *a* and *b*. CCSS SMP 7

b. **INTERPRET PROBLEMS** Solve the problem. Explain how you found your answer. CCSS SMP 1

EXAMPLE 4 Using the Difference of Squares for Mental Calculation CCSS A.SSE.3a

The difference of squares formula can be used for some math calculations that seem difficult at first.

a. **USE STRUCTURE** Consider the value of $99^2 - 1$. What is the value of *a*? What is the value of *b*? Factor this expression as the difference of squares. CCSS SMP 7

b. **CALCULATE ACCURATELY** Using the factorization, find the value of $99^2 - 1$. CCSS SMP 6

c. **USE STRUCTURE** Why is determining the value of $99^2 - 16$ much harder when factoring as the difference of squares? CCSS SMP 7

d. **CALCULATE ACCURATELY** Instead of factoring as the first step, use **parts a** and **b** to find the value of $99^2 - 16$ without a calculator. CCSS SMP 6

e. **CALCULATE ACCURATELY** Find the value of $1001^2 - 1$ using the difference of squares. CCSS SMP 6

PRACTICE

1. A company that manufactures cardboard boxes sells three sizes of boxes. The volume of the small box is represented by $n^3 + 4n^2 - 16n - 64$ in^3. The volume of the medium box is represented by $n^3 + 6n^2 - 36n - 216$ in^3. The volume of the large box is represented by $n^3 + 8n^2 - 64n - 512$ in^3. The company wants to start selling an extra-large box. Predict the dimensions and the volume of the extra-large box. CCSS A.SSE.3a

   a. **REASON ABSTRACTLY** Find the dimensions of the small, medium, and large boxes. Explain the method you used. CCSS SMP 2

   b. **FIND A PATTERN** Look for a pattern in the areas. What pattern is represented by the areas of the small, medium, and large boxes? Use this pattern to predict the dimensions of the extra-large box. CCSS SMP 7

   c. **REASON ABSTRACTLY** Use your prediction in **part b** to find the volume of the extra-large box. CCSS SMP 2

2. **REASON ABSTRACTLY** Factor $x^4 - 16$ completely and use the factorization to determine any real solutions to the equation $x^4 - 16 = 0$. CCSS A.REI.4b, SMP 2

3. **CRITIQUE REASONING** Janelle factored an expression using these steps: $25x^2 - 16 = 5^2x^2 - 4^2 = (5x - 4)^2 = (5x - 4)(5x - 4)$. Explain why Janelle is correct or incorrect. If she is not correct, factor the expression correctly. CCSS A.SSE.3a, SMP 3

4. **CALCULATE ACCURATELY** Use the difference of squares to factor and simplify the following expressions. CCSS A.SSE.3a, SMP 6

   a. $121x^2y^6z^4 - 16y^2z^2$

   b. $991^2 - 81$

9.5 Solving by Factoring: Perfect Squares

Objectives

- Identify and factor perfect square trinomials.
- Use factors to identify the zeros of a quadratic function.
- Solve quadratic equations involving perfect squares.

Content: A.SSE.3a, A.REI.4b
Practices: 1, 2, 3, 4, 7, 8
Use with Lesson 8–9

EXAMPLE 1 Recognize and Factor Perfect Square Trinomials

You already know how to expand $(x + y)^2$ to get $x^2 + 2xy + y^2$. In this section you will learn how to go in the other direction and factor a perfect square trinomial, a trinomial that can be factored into the square of a binomial. You will learn to recognize a square trinomial as a trinomial whose first and last terms are perfect squares and whose middle term is twice the product of the square roots of the first and last terms. CCSS A.SSE.3a

EXPLORE Examine $f(x) = 4x^2 + 20x + 25$, $g(x) = 6x^2 + 30x + 25$, $h(x) = 4x^2 + 10x + 25$, and $m(x) = 4x^2 - 20x + 25$.

a. REASON QUANTITATIVELY Two of the functions above do not match the description provided in the description of perfect square trinomials. Which are not perfect square trinomials? Explain. CCSS SMP 2

b. REASON QUANTITATIVELY For the two remaining expressions, find the square root of the first term and the last term. Then consider how the middle term is related to the square roots. CCSS SMP 2

c. USE A MODEL Model each expression using algebra tiles. Describe how the models are similar and how they are different. CCSS SMP 4

d. USE STRUCTURE Factor each expression. Find the zeros of each function. What do you notice about the zeros of perfect square trinomials? What does that tell you about the graph? CCSS SMP 2

e. MAKE A CONJECTURE Use the rules for factoring a trinomial in the form $ax^2 + bx + c$ to write a general rule for factoring of a perfect square trinomial in the form $x^2 + 2xy + y^2$. How does the rule change when y is negative? CCSS SMP 3

Combining the ability to recognize and factor perfect trinomial squares with a tool called the Square Root Property provides an efficient method for solving a variety of quadratic equations. To understand the Square Root Property, look at a problem like $x^2 - 25 = 0$, which you have already solved by factoring. You can find the solutions of the equation by using the definition of square root.

$x^2 - 25 = 0$	Original Equation
$x^2 = 25$	Add 25 to each side.
$x = \pm\sqrt{25}$	Take the square root of each side.

The square roots of 25 are 5 and −5. You can write the solution as ± 5.

The Square Root Property can help solve problems whose answers are not as obvious as the problem above.

$9x^2 + 30x + 25 = 100$	Original Equation
$(3x + 5)^2 = 10^2$	Factor
$3x + 5 = \pm 10$	Take the square root of each side
$x = -5$ and $x = \frac{5}{3}$	Solve for the positive and negative square root

KEY CONCEPT Square Root Property

Use the definition of square root to complete the following.

To solve a quadratic equation in the form of $x^2 = n$, when n is greater than or equal to 0, take the

__________ of each side.

Example: $x^2 = 100$; $x =$ _____ $\sqrt{100} = \pm$ _____

EXAMPLE 2 Using the Square Root Property CCSS A.REI.4b

Follow these steps to solve $(x - 3)^2 = 81$ and then $x^2 + 10x + 25 = 64$.

a. USE STRUCTURE Solve the first equation using the Square Root Property. Explain. CCSS SMP 7

b. USE STRUCTURE Rewrite the second equation using perfect squares. Explain how you identified the perfect squares. CCSS SMP 7

c. REASON QUANTITATIVELY Apply the Square Root Property. What is the result? CCSS SMP 2

d. INTERPRET PROBLEMS After applying the Square Root Property, explain why there are two equations you have to solve. Then solve for x. CCSS SMP 1

e. MAKE A CONJECTURE Can you solve this equation by factoring and using the Zero Product Property? Explain. CCSS SMP 3

f. USE STRUCTURE Consider an equation that does not have a perfect square as an answer, $x^2 + 16x + 64 = 19$. Use the same steps you have in the problems above, but leave terms that are not perfect squares in radical form. What does this tell you about where the parabola would cross the x-axis. CCSS SMP 7

g. MAKE A CONJECTURE Can you solve the equation in **part f** by factoring and using the Zero Product Property? Explain. CCSS SMP 3

EXAMPLE 3 Solve Problems with Perfect Squares CCSS A.SSE.3a

The rectangle prism at the right has a square base with an area of 1521 square centimeters and a height of x centimeters. The perimeter of the base is $48x + 12$ centimeters.

x cm

a. PLAN A SOLUTION How can you use the perimeter of the square base to write an equation of its area in terms of x? CCSS SMP 1

b. **REASON ABSTRACTLY** Explain how to solve for x and then use its value to find the volume of the prism. CCSS SMP 2

c. **INTERPRET PROBLEMS** Find the volume of the prism. Explain your reasoning. CCSS SMP 1

PRACTICE

1. **CRITIQUE REASONING** Elly claims the expression $9x^2 + 50x + 25$ is a perfect square trinomial. Is she correct? If she is incorrect, show how the expression can be changed so that it is a perfect square. CCSS A.SSE.3a, SMP3

2. **CRITIQUE REASONING** Paul suggests that to find the zeros of $y = x^4 - 81$, you can use Difference of Squares twice. He says that the result will have the same zeros as $y = x^2 - 9$. Prove or disprove his statement. CCSS A.SSE.3a, SMP 3

3. **USE STRUCTURE** Find a value for m that will make the expression $4x^4 - 44x^2 + m$ a perfect square. Then use the value to factor the expression completely. CCSS A.SSE.3a, SMP 7

4. **DESCRIBE A METHOD** Solve the equation $25b^2 - 90b + 81 = 441$. CCSS A.REI.4b, SMP8

   a. Can you find a value of b by factoring and using the Zero Product Property? Explain.

   b. Explain another way to find the value of b. Then solve the equation.

9.6 Solving by Graphing

STANDARDS

Content: A.REI.4b, F.IF.7a, F.IF.8a
Practices: 1, 2, 3, 4, 5, 6, 7, 8
Use with Lesson 9-2

Objectives

- Use a graph to find solutions to quadratic equations.
- Use a graph to determine whether a quadratic equation has no real solutions, one real solution, or two real solutions.

Graphs of quadratic functions can be used to find the solutions of the related equation. The solutions (also known as roots) of an equation can be determined by making one side of the equation equal to zero, replacing the zero with the variable y, graphing the result and examining the x-intercepts (the zeros) of its related graph.

EXAMPLE 1 Understand Graphs of Quadratic Functions CCSS A.REI.4b, F.IF.7a

a. DESCRIBE A METHOD Solve the quadratic equation $4x^2 = 16$. Describe the solution(s) and how you solved the equation. CCSS SMP 8

b. DESCRIBE A METHOD Solve the quadratic equation $2x^2 - 13x - 7 = 0$. Describe the solution(s) and how you solved the equation. CCSS SMP 8

c. DESCRIBE A METHOD Solve the quadratic equation $4x^2 = 4x - 1$. Describe the solution(s) and how you solved the equation. CCSS SMP 8

d. REASON ABSTRACTLY The graph of a quadratic function is a U-shaped curve called a parabola. The graphs of a–c are shown below. Describe where the graphs of each of the functions cross the x-axis. Consider how these intercepts relate to the work you did in a–c. CCSS SMP 2

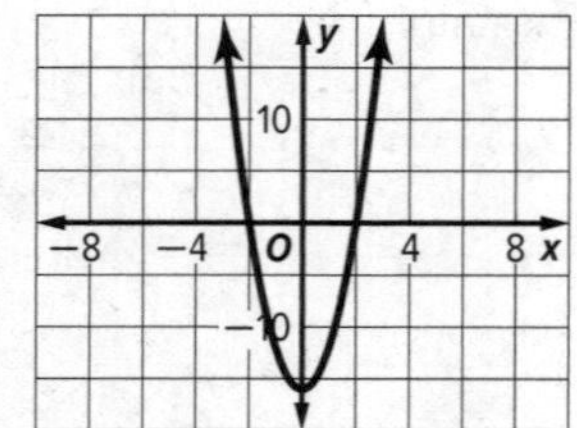

a

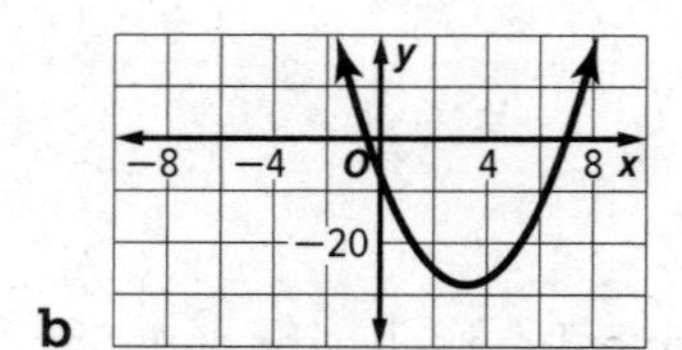

b

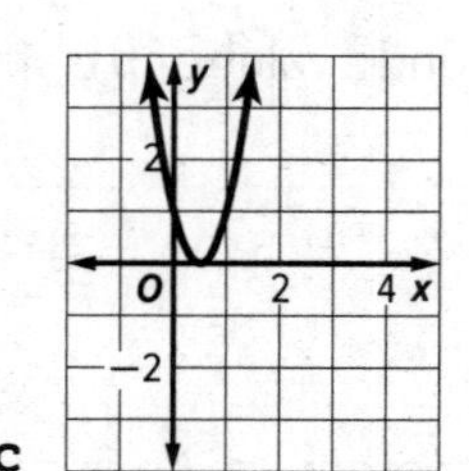

c

e. REASON ABSTRACTLY Suppose the graph of a quadratic function never crosses the x-axis. What conclusions can you draw about the solutions of the function? CCSS SMP 2

f. USE STRUCTURE Label the number and type of solutions for each graph below. CCSS SMP 7

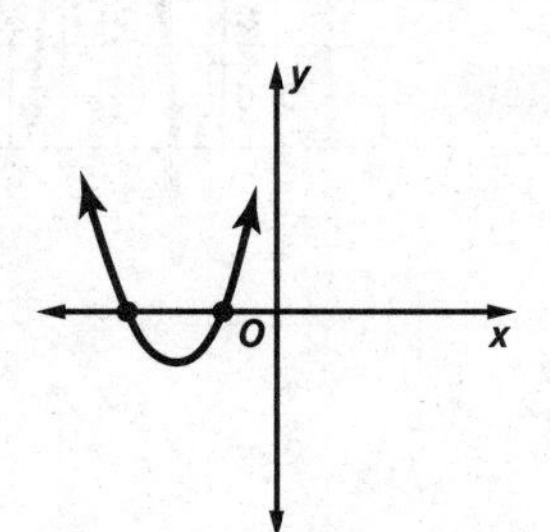

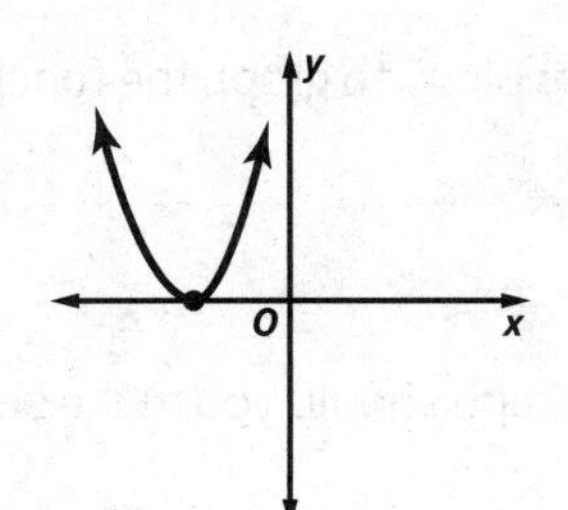

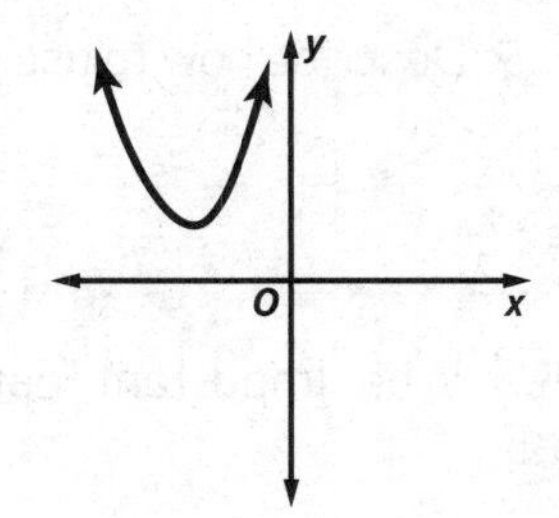

EXAMPLE 2 Factor to Find Solutions CCSS F.IF.8a

The graphs of different quadratic equations are shown.

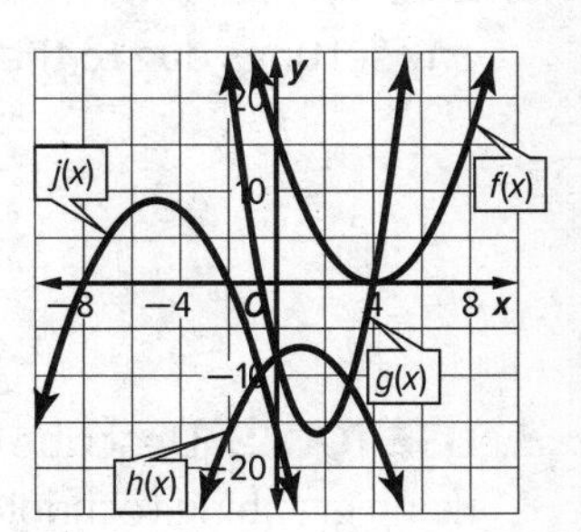

a. REASON QUANTITATIVELY How can you determine which curve is the graph of $3x^2 - 10x - 8 = 0$? CCSS SMP 2

b. USE A MODEL Describe the number and type of solutions for each function shown. CCSS SMP 4

c. REASON QUANTITATIVELY Match these two equations to the remaining functions on the graph: $-x^2 - 10x - 16 = 0$ and $x^2 + 16 = 8x$. Explain your choices. CCSS SMP 2

d. REASON QUANTITATIVELY The remaining graph is the graph of the related equation $-x^2 + 2x - 8 = 0$. Explain how you know this function has no x-intercepts. CCSS SMP 2

e. REASON QUANTITATIVELY Solve the equation $2x^2 - 6x = 20$ by factoring. Write a function $f(x)$ that could be graphed and used to solve the equation. Based on the solutions to the equation, what do you expect to find on the graph of $f(x)$? CCSS SMP 2

Graphing technology provides another means for solving quadratic functions by analyzing graphs and their intercepts.

EXAMPLE 3 Use Graphing Technology CCSS F.IF.8a

Peter is designing and selling a computer game. The amount of profit he will make is dependent on how much he charges for the game. The profit can be represented by the function $f(x) = -150x^2 + 3400x - 15{,}000$. Reproduce the graph shown using graphing technology.

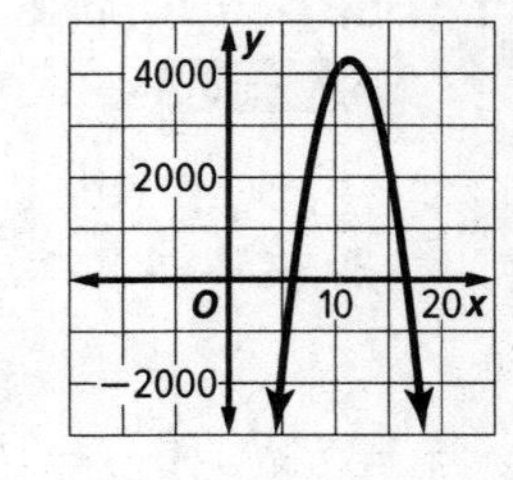

a. **USE TOOLS** Describe how to use graphing technology to graph the function. CCSS SMP 5

b. **USE TOOLS** What important features of the graph should you make sure are shown? CCSS SMP 5

c. **USE STRUCTURE** What are the domain and range of the function? Should you change any settings due to the size of the domain and range? CCSS SMP 7

d. **USE TOOLS** Describe how using the graph on paper and using the graph created by using graphing technology are similar and different. CCSS SMP 5

e. **INTERPRET PROBLEMS** Find the *x*-intercepts. What do the *x*-intercepts tell you in regards to the context of the problem? CCSS SMP 1

f. **INTERPRET PROBLEMS** What prices do you think Peter should sell the game for? CCSS SMP 1

g. **CALCULATE ACCURATELY** Find the *x*-intercepts algebraically. CCSS SMP 6

h. **INTERPRET PROBLEMS** Write an equation to find how much Peter should charge for the game to make a profit of \$2000. How can this equation be solved graphically? CCSS SMP 1

EXAMPLE 4 Solving Equations Graphically CCSS F.IF.7a

A rock is thrown into the air and its height in feet is modeled by the function $h(x) = -16x^2 + 20x + 5$ where x is the time in seconds since the rock was thrown.

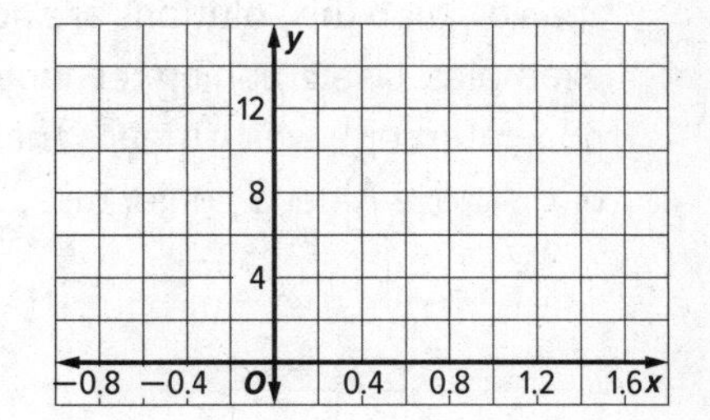

a. USE TOOLS Describe how to use graphing technology to graph the function and provide a sketch of the graph. CCSS SMP 5

b. REASON QUANTITATIVELY Considering the context of the problem, are negative x values included in the domain of $h(x)$? Explain your reasoning. CCSS SMP 2

c. COMMUNICATE PRECISELY Find any x-intercepts. What does each x-intercept mean in the context of the problem? CCSS SMP 6

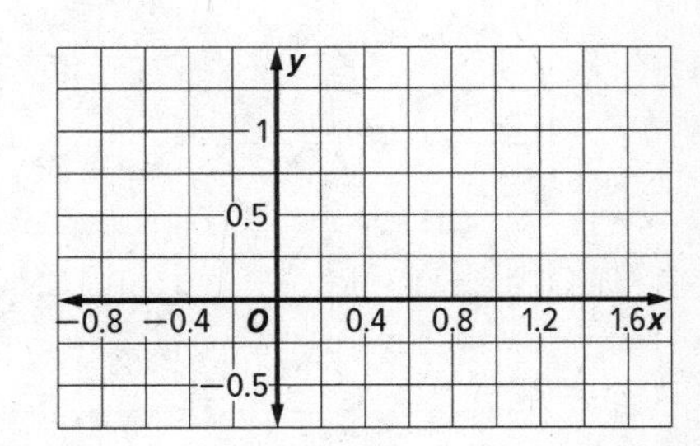

d. INTERPRET PROBLEMS Write an equation to solve for the time x at which the rock reaches a height of 10 feet. Use the method outlined at the beginning of the lesson to solve the equation graphically and provide a sketch of the graph. CCSS SMP 1

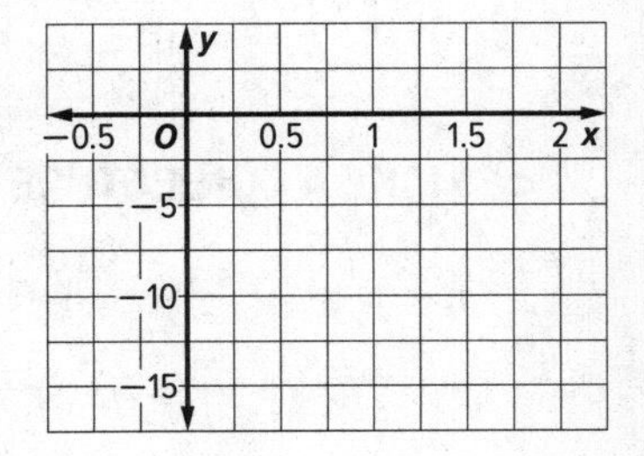

e. INTERPRET PROBLEMS Write an equation to solve the time x at which the rock reaches a height of 15 feet. Sketch the graph of the function relating to this equation. What can you conclude about the solutions to the equation? What does this mean in the context of the problem? CCSS SMP 1

PRACTICE

1. **CRITIQUE REASONING** For a quadratic equation $ax^2 + bx + c = 0$, where a, b, and c are all real numbers, $a \neq 0$, Aiden suggests that if the product of 4, a, and c is greater than b^2, then the solutions are not real numbers. Test his conjecture with three examples. Use graphing technology to graph each example to check that they have no x-intercept, which means they have no real solutions. Do your examples support or disprove Aiden's conjecture? Explain your reasoning. CCSS F.IF.7a, SMP 3

2. Which values of m and n will result in these three equations representing the related functions shown on the graph? CCSS F.IF.8a

$y = x^2 + 12x + m; y = 2x^2 - nx + 72; y = -x^2 + 3x - 10$

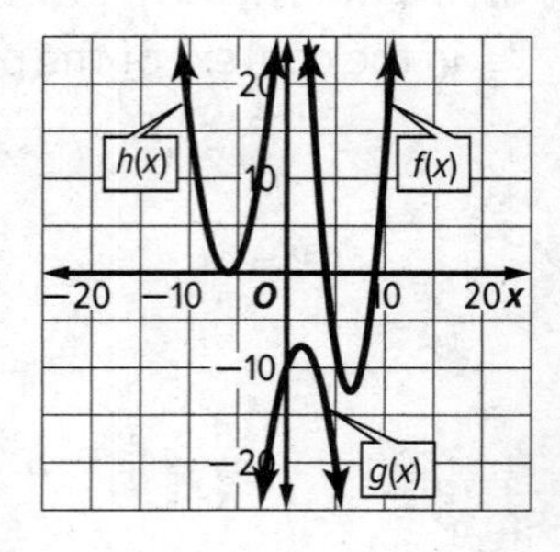

a. REASON QUANTITATIVELY Describe the solutions for each function. Then, match each graph to its related equation. Explain your reasoning. CCSS SMP 2

b. MAKE A CONJECTURE How could you choose an appropriate value for m? CCSS SMP 3

c. MAKE A CONJECTURE How could you choose an appropriate value for n? CCSS SMP 3

3. The height, *h* of a model rocket launched from ground level into the air after *t* seconds can be modeled by the equation $h = -16t^2 + 160t$. The equation is graphed on the coordinate grid below. CCSS F.IF.7a

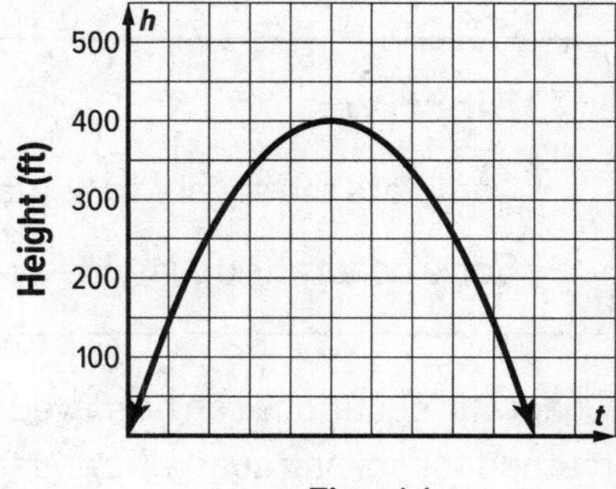

a. USE A MODEL Where does the graph intersect the *x*-axis? What do these points represent? CCSS SMP 4

b. INTERPRET PROBLEMS How long does it take the rocket to reach its maximum height? Explain your reasoning. CCSS SMP 1

4. USE A MODEL Through market research a company finds that its profit in dollars can be modeled by the function $P(x) = -50{,}000x^2 + 300{,}000x - 250{,}000$ where *x* is the price in dollars at which they sell their product. CCSS F.IF.7a, SMP 4

a. Describe how to use graphing technology to graph the function, and then provide a sketch of the graph.

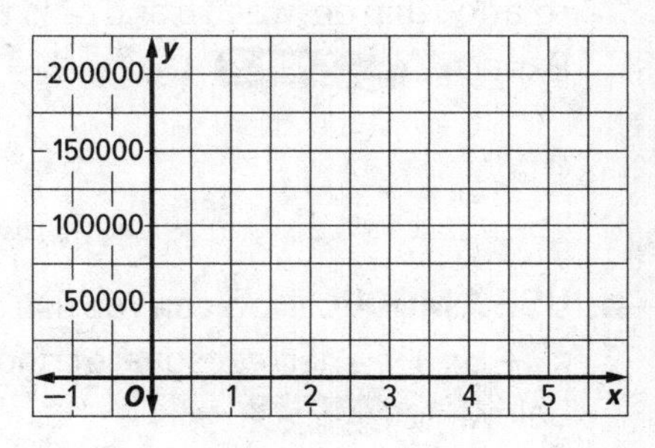

b. Find the *x*-intercepts. What do these signify in the context of the problem?

c. Write an equation for the price at which the company should sell their product to make a profit of $150,000. Sketch a graph of the function relating to this equation and use it to solve the equation graphically.

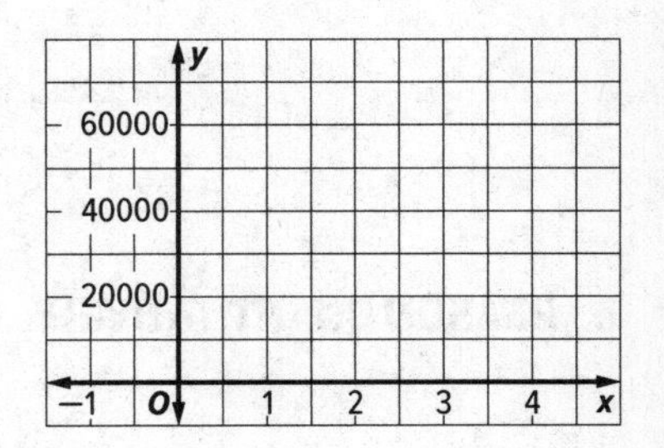

d. Is there a price at which the company can sell their product to make a profit of $300,000? Explain your reasoning and use a graph to support your answer.

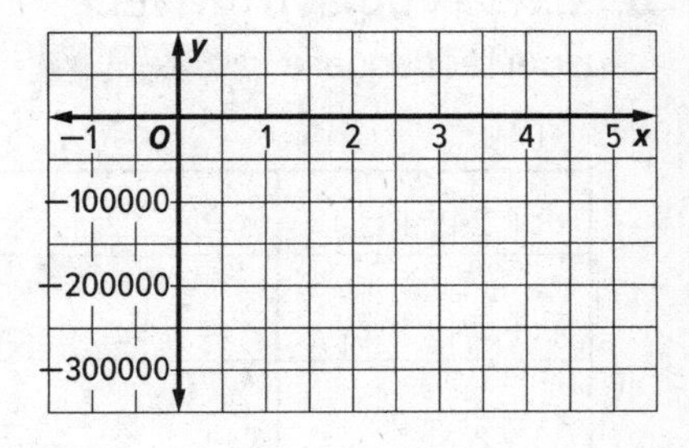

9.7 Solving by Completing the Square

Objectives

- Complete the square to write perfect square trinomials.
- Solve quadratic equations by completing the square.

Content: A.REI.4a, F.IF.8a, A.SSE.3b
Practices: 1, 2, 3, 4, 7, 8
Use with Lesson 9-4

Quadratic equations can be solved by inspection, factoring, and using a graph. Another method for solving quadratic equations is called completing the square. To **complete the square**, add a constant c to the expression $x^2 + bx$ to form a perfect square trinomial $x^2 + bx + c$. Understanding the geometric model that the algebra describes is an important first step in mastering this method.

EXAMPLE 1 Understand Completing the Square CCSS A.REI.4a

EXPLORE Find the value of c that makes $x^2 + bx + c$ a perfect square.

a. REASON ABSTRACTLY Suppose $x^2 + bx + c$ is a perfect square. What is the relationship between b and c? Is this relationship the same when $a \neq 1$? Explain. CCSS SMP 2

b. USE A MODEL How can you use algebra tiles to find the value of c that makes $x^2 + 8x + c$ a perfect square trinomial? Explain. Then sketch the arrangement below. CCSS SMP 4

c. REASON QUANTITATIVELY If b is an odd number, what is always true about c? CCSS SMP 2

d. REASON QUANTITATIVELY Find the values of c and b that make each trinomial a perfect square. CCSS SMP 2

$x^2 + 24x + c$	$c =$ ______
$x^2 + 9x + c$	$c =$ ______
$x^2 - \frac{1}{2}x + c$	$c =$ ______

$x^2 - bx + 81$	$b =$ ______
$x^2 + bx + \frac{25}{4}$	$b =$ ______
$x^2 + bx + \frac{1}{9}$	$b =$ ______

KEY CONCEPT How to Complete the Square

Complete the following to develop an algorithm to complete the square for an expression of the form $x^2 + bx$.

STEP 1: Divide b by ________________.

STEP 2: ________________ the quotient.

STEP 3: ________________ the result of Step 2 to ________________.

EXAMPLE 2 Solve an Equation by Completing the Square CCSS A.SSE.3b, F.IF.8a

Solve the equation $x^2 - 12x - 10 = 3$ by completing the square.

a. **USE STRUCTURE** How can you write an equivalent equation so that the left side is in a form for completing the square? CCSS SMP 7

b. **REASON QUANTITATIVELY** What value is added to the left side of the equation to complete the square? Since this is an equation, not an expression, what else must be done to the equation? What is the equivalent equation? CCSS SMP 2

c. **DESCRIBE A METHOD** Solve the equation. Explain your steps. CCSS SMP 8

d. **DESCRIBE A METHOD** Alana is considering the graph of $y = -3x^2 + 18x - 24$. Show how substituting 0 for y and completing the square can provide information about where the graph will cross the x-axis. CCSS SMP 8

e. **DESCRIBE A METHOD** Now that Alana knows the x-intercepts of the graph of $y = -3x^2 + 18x - 24$, she wants to find its maximum point. Explain how this can be done. CCSS SMP 8

EXAMPLE 3 Solve Problems by Completing the Square CCSS A.REI.4A, F.IF.8a

The value of a share of a stock can be modeled by the quadratic equation $v = 4t^2 - 20t$, where t represents the number of days after the stock has been listed. Mrs. Lopez bought 5 shares of the stock on the 6th day after it was listed and sold it when each share had a value of \$144. How much was Mrs. Lopez's profit or loss? How many days did she own the stock?

a. **REASON QUANTITATIVELY** Complete the square to find the value of t when $v = \$144$. CCSS SMP 2

b. **INTERPRET PROBLEMS** Interpret the results from **part a** to find the number of days Mrs. Lopez owned the stock. CCSS SMP 1

c. **PLAN A SOLUTION** What information do you need to know to find Mrs. Lopez's profit or loss? Find the profit or loss. CCSS SMP 1

d. **INTERPRET PROBLEMS** Complete the square in the original problem to find the stock's lowest price. What does that suggest about the equation? CCSS SMP 1

PRACTICE

1. **REASON QUANTITATIVELY** Find the value of q that makes $0.5x^2 + 0.5qx + 72$ a perfect square trinomial. Show that the same value of q makes $4x^2 + 24x + 1.5q$ a perfect square trinomial. CCSS A.REI.4a, SMP 2

2. REASON QUANTITATIVELY The area of the rectangle shown is 352 square inches. What are the dimensions of the rectangle? CCSS A.REI.4a, SMP 2

$4x$ in.

$(x + 3)$ in.

3. CONSTRUCT ARGUMENTS Use completing the square to show that no two consecutive positive even integers can have a product of 27. CCSS A.REI.4a, SMP 3

4. REASON ABSTRACTLY Jeremy writes a trinomial that can be solved by completing the square and cannot be solved by factoring. One of the solutions of his equation is between 4 and 5. CCSS A.REI.4a

a. Write a possible equation. CCSS SMP 2

b. What are the solutions to the equation? CCSS SMP 2

c. EVALUATE REASONABLENESS Explain why your equation meets Jeremy's criteria. CCSS SMP 8

5. USE STRUCTURE The quadratic function $f(t) = -5000t^2 + 70000t + 5000$ is used to model the number of sales of a company's product in years t since the product was released. CCSS SMP 7

a. Complete the square to find the zeros of $f(t)$ and interpret each in the context of the problem. CCSS F.IF.8a

b. Use the answer from **part a** to find the maximum number of sales for the product. CCSS A.SSE.3b

9.8 Solving by Using the Quadratic Formula

Objectives

- Derive the Quadratic Formula.
- Use the Quadratic Formula to solve quadratic equations.
- Write and solve quadratic equations to model real-world problems.

STANDARDS

Content: A.REI.4b, A.REI.4a, A.CED.1

Practices: 1, 2, 3, 4, 6, 7, 8

Use with Lesson 9-5

EXAMPLE 1 Derive the Quadratic Formula CCSS A.CED.1

The general quadratic equation can be written in the form $ax^2 + bx + c = 0$, where a, b, and c are any numbers and $a \neq 0$. If you complete the square using the general quadratic equation, you come up with an equation called the quadratic formula. As you will see in this section, this formula is a tool you can use to solve quadratic equations.

EXPLORE You can use completing the square to derive a formula for solving any quadratic equation. Follow these steps to derive the Quadratic Formula.

a. DESCRIBE A METHOD Use what you know regarding how to complete the square of a specific equation to consider completing the square of the general quadratic equation. Consider $4x^2 - 2x - 5 = 0$ and $ax^2 + bx + c = 0$. Apply the first step of completing the square to both equations. CCSS SMP 8

____

b. DESCRIBE A METHOD Take the equations from **part a** and continue completing the square by dividing through by the constant in front of the quadratic term. CCSS SMP 8

$x^2 - \frac{\square}{\square}x = \frac{\square}{\square}$ and $x^2 + \frac{\square}{\square}x = -\frac{\square}{\square}$

c. DESCRIBE A METHOD Recall the next step in completing the square. Explain that step. Apply to both the specific and general cases. CCSS SMP 8

____

____

The general case is $x^2 + \frac{\square}{\square}x + \frac{\square^{\square}}{\square\square^{\square}} = -\frac{\square}{\square} + \frac{\square^{\square}}{\square\square^{\square}}$

d. DESCRIBE A METHOD You should have $x^2 - \frac{1}{2}x + \frac{1}{16} = \frac{5}{4} + \frac{1}{16}$ and $x^2 + \frac{b}{a}x + \frac{b^2}{(4a)^2} = -\frac{c}{a} + \frac{b^2}{(4a)^2}$. Factor the left side of each equation and simplify the right side. CCSS SMP 8

____

e. USE STRUCTURE The final two steps in solving a quadratic equation by completing the square are taking the square root of both sides of the equation and then isolating x. Apply those two steps to each equation. CCSS SMP 7

KEY CONCEPT The Quadratic Formula: $x = \dfrac{-b \pm \sqrt{b^2 - 4ac}}{2a}$

The equation derived from completing the square of the general form of a quadratic equation is called the quadratic formula. Where factoring allows you to find solutions to quadratic equations that are factorable and completing the square can be cumbersome, using the quadratic formula is an efficient method of finding the roots to any quadratic equation.

EXAMPLE 2 Use the Quadratic Formula CCSS A.REI.4b

Solve $2x^2 + 3x = 6$.

a. CONSTRUCT ARGUMENTS Explain why factoring is not a viable method for solving this quadratic equation. CCSS SMP 3

b. CALCULATE ACCURATELY Solve the equation in two ways. Round to the nearest tenth, if necessary. For this problem, which method do you prefer using? Explain. CCSS SMP 6

Completing the Square:	Quadratic Formula:

c. USE STRUCTURE Use the Quadratic Formula to verify that all equations of the form $k(2x^2 + 3x - 6) = 0$ for $k \neq 0$ have the same solutions. CCSS SMP 7

EXAMPLE 3 Solve a Quadratic Equation with No Real Solutions CCSS A.REI.4b

Solve $2x^2 - 5x + 20 = 0$.

a. **PLAN A SOLUTION** Which method would you use to solve the equation? Explain. CCSS SMP 1

b. **CALCULATE ACCURATELY** Solve the equation using the Quadratic Formula. Leave your answer in radical form. CCSS SMP 6

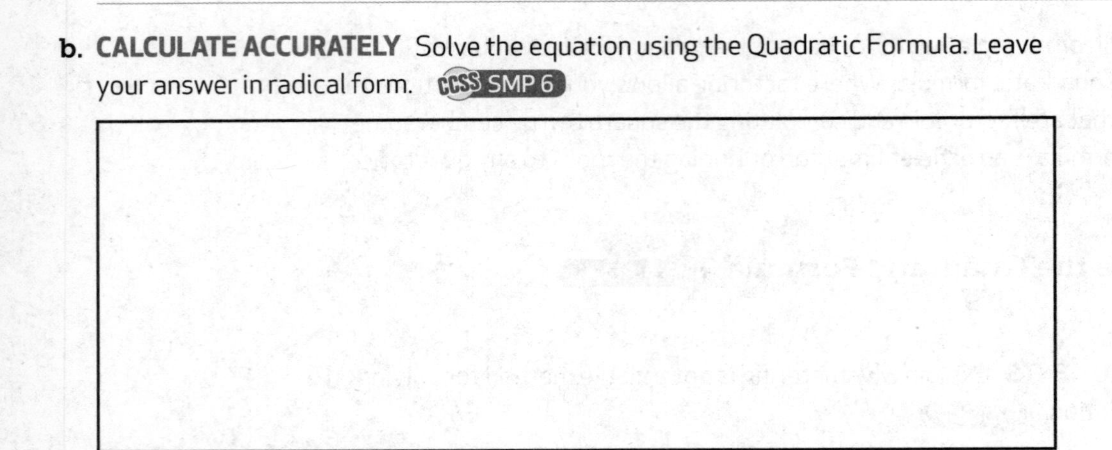

c. **REASON QUANTITATIVELY** How do you know that the equation has no real solutions?

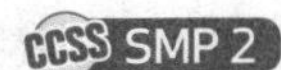

d. **MAKE A CONJECTURE** If all you needed to know regarding a quadratic equation is whether or not it had any real answers, which part of the quadratic formula would you need to consider? Explain. CCSS SMP 3

e. **REASON QUANTITATIVELY** The portion of the quadratic formula that tells you how many solutions there are to a quadratic equation is called the **discriminant**. Use the discriminant to determine how many solutions there are to the equation $3x^2 - 11x = 31$. What can you say about the graph of its related function? CCSS SMP 7

f. **MAKE A CONJECTURE** What would you know about a quadratic equation if the result you get when substituting values into the discriminant is 0? Explain. CCSS SMP 3

g. **INTERPRET PROBLEMS** Consider the quadratic equation $3x^2 - 4x + k = 0$. Give the values of k for which the equation has no real solutions, one real solution, and two real solutions. CCSS SMP 1

EXAMPLE 4 Create an Equation to Solve a Real-World Problem

A rectangular fishpond is surrounded by a walkway. The dimensions of the pond and walkway are shown. The area of the surface of the pond is 247 square feet. Find *w*, the width of the walkway.

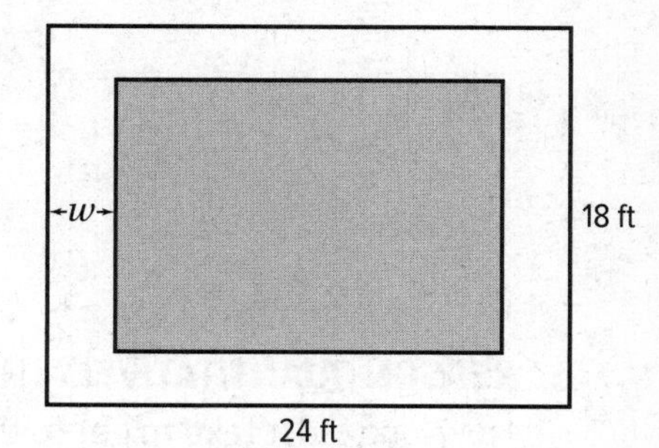

a. INTERPRET PROBLEMS What are the length and width of the fishpond in terms of *w*? CCSS SMP 1

b. REASON QUANTITATIVELY Write an expression that represents the area of the fishpond in terms of *w*. How can you use this expression to find *w*? CCSS SMP 2

c. CALCULATE ACCURATELY Solve the equation from **part b**. Use your solutions to describe the graph of the function. CCSS SMP 6

d. REASON QUANTITATIVELY Explain why one of the roots does not apply to the problem you just solved. CCSS SMP 3

PRACTICE

1. **REASON QUANTITATIVELY** For the equation $3x^2 + 2x + q = 0$, find all values of q so that there are two real solutions to the equation. Then find all values of q so that there are two complex solutions for the equation. Explain. CCSS A.REI.4b, SMP 2

2. New City's public works is putting together a fireworks presentation to celebrate the fourth of July. They will be firing rockets from the roofs of buildings at different heights. The path of one of the rockets is defined by the equation $h_1 = -16t^2 + 90t + 120$ with t in seconds and h_1 in feet. CCSS A.REI.4b

   a. INTERPRET PROBLEMS The equation provides the height the rocket is above the ground at any time between when it is launched and when it hits the ground. What do each of the coefficients mean? CCSS SMP 1

b. REASON QUANTITATIVELY Write the equation that models when the rocket will hit the ground. Solve the equation. CCSS SMP 2

c. REASON QUANTITATIVELY How high is the rocket when it is at its highest point? How long did it take to get to this point? Explain. CCSS SMP 2

d. USE STRUCTURE Without actually graphing the equation, use the information from **parts a, b** and **c** to describe what the graph would look like. Explain. CCSS SMP 7

e. USE STRUCTURE Write and solve an equation for the times at which the rocket reaches a height of 200 feet. CCSS SMP 7

3. REASON QUANTITATIVELY A cylinder is filled completely with water. Its base has a radius of x cm, and its height is 8 cm. After Ella removes $40x$ cm^3 of water, the volume of the remaining water is 250 cm^3. What is the radius of the cylinder? Round to the nearest whole number, if necessary. CCSS A.REI.4b, SMP 1

4. REASON ABSTRACTLY Solve the equation $x^2 + bx + c = 0$ for x in terms of b and c. Then use *sometimes, always,* or *never* to complete the sentences. CCSS A.REI.4b, SMP 2

If both b and c are negative, there will ____________ be at least one real solution.

If both b and c are positive, there will ____________ be complex nonreal solutions

5. **USE A MODEL** Braden wrote three quadratic equations, $2x^2 - 3x + 8 = 0$, $x^2 - 6x - 4 = 0$, and $-2x^2 + 4x + 3 = 0$. He graphs one of the equations as shown. He shows Ava the graph and the three equations and challenges her to match the correct equation with the graph. Does Ava need to solve the equations to do the matching? Explain why or why not. CCSS A.REI.4b, A.CED.1, SMP 4

6. **USE A MODEL** A rectangular area is to be enclosed with its length 20 feet less than its width as shown. CCSS A.REI.4b, A.CED.1, SMP 4

$w - 20$

w

 a. What should the width be in order for the enclosure to have an area of 100 ft^2?

 b. Find the width for which the area enclosed is R ft^2 for $R > 0$?

7. **USE STRUCTURE** A frame is made with width w inches and length $15 - w$ inches as shown in the figure. CCSS A.REI.4b, A.CED.1, SMP 7

$15 - w$

w

 a. Is there any width for which the frame will have an area of 150 in^2? Explain your reasoning.

 b. Find maximum area of the frame and the dimensions that produce the maximum area.

Performance Task

Building an Aquarium

Provide a clear solution to the problem. Be sure to show all of your work, include all relevant drawings, and justify your answers.

An aquarium in the shape of a rectangular prism is being designed. The following conditions must be met:

- The height of the interior must be 4 feet.
- The width of the interior must be 5 feet longer than the length of the interior.
- The aquarium must be 75% full, and the weight of the water cannot exceed 9300 pounds.

Part A

Density is the ratio of a substance's weight to its volume. The density of water is about 62 pounds per cubic foot.

Use this information to write a quadratic equation that can be used to find the interior dimensions of the aquarium assuming the maximum amount of water is used.

Part B

Use the equation from **Part A** to determine the length and width of the aquarium.

Part C

When the aquarium is built, it will weigh 500 pounds. Suppose the total weight of the aquarium and the water may not exceed 9600 pounds. Write and solve an equation to find the length and width of the aquarium so that it uses the maximum amount of water. Round your answers to the nearest hundredth.

Part D

The designers are building an aquarium in the shape of a cylinder that will meet the first and third conditions for the aquarium in **Part A**. Write and solve an equation to determine the area of the base for an aquarium that uses the maximum amount of water.

Performance Task

Stopping Distance

Provide a clear solution to the problem. Be sure to show all of your work, include all relevant drawings, and justify your answers.

The diagram below illustrates that the faster someone drives, the farther it takes to stop.

fast

longer stopping distance

slow

shorter stopping distance

This table shows approximate stopping distances for low speeds. As speed increases, the stopping distance increases.

Speed s (miles per hour)	5	10	15	20
Stopping Distance d (feet)	6.25	15.00	26.25	40.00

Part A

Based on data in the table, Terry conjectured that the formula $d = 1.75s - 2.5$ could be used to predict stopping distance d for a given speed s. Doug conjectured that the formula $d = 0.05s^2 + s$ would be a better fit. Which formula predicts stopping distance for a given speed? Explain your reasoning.

Part B

It takes a motorist 200 feet to stop. Using Doug's equation from **Part A**, write and solve a quadratic equation to find the speed the motorist was traveling. Round to the nearest whole number.

Part C

Terry has decided to use Doug's formula to estimate speed given stopping distance. She graphed $d = 0.05s^2 + s$, as shown below.

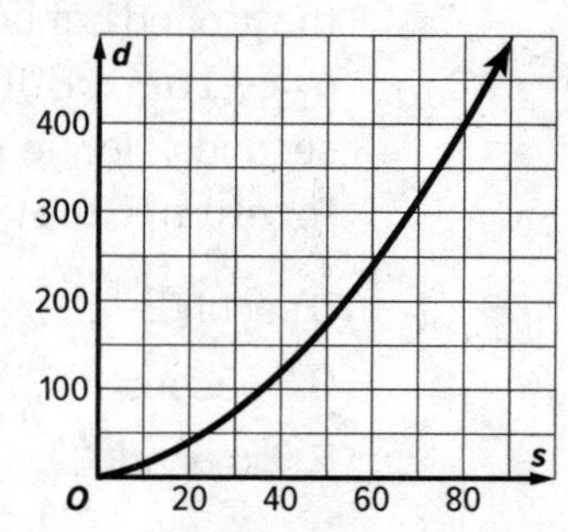

Estimate the speed at which a motorist is driving if it takes 250 feet to stop using Terry's graph. Then find the actual speed by evaluating $d = 0.05s^2 + s$. Is your estimated speed greater than, less than, or equal to the actual speed?

Part D

Terry claims that if a stopping of 100 feet is doubled, then the motorist's speed when the brakes are applied is also doubled. Doug was not convinced. Prove or disprove Terry's claim.

Standardized Test Practice

1. Solve by factoring. CCSS A.SSE.3a

$x^2 - 3x + 15 = -5(x - 10)$

2. The graph of the function $y = x^2 + x - 6$ is shown below.

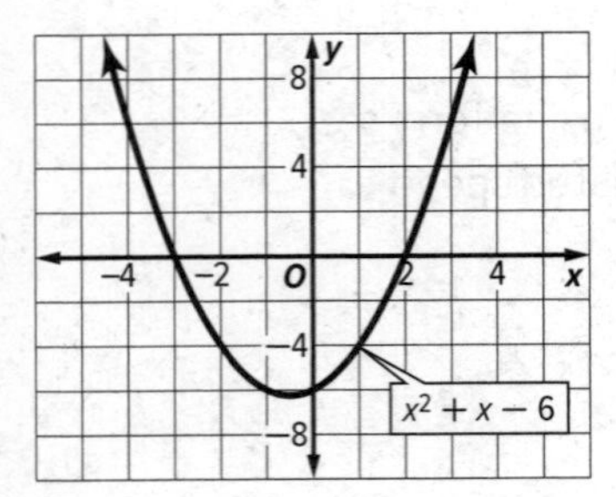

What are the solutions to the related equation $x^2 + x - 6 = 0$? CCSS A.REI.4b

3. Which of the following are solutions to the equation? Solve by factoring. CCSS A.SSE.3a

$6x^2 = 24x$

$x = -4$ | $x = 0$

$x = 2$ | $x = 4$

4. The legs of a right triangle are $2x + 1$ and $7x - 2$ inches, and the hypotenuse is $5x + 3$ inches. In standard form, what equation can be used to find the value of x for this triangle? CCSS A.SSE.3a

Solve by factoring. What is the value of x?

$x =$ ☐ in.

5. What is the sum of the solutions of the equation $(x + 3)^2 = 6x + 18$? CCSS A.REI.4b

−6 | 3

0 | 6

6. The height in feet of a golf ball hit from the ground is modeled by the equation $h = -16t^2 + 80t$, where t is the time in seconds. How long will it take for the golf ball to return to the ground? CCSS A.SSE.3a

0 seconds

0.5 second

2.2 seconds

5 seconds

7. The triangle shown in the diagram below has an area of 13 square inches. CCSS A.REI.4b

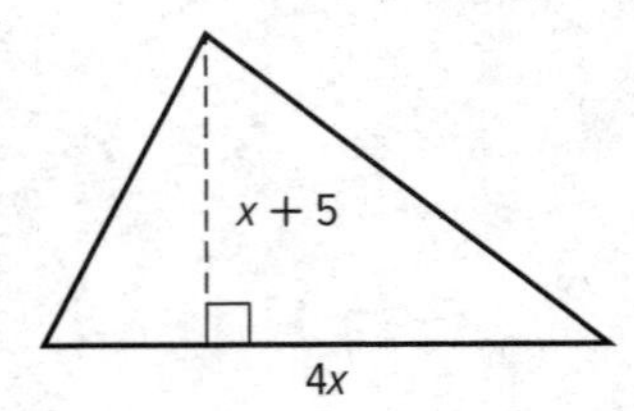

In standard form, what quadratic equation can be used to find the value of x?

Use the quadratic formula to solve. What are the base and height of the triangle, to the nearest tenth of an inch?

Base: ☐ in. Height: ☐ in.

8. Solve by factoring. CCSS A.SSE.3a

$3x^2 - 19x - 40 = 0$

9. Solve using the quadratic formula. CCSS A.REI.4b

$3x^2 + x - 8 = 0$

10. After completing the square for the equation $x^2 + 8x - 9 = 0$, what is the value of the right side of the equation? CCSS A.REI.4a

−9 | 16

9 | 25

11. Consider each quadratic equation shown in the table below. For each equation, find the value of the discriminant, and tell how many real solutions the equation has. CCSS A.REI.4b

Equation	Value of Discriminant	Number of Real Solutions
$3x^2 - 11x + 5 = 0$		
$2x^2 - x + 5 = 2x - 7$		
$4x^2 + 28x + 49 = 0$		

12. The area of the shaded portion of the figure below is 99 square feet. What is *x*? Describe your solution process. CCSS A.SSE.3a

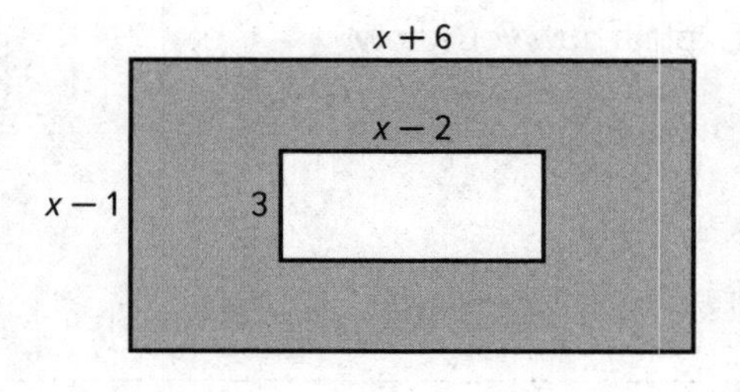

13. The area of a trapezoid is given by the formula $A = \frac{1}{2}(b_1 + b_2)h$, where b_1 and b_2 are the lengths of the bases and *h* is the height. A trapezoid has a height equal to its shorter base and a longer base that is 24 cm longer than the shorter base. The area of the trapezoid is 45 square centimeters. Describe how to find the height of the trapezoid by completing the square. What is the height of the trapezoid? CCSS A.REI.4a

14. The side length of a larger square is five times the side length of a smaller square. If the combined area of the two squares is 650 square inches, what is the side length of the larger square? Show your work. CCSS A.SSE.3a

15. Write a quadratic equation that has no real solutions. In simplified form your equation must have a linear term and a constant term. Explain how you know that the equation has no real solutions. CCSS A.REI.4b

10 Quadratic Functions

CHAPTER FOCUS Learn about some of the Common Core State Standards that you will explore in this chapter. Answer the preview questions. As you complete each lesson, return to these pages to check your work.

What You Will Learn	Preview Question
Lesson 10.1: Graphing Quadratic Functions	
CCSS F.IF.4 For a function that models a relationship between two quantities, interpret key features of graphs and tables in terms of the quantities, and sketch graphs showing key features given a verbal description of the relationship. **Also addresses:** F.IF.7a, F.IF.2, F.IF.5, F.IF.9, N.Q.1, A.CED.2	CCSS SMP 7 The graph of the function $y = 2x^2 + 4x + 3$ has a vertex at $(-1, 1)$. Is the vertex a maximum or a minimum? Explain how you know.
Lesson 10.2: Transforming Quadratic Functions	
CCSS F.BF.3 Identify the effect on the graph of replacing $f(x)$ by $f(x) + k$, $k\,f(x)$, $f(kx)$, and $f(x + k)$ for specific values of k (both positive and negative); find the value of k given the graphs. Experiment with cases and illustrate an explanation of the effects on the graph using technology.	CCSS SMP 7 Describe the transformation performed on the graph of $f(x)$ to obtain the graph of $g(x)$. $f(x) = x^2 - 16$ $g(x) = x^2 + 6x + 4$
Lesson 10.3: Modeling: Quadratic Functions	
CCSS N.Q.2 Define appropriate quantities for the purpose of descriptive modeling. CCSS N.Q.3 Choose a level of accuracy appropriate to limitations on measurement when reporting quantities. **Also addresses:** A.SS2.1a, A.SSE.1b, F.BF.1a, F.BF.1b, F.IF.4, F.IF.7a, A.CED.1, A.CED.2	CCSS SMP 2 Jason is standing on top of a building that is 100 feet high. He throws a ball upwards. Explain why the function $f(t) = t(-16t + 45)$ cannot model the height of the ball after t seconds.
Lesson 10.4: Solving Linear-Quadratic Systems	
CCSS A.REI.7 Solve a simple system consisting of a linear equation and a quadratic equation in two variables algebraically and graphically. **Also addresses:** A.CED.2	CCSS SMP 6 What is the solution to the system? $y = -x + 7$ $y = x^2 - 7x + 12$

What You Will Learn	Preview Question
Lesson 10.5: Function Intersections	
CCSS A.REI.11 Explain why the *x*-coordinates of the points where the graphs of the equations $y = f(x)$ and $y = g(x)$ intersect are the solutions of the equation $f(x) = g(x)$; find the solutions approximately, e.g., using technology to graph the functions, make tables of values, or find successive approximations. Include cases where $f(x)$ and/or $g(x)$ are linear, polynomial, rational, absolute value, exponential, and logarithmic functions.	CCSS SMP 7 Use the graph to solve the related equation $x^2 - 4 = -2x^2 - 1$. 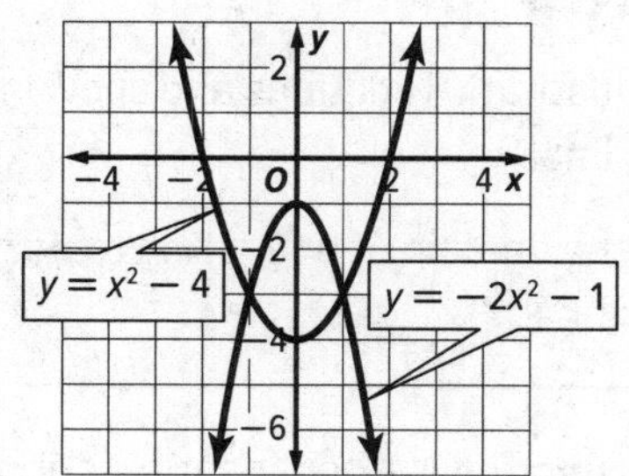
Lesson 10.6: Combining Functions	
CCSS F.BF.1b Combine standard function types using arithmetic operations. **Also addresses:** F.BF.1a	CCSS SMP 1 Let $f(x) = 3x^2 + 5x - 2$ and $g(x) = 2x - 6$. Find $f(x) + g(x)$. What type of function is the sum?
Lesson 10.7: Analyzing Functions with Successive Differences	
CCSS F.LE.2 Construct linear and exponential functions, including arithmetic and geometric sequences, given a graph, a description of a relationship, or two input-output pairs (include reading these from a table). **Also addresses:** F.IF.6, F.LE.1b, F.LE.1c, F.LE.3	CCSS SMP 7 Write an equation for a function that models the data in the table. Explain how you determined what type of function models the data. *x*: 0, 1, 2, 3, 4 *y*: 0, −1.5, −6, −13.5, −24
Lesson 10.8: Special Functions	
CCSS F.IF.4 For a function that models a relationship between two quantities, interpret key features of graphs and tables in terms of the quantities, and sketch graphs showing key features given a verbal description of the relationship. CCSS F.IF.7b Graph square root cube root, and piecewise-defined functions, including step functions and absolute value functions. **Also addresses:** F.BF.3, F.IF.4a, F.IF.7a	CCSS SMP 4 For orders of less than 10 cartons of eggs, a farmer charges \$2.25 per carton. Orders of 10 to 100 cartons cost \$2.00 per carton. For orders over 100, the cost is \$1.50 per carton. Sketch the graph of a piecewise function that models this problem. 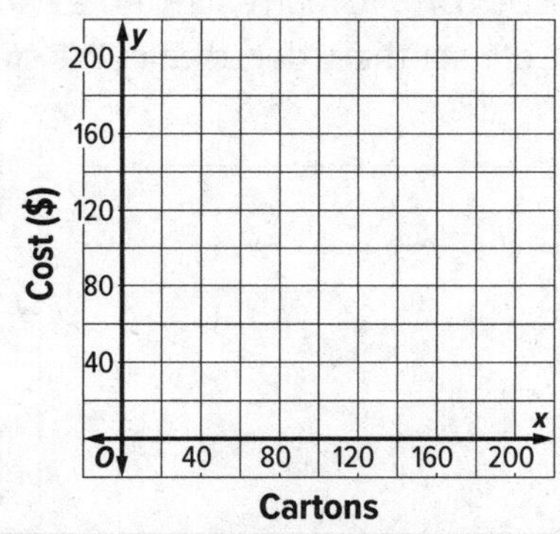

10.1 Graphing Quadratic Functions

Objectives

- Graph quadratic functions and show intercepts, maxima, and minima.
- Relate the domain of a function to its graph and the relationship it describes.

Content: F.IF.2, F.IF.4, F.IF.5, F.IF.7a, F.IF.9, N.Q.1, A.CED.2
Practices: 2, 3, 4, 6, 7
Use with Lesson 9–1

A **quadratic function** is a nonlinear function that can be written in the standard form $f(x) = ax^2 + bx + c$, where $a \neq 0$. The graph of a quadratic function is called a **parabola** and has a maximum or minimum point called the **vertex**. The graph may have x-intercepts where the graph crosses the x-axis. It has a y-intercept where the graph crosses the y-axis.

EXAMPLE 1 Explore Quadratic Functions CCSS F.IF.7a

EXPLORE Complete the tables for each of the following quadratic functions.

a. $y = x^2 - 4x - 4$

−2	
−1	
0	
1	
2	
3	
4	
5	

b. $y = -x^2 - 6x - 3$

−6	
−5	
−4	
−3	
−2	
−1	
0	
1	

c. REASON ANALYTICALLY What do you notice about the two tables? What happened to the values of y as you entered the given values of x? Why do you think this might have happened? CCSS SMP 2

d. USE STRUCTURE Graph each of the two sets of ordered on separate coordinate planes. Connect the ordered pairs in a smooth curve. Describe the result. CCSS SMP 7

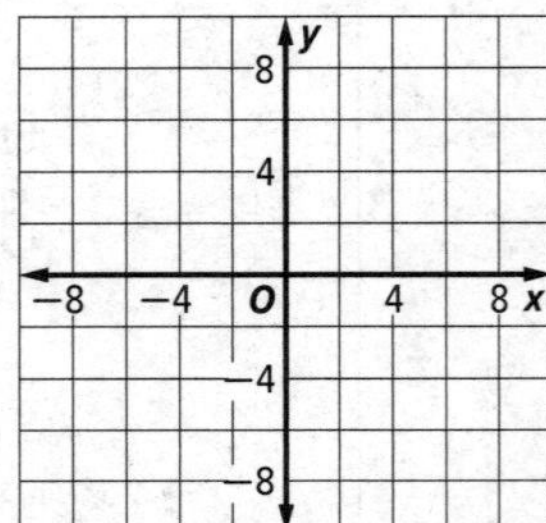

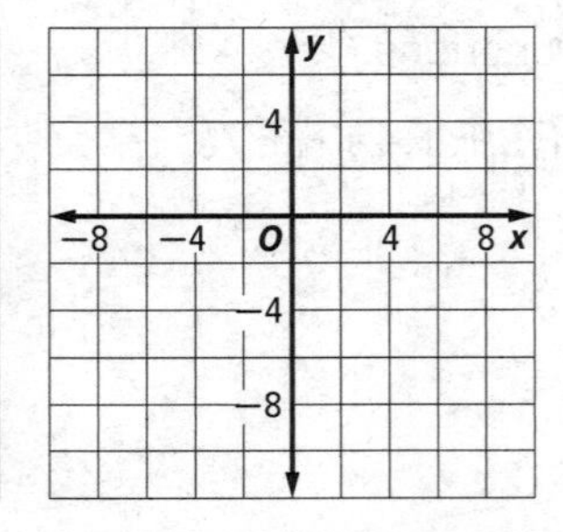

e. CALCULATE ACCURATELY Use the graph of the first function to determine the coordinates of the vertex, the y-intercept, and the approximate coordinates of the x-intercepts. CCSS SMP 6

____

f. REASON QUANTITATIVELY Use the table of the second function to determine the coordinates of the vertex, the y-intercept, and the approximate coordinates of the x-intercepts. Explain how you can find these in the table. CCSS SMP 2

____

____

____

EXAMPLE 2 Graph Quadratic Functions CCSS F.IF.7a, SMP 7

Graph each quadratic function. Plot a point at every intercept and at the maximum or minimum. List the intercepts and the maximum value or minimum value of the function.

a. $y = x^2 - 2x - 3$

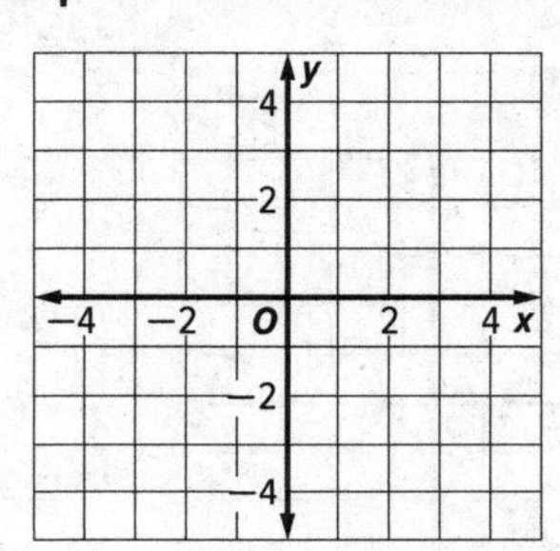

Intercepts: ____________________

vertex: ____________________

b. $y = -x^2 + 2x - 1$

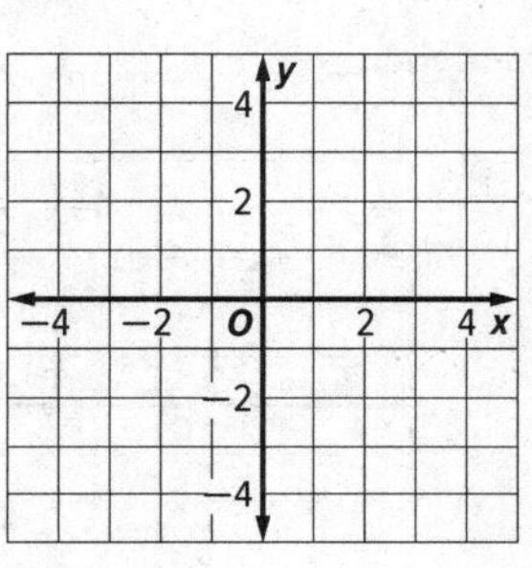

Intercepts: ____________________

vertex: ____________________

c. $y = -x^2 - 4x - 3$

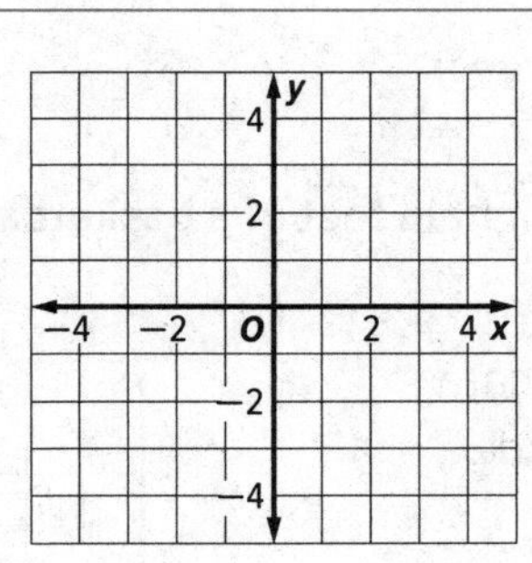

Intercepts: ____________________

vertex: ____________________

d. $y = x^2 + 2$

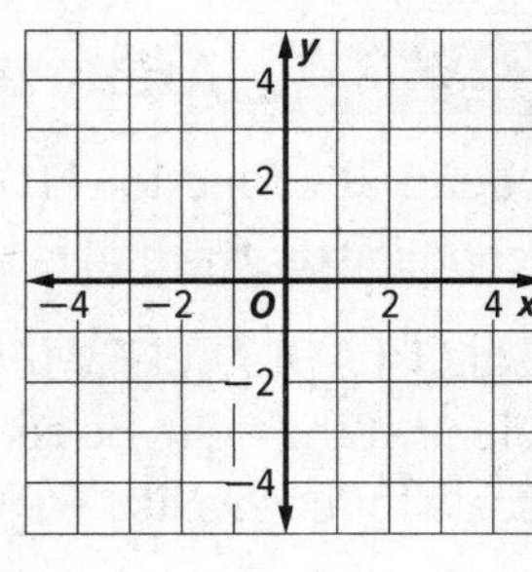

Intercepts: ____________________

vertex: ____________________

e. $y = -x^2 + x + 2$

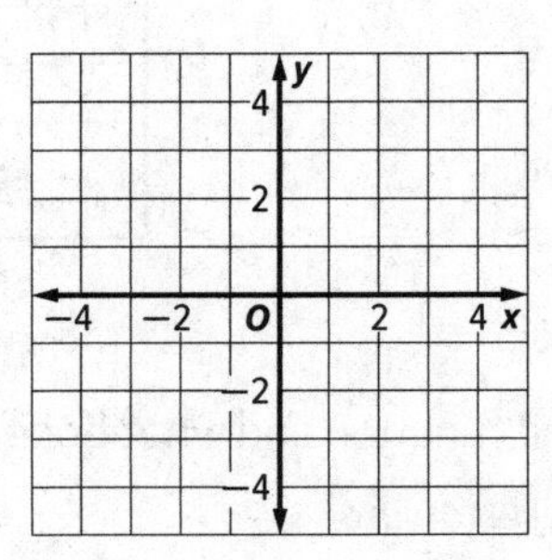

Intercepts: ____________________

vertex: ____________________

f. $y = x^2 + 3x + 2$

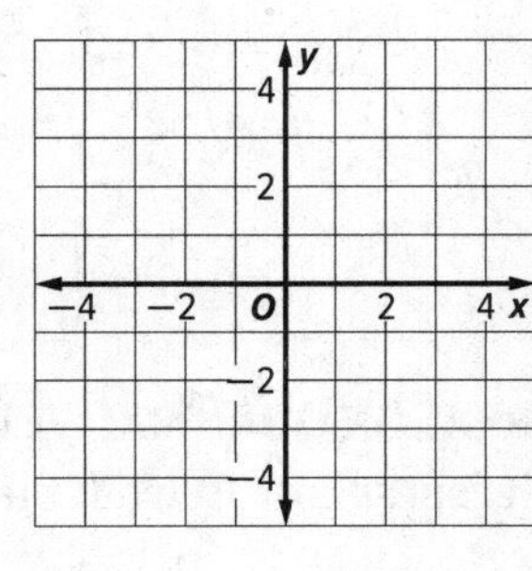

Intercepts: ____________________

vertex: ____________________

g. **USE STRUCTURE** Is there a connection between the structure of a quadratic equation and whether the graph opens upward or downward? What about whether the function has a maximum value or a minimum value? Explain. CCSS SMP 7

____

____

____

KEY CONCEPT Graphs of Quadratic Functions

Complete the following. Give an example of each type of quadratic function and sketch the graph.

$a > 0$	$a < 0$
When $a > 0$, the graph of $y = ax^2 + bx + c$ opens ____________. The lowest point of the graph is the ____________.	When $a < 0$, the graph of $y = ax^2 + bx + c$ opens ____________. The highest point of the graph is the ____________.
Example: ____________	Example: ____________
Graph:	Graph:

EXAMPLE 3 Analyze a Quadratic Model

The quadratic function $h(t) = -16t^2 + 25t + 5$ models the height in feet of a basketball t seconds after the ball is thrown.

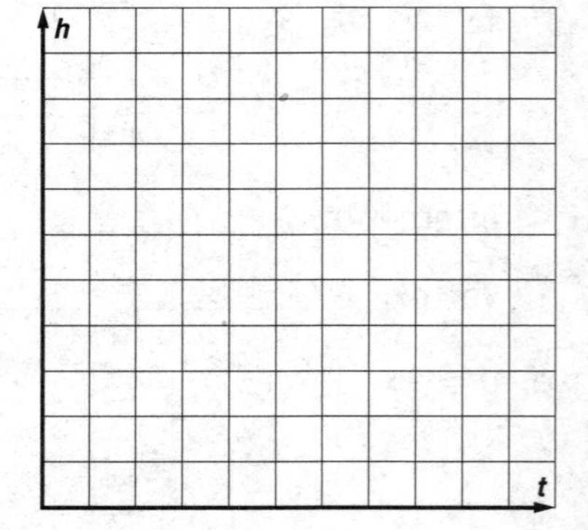

a. **CALCULATE ACCURATELY** Graph the function on the coordinate plane at the right. Be sure to label the x- and y-axes and provide a scale for each axis. CCSS N.Q.1, SMP 6

b. **REASON QUANTITATIVELY** Find the intercepts of the graph. Describe what they represent. CCSS F.IF.4, SMP 2

____

____

____

____

c. **REASON QUANTIATIVELY** What is the maximum value of the function and what does it represent? At what time is the maximum reached? CCSS F.IF.4, SMP 2

____

____

d. USE A MODEL What are the values of $h(0)$ and $h(1)$? What do these represent? CCSS F.IF.2, SMP 4

e. USE A MODEL State a reasonable domain for this function. What does the domain represent? CCSS F.IF.5, SMP 4

f. CRITIQUE REASONING A student claims that the graph of the parabola you drew in **part a** represents the physical path of the basketball as it moves through the air. Do you agree? Explain. CCSS F.IF.4, SMP 3

PRACTICE

USE STRUCTURE Make a table for each quadratic function. Use the table to create the graph. Show each intercept and maximum or minimum. Then list the intercepts and give the maximum value or minimum value of the function. CCSS F.IF.7a, SMP 7

1. $y = x^2 + x + 3$

−3	
−2	
−1	
0	
1	
2	

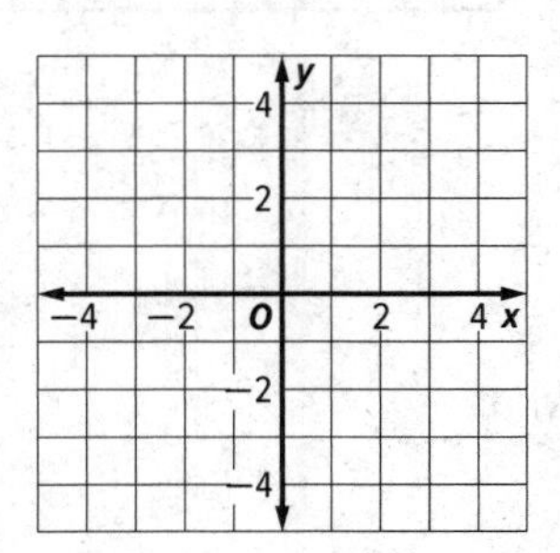

Intercepts: ______

Max or min: ______

2. $y = -x^2 - 3x - 2$

−4	
−3	
−2	
−1	
0	
1	

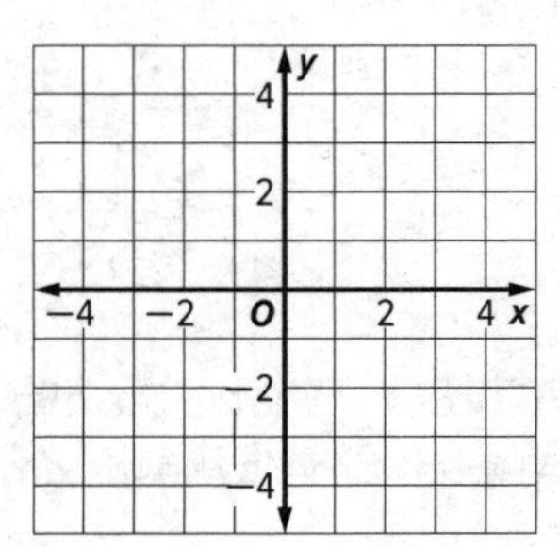

Intercepts: ______

Max or min: ______

USE STRUCTURE **Graph each quadratic function. Use the intercepts and maximum or minimum. Then list the intercepts and give the maximum value or minimum value of the function.** CCSS F.IF.7a, SMP 7

3. $y = -2x^2 + 2$

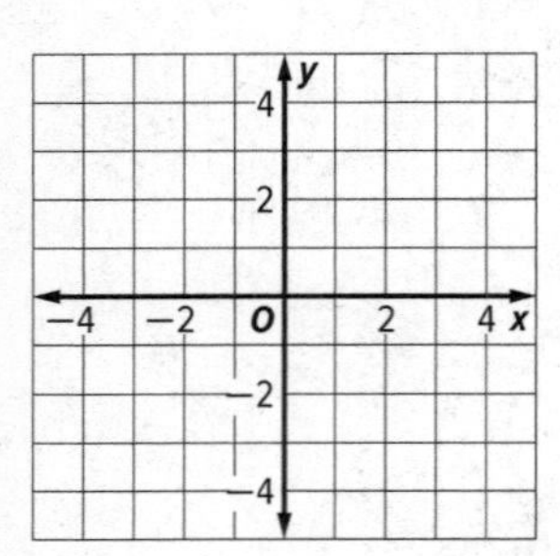

Intercepts: ______________________

vertex: ______________________

4. $y = \frac{1}{2}x^2 + 2x$

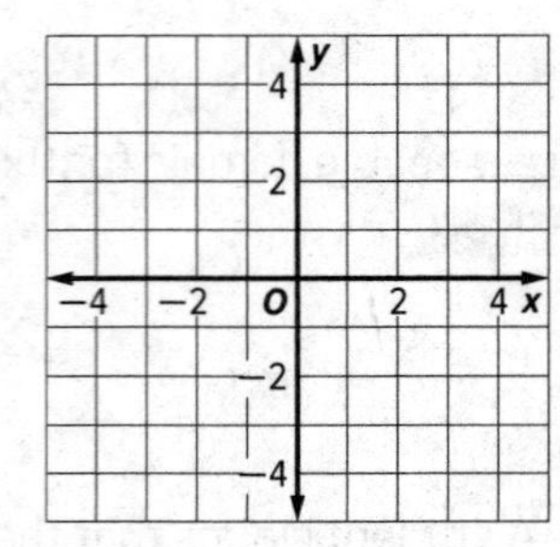

Intercepts: ______________________

vertex: ______________________

5. **CRITIQUE REASONING** Marissa said she graphed a quadratic function for which the maximum or minimum value is the same as both the x-intercept and the y-intercept. Is this possible? If so, give an example of such a function and draw its graph on the coordinate plane at the right. If it is not possible, explain why not. CCSS F.IF.7a, SMP 3

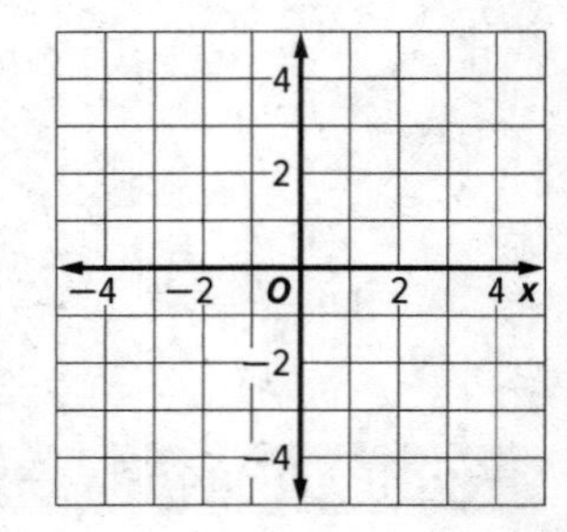

____

____

6. DeMarcus is sitting in a lifeguard's chair at the beach. He tosses a beanbag into the air and lets it lands on the beach below him. The function $h(t) = -16t^2 + 8t + 12$ models the beanbag's height in feet, t seconds after DeMarcus tosses it into the air.

a. **CALCULATE ACCURATELY** Graph the function on the coordinate plane at the right. Be sure to label the x- and y-axes and provide a scale for each axis. CCSS N.Q.1, SMP 6

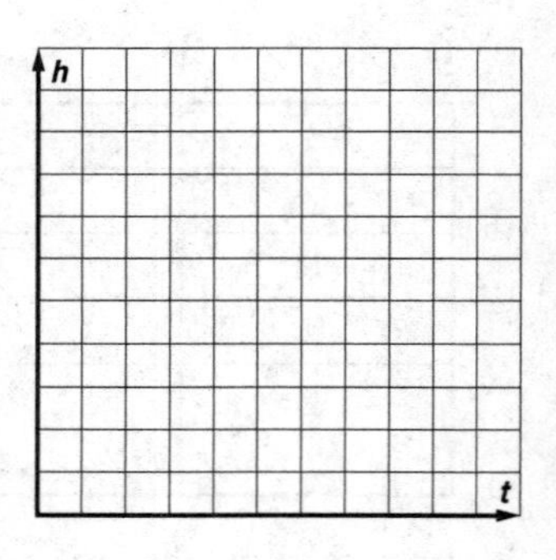

b. **REASON QUANTITATIVELY** Find the intercepts of the graph. Describe what they represent. CCSS F.IF.4, SMP 2

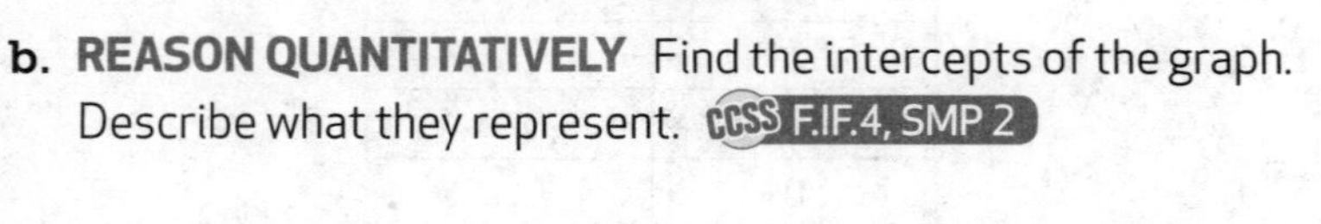

____

____

____

____

____

c. **REASON QUANTITATIVELY** What is the maximum value of the function and what does it represent? At what time is the maximum reached? CCSS F.IF.4, SMP 2

____

____

____

d. USE A MODEL What is the value of $h(0.5)$ What does this tell you? 

e. USE A MODEL State a reasonable domain for this function. What does the domain tell you? CCSS F.IF.5, SMP 4

REASON ABSTRACTLY Determine which quadratic function has the greater maximum or smaller minimum. CCSS F.IF.9, SMP 2

7. Function 1: $y = x^2 + 2x + 1$
Function 2: Shown in graph below.

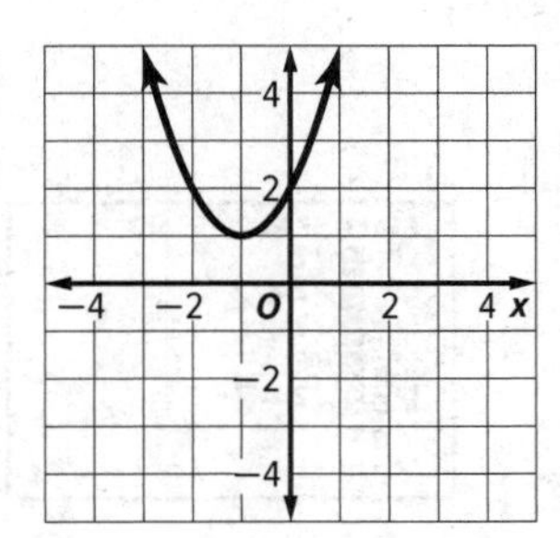

8. Function 1: $y = -x^2 + 3x - 2$
Function 2: Shown in graph below.

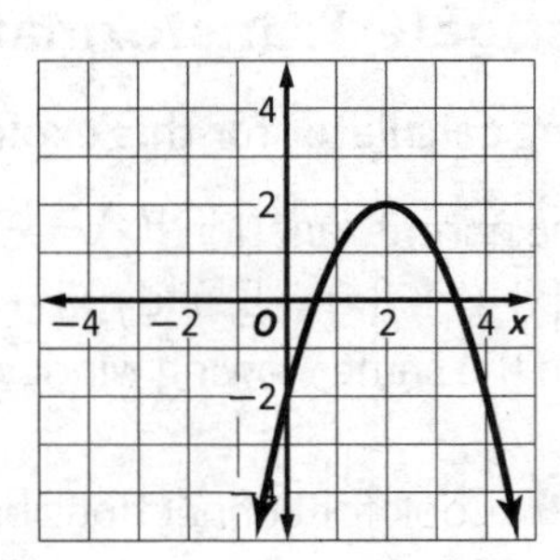

9. Function 1: $y = -x^2 + 5x - 2$
Function 2: Shown in graph below.

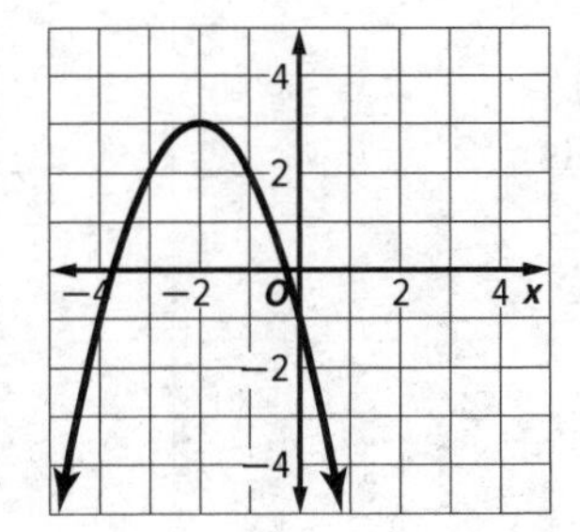

10. Function 1: $y = x^2 - 6x + 8$
Function 2: Shown in graph below.

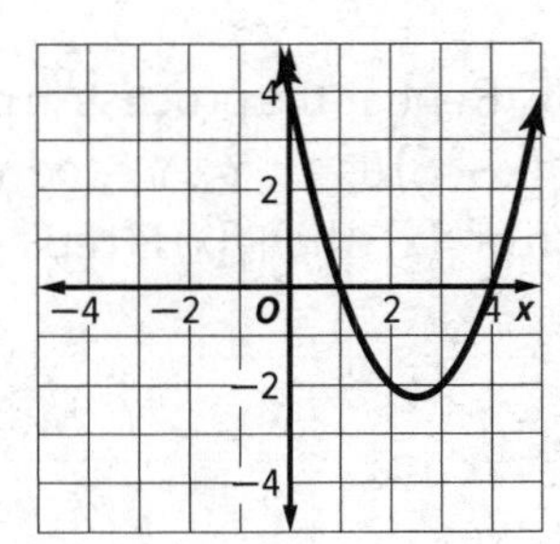

11. COMMUNICATE PRECISELY Write the equation for a quadratic function that has a y-intercept of 0 and a minimum value at $x = 2$. Explain the steps you used to write the equation and graph the function on the coordinate plane at the right. CCSS F.IF.7a, A.CED.2, SMP 6

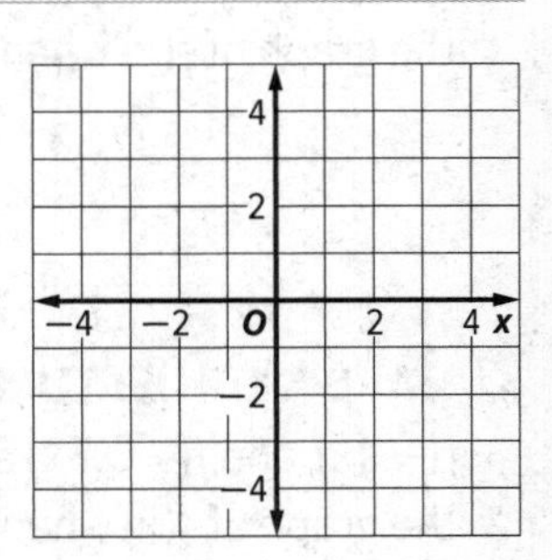

10.2 Transforming Quadratic Functions

Content: F.BF.3
Practices: 1, 2, 3, 4, 5, 6, 7, 8
Use with Lesson 9-3

Objectives

- Identify the effect on the graph of $f(x)$ under the transformations $f(x) + k$, $kf(x)$, $f(kx)$, and $f(x + k)$.
- Find the value of k given the graph of a quadratic function.

A parent function is the most basic function in a family of functions. The parent quadratic function is the function $f(x) = x^2$. Other functions in the family are transformations of the parent function.

EXAMPLE 1 Investigate Transformations of $f(x) = x^2$ CCSS F.BF.3

EXPLORE Use a graphing calculator for this exploration.

a. USE TOOLS Enter the parent function $f(x) = x^2$ in your graphing calculator as Y_1. Then enter $f(x) + 2$, $f(x) + 4$, $f(x) - 5$, and $f(x) - 7$ as Y_2, Y_3, Y_4, and Y_5, as shown. Graph the functions in the same viewing window. CCSS SMP 5

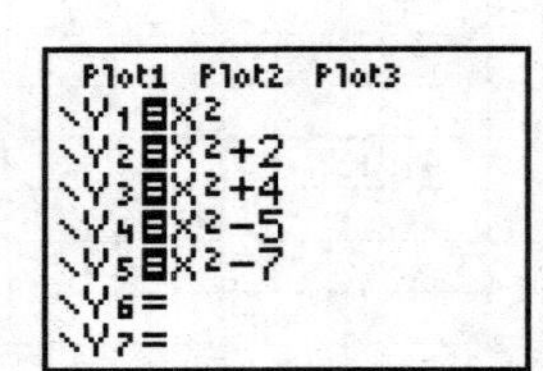

b. MAKE A CONJECTURE Look for connections between the function rules and the graphs. Then make a conjecture. What is the effect on the graph of $f(x)$ when $f(x)$ is replaced by $f(x) + k$? CCSS SMP 3

c. MAKE A CONJECTURE Repeat the process in **part a,** but this time enter $f(x + 3)$, $f(x + 6)$, $f(x - 4)$, and $f(x - 8)$ as Y_2, Y_3, Y_4, and Y_5. Then make a conjecture. What is the effect on the graph of $f(x)$ when $f(x)$ is replaced by $f(x + k)$? CCSS SMP 3

d. MAKE A CONJECTURE Repeat the process in **part a,** but this time enter $2f(x)$, $0.5f(x)$, $-3f(x)$, and $-0.2f(x)$ as Y_2, Y_3, Y_4, and Y_5. Then make a conjecture. What is the effect on the graph of $f(x)$ when $f(x)$ is replaced by $kf(x)$? CCSS SMP 3

e. MAKE A CONJECTURE Repeat the process in **part a,** but this time enter $f(3x)$, $f(0.2x)$, $f(-4x)$, and $f(-0.5x)$ as Y_2, Y_3, Y_4, and Y_5. Then make a conjecture. What is the effect on the graph of $f(x)$ when $f(x)$ is replaced by $f(kx)$? CCSS SMP 3

KEY CONCEPT Transformations of Quadratic Functions

Complete the following by describing the effect on the graph of $f(x) = x^2$ when $f(x)$ is replaced by each function. Then give an example of each transformation by drawing the graph for the given function.

$f(x) + k$	$f(x + k)$
$k > 0$: ______	$k > 0$: ______
$k < 0$: ______	$k < 0$: ______
Example: $y = x^2$; $y = x^2 + 2$	Example: $y = x^2$; $y = (x - 3)^2$

$kf(x)$	$f(kx)$
$0 < \lvert k \rvert < 1$: ______	$0 < \lvert k \rvert < 1$: ______
$\lvert k \rvert > 1$: ______	$\lvert k \rvert > 1$: ______
$k < 0$: ______	
Example: $y = x^2$; $y = \frac{1}{10}x^2$	Example: $y = x^2$; $y = (0.2x)^2$

EXAMPLE 2 Use Transformations to Graph a Quadratic Function CCSS F.BF.3

Follow these steps to graph $g(x) = 2(x + 3)^2 - 1$.

a. PLAN A SOLUTION Explain how you can graph $g(x)$ by performing a series of transformations on the graph of the parent function. How did you perform the compression or stretch? CCSS SMP 1

______

______

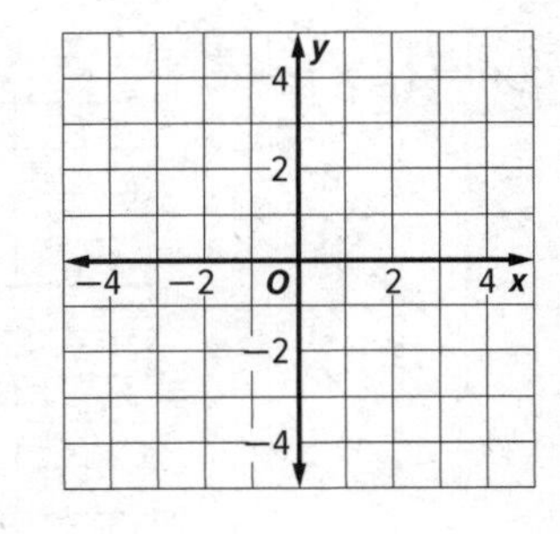

b. USE STRUCTURE Use the transformations you identified to graph $g(x)$ on the coordinate plane at the right. CCSS SMP 7

EXAMPLE 3 Find Values of k Given a Graph CCSS F.BF.3

An artist is designing an arch for a sculpture garden. The arch is based on a parabola, as shown in the figure. The artist wants to write an equation that her crew can use to construct the arch.

a. USE STRUCTURE How do you determine the compression or stretch factor? 

b. USE STRUCTURE Explain how to use transformations to write an equation for the arch. CCSS SMP 7

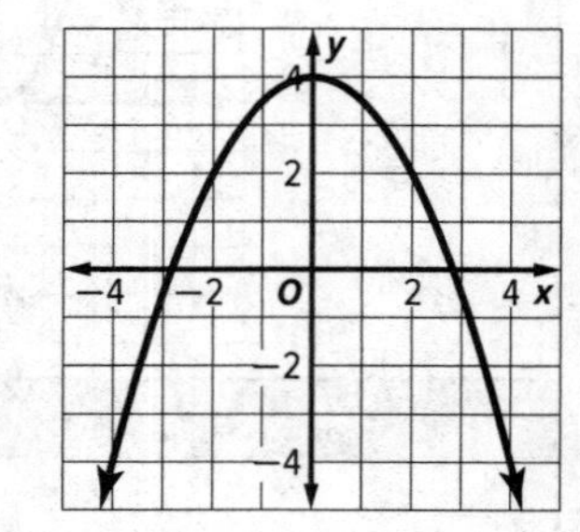

c. EVALUATE REASONABLENESS Explain how you can check that the equation you wrote in **part b** is correct. CCSS SMP 8

PRACTICE

USE STRUCTURE Use transformations to graph each quadratic function. CCSS F.BF.3, SMP 7

1. $g(x) = 3(x - 1)^2$

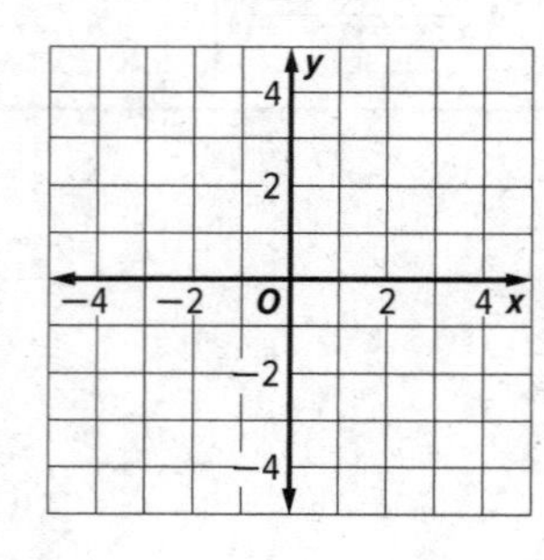

2. $h(x) = -2(x + 2)^2 + 2$

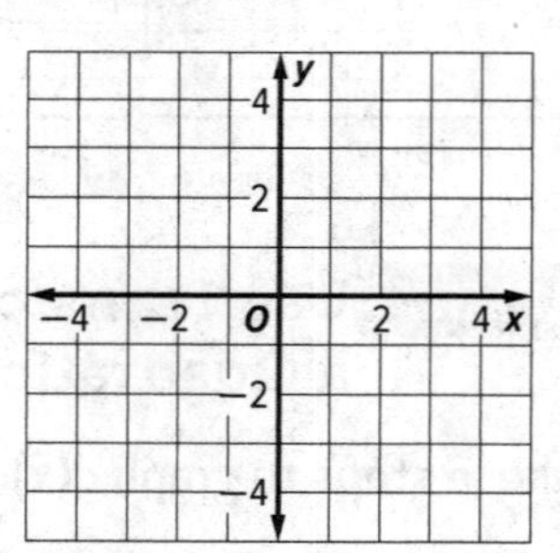

3. $g(x) = \left(\frac{1}{2}x\right)^2 - 3$

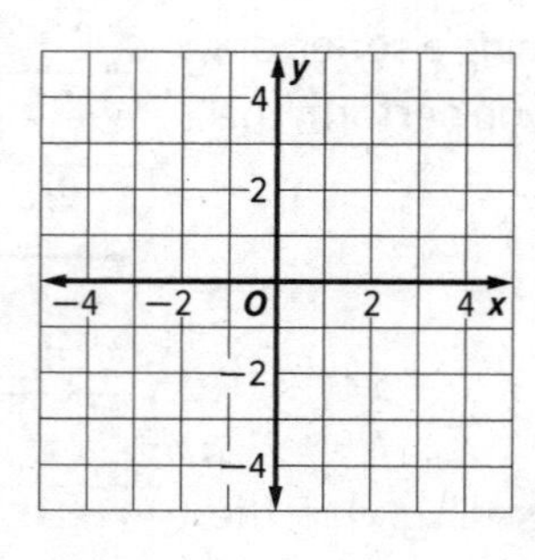

4. $h(x) = -\frac{1}{4}(x - 1)^2 + 4$

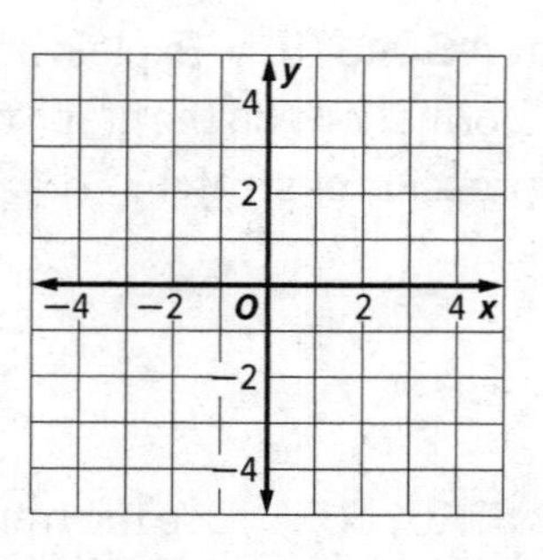

USE STRUCTURE Describe the transformations done to the parent functions to create the graphs in Exercises 1–4. CCSS F.BF.3, SMP 7

5. $g(x) = 3(x - 1)^2$ ____________________

6. $h(x) = -2(x + 2)^2 + 2$ ____________________

7. $g(x) = \left(\frac{1}{2}x\right)^2 - 3$ ____________________

8. $h(x) = -\frac{1}{4}(x - 1)^2 + 4$ ____________________

9. **COMMUNICATE PRECISELY** Write a set of instructions that a classmate could use to transform the graph of $f(x) = x^2$ and end up with the graph of $g(x) = -10(x - 17)^2$. Explain how to do the vertical stretch. CCSS F.BF.3, SMP 6

____

USE STRUCTURE Use transformations to write an equation for each graph. 

10.

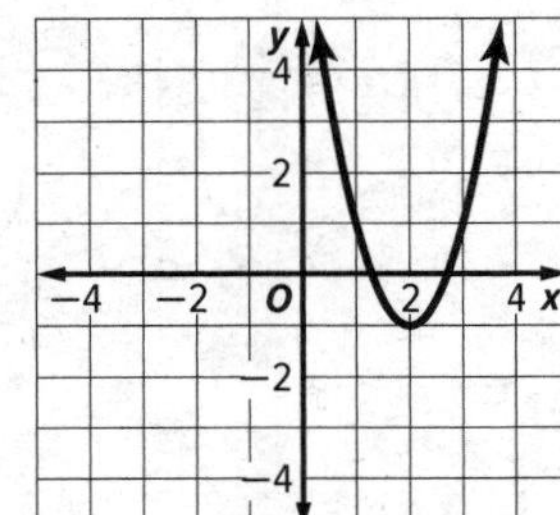

11.

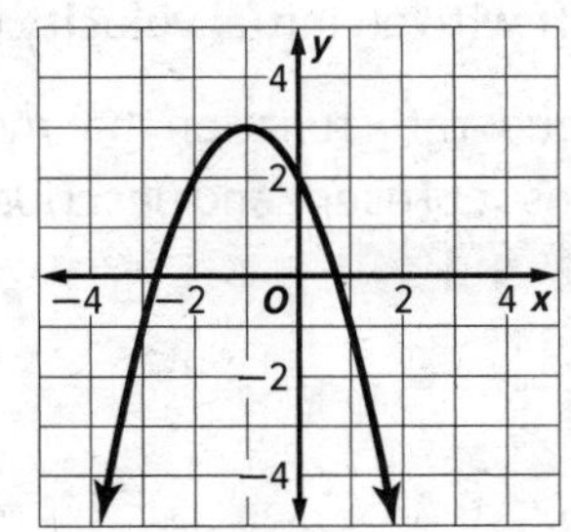

____

12. An animator is using a coordinate plane to design a scene in a movie. In the scene, a comet enters the screen in Quadrant II, moves around the screen in a parabolic path, and leaves the screen in Quadrant I.

    a. **USE A MODEL** The figure shows the path of the comet. What equation represents the path? CCSS F.BF.3, SMP 4

    ____

    b. **REASON ABSTRACTLY** The animator decides to translate the path of the comet 8 units up and 7 units left. What equation represents the new path? CCSS F.BF.3, SMP 2

    ____

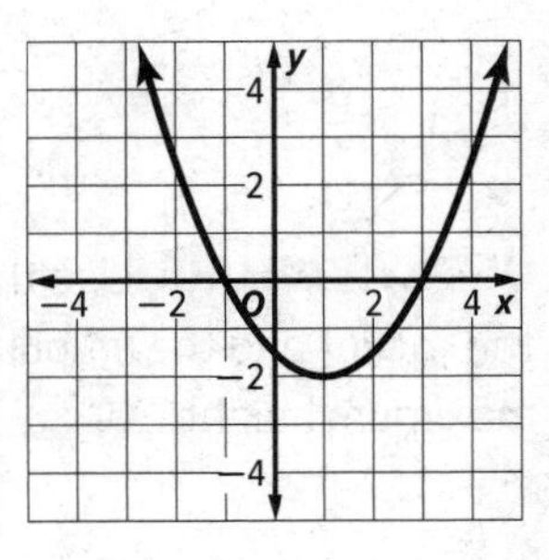

13. **REASON ABSTRACTLY** A function f is an *even function* if $f(-x) = f(x)$ for all x in the domain of the function. A function f is an *odd function* if $f(-x) = -f(x)$ for all x in the domain of the function. Is the parent quadratic function an even function or an odd function? Justify your answers. CCSS F.BF.3, SMP 2

____

10.3 Modeling: Quadratic Functions

Objectives

- Use quadratic functions to model real-world phenomena.
- Write and graph quadratic functions to solve problems.

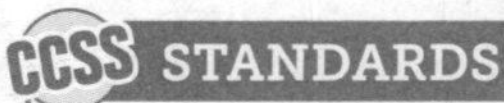

Content: N.Q.2, N.Q.3, A.SSE.1a, A.SSE.1b, F.BF.1a, F.BF.1b, F.IF.4, F.IF.7a, A.CED.1, A.CED.2
Practices: 2, 4, 5, 6, 7, 8
Use with Lesson 9-6

A *projectile* is any object that is thrown or launched into the air. You can use a quadratic function to model the motion of a projectile. The function $h(t) = \frac{1}{2}gt^2 + v_0t + h_0$ represents the height *h* of the projectile after *t* seconds. The coefficients and constant in the equation are as follows.

g acceleration due to gravity (-32 ft/s^2 or -9.8 m/s^2)
v_0 initial vertical velocity
h_0 initial height

EXAMPLE 1 Model Projectile Motion

EXPLORE Miguel launches a model rocket from a platform that is 5 feet above the ground. The rocket takes off with an initial velocity of 50 feet per second.

a. USE A MODEL Write a quadratic function that models the motion of the rocket. Explain what the variables represent and describe the units associated with each variable. CCSS N.Q.2, F.BF.1a, SMP 4

b. USE STRUCTURE Graph the function on the coordinate plane at the right. CCSS A.CED.2, SMP 7

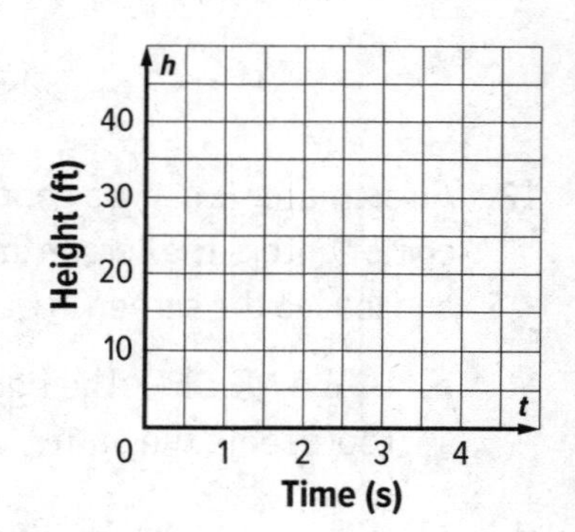

c. REASON QUANTITATIVELY Write and solve an equation to determine the length of time the rocket is in the air. Explain your steps. CCSS A.CED.1, SMP 2

d. USE STRUCTURE Use the graph to estimate the maximum height reached by the rocket and the amount of time it will take the rocket to reach the maximum height. CCSS SMP 7

e. REASON QUANTITATIVELY Compare your answers to **parts c** and **d**. When considered together, do the answers make sense? Explain. CCSS SMP 2

EXAMPLE 2 Develop a Revenue Model

The owner of a community theater keeps track of the price of tickets and the number of tickets sold. She finds that when the price is $12, she can sell 400 tickets. For each $1 increase in the price, she sells 20 fewer tickets. The owner would like to determine the ticket price that maximizes her revenue from ticket sales.

a. USE A MODEL Let x represent the number of one-dollar increases in the price of a ticket. Write a function for the new price P of a ticket in terms of x, and a function for the number of tickets N that will be sold at the new price. CCSS F.BF.1a, SMP 4

b. USE A MODEL Write a quadratic function that models the total revenue R from ticket sales as a function of x. Write the function as a product of two factors and tell what each factor represents. CCSS F.BF.1b, SMP 4

c. USE STRUCTURE Graph the function on the coordinate plane at the right. CCSS A.CED.2, SMP 7

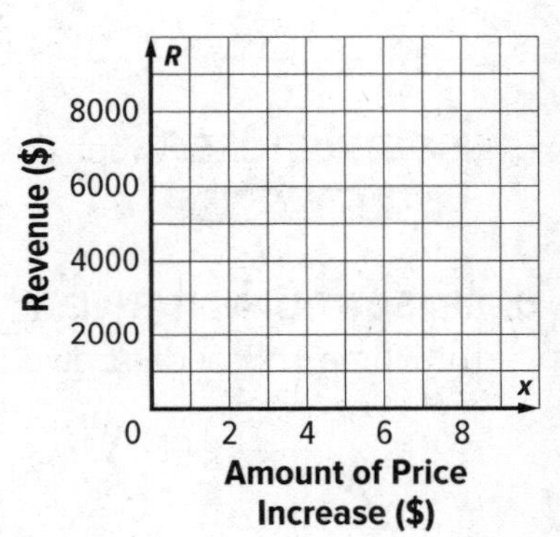

d. REASON QUANTITATIVELY Based on the graph, should the owner of the theater increase ticket prices? How do you know? CCSS F.IF.7a, SMP 2

e. USE TOOLS Determine the value of x that corresponds to the maximum value of the revenue function. Explain the steps you used. CCSS A.SSE.1b, SMP 5

f. COMMUNICATE PRECISELY Explain to the theater owner the new price she should charge for tickets, the number of tickets she can expect to sell, and the total revenue she can expect at this price. CCSS A.SSE.1b, SMP 6

g. REASON QUANTITATIVELY What is the value of $R(-12)$? What does it mean in the context of this problem? CCSS SMP 2

h. REASON QUANTITATIVELY What is the value of $R(20)$? What does it mean in the context of this problem? CCSS SMP 2

EXAMPLE 3 Develop a Model from a Table

At a diving competition, special equipment collects data on the height of each diver above the surface of the pool at different times during his or her dive. The table shows the data that was collected during one of Brendan's dives.

Time Since Start of Dive (s)	Height of Diver Above Pool (ft)
0.75	45
1.00	40
1.25	33

a. **USE TOOLS** Enter the data from table in two lists in your calculator. Then use the calculator's quadratic regression tool to write a function that models Brendan's dive. Tell what each variable represents. CCSS N.Q.2, A.CED.2, SMP 5

b. **USE A MODEL** What do the coefficients and constant in your function tell you about Brendan's dive? Explain. CCSS A.SSE.1a, SMP 4

c. **USE STRUCTURE** Graph the function on the coordinate plane at the right. CCSS A.CED.2, SMP 7

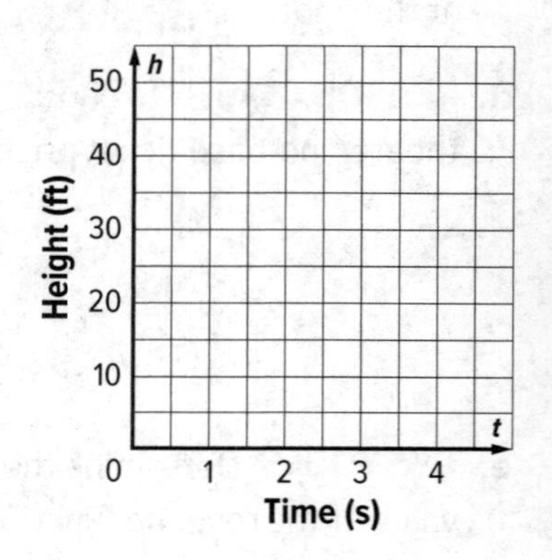

d. **REASON QUANTITATIVELY** Write and solve an equation to determine the total time of Brendan's dive. Explain your steps. CCSS A.CED.1, SMP 2

e. **REASON QUANTITATIVELY** What is the maximum height that Brendan reached during his dive? At what time did this occur? Explain how you found your answer. CCSS A.CED.2, SMP 2

PRACTICE

1. Alana kicks a soccer ball with an initial velocity of 10 m/s when the ball is 0.5 meter above the ground.

   a. **USE A MODEL** Write a quadratic function that models the motion of the soccer ball. Explain what the variables represent and describe the units associated with each variable. CCSS N.Q.2, F.BF.1a, SMP 4

b. USE STRUCTURE Graph the function on the coordinate plane at the right. CCSS A.CED.2, SMP 7

c. REASON QUANTITATIVELY Based on the graph, when is the ball moving most rapidly? When is it not moving at all? CCSS A.CED.2, SMP 2

Height (m)

h

8

6

4

2

0

t

1 2 3 4

Time (s)

d. REASON QUANTITATIVELY Write and solve an equation to determine the length of time the ball is in the air. Explain your steps. CCSS N.Q.3, A.CED.1, SMP 2

e. EVALUATE REASONABLENESS Explain how you know your answer to **part d** is reasonable. CCSS A.CED.1, SMP 8

f. REASON QUANTITATIVELY At approximately what times is the soccer ball at a height of 3 meters? Why does it make sense that there are two answers? CCSS A.CED.1, SMP 2

g. USE A MODEL A machine that the soccer team uses for practice can propel the ball upward with the same initial velocity of Alana's kick, but it can be done so from ground level. What equation would model the height of the soccer ball when the machine mimics Alana's kick? CCSS SMP 4

h. REASON QUANTITATIVELY Write and solve an equation to determine the length of time the ball is in the air. CCSS SMP 2

i. REASON ABSTRACTLY Does your answer to **part h** seem reasonable when compared to your answer to **parts d** and **e**?

2. Casey owns a goat farm and she is planning to build a new rectangular pen for 8 of her goats. She will use an existing straight wall for one side of the pen. She will use 90 feet of fencing to building the other three sides of the pen, as shown in the figure.

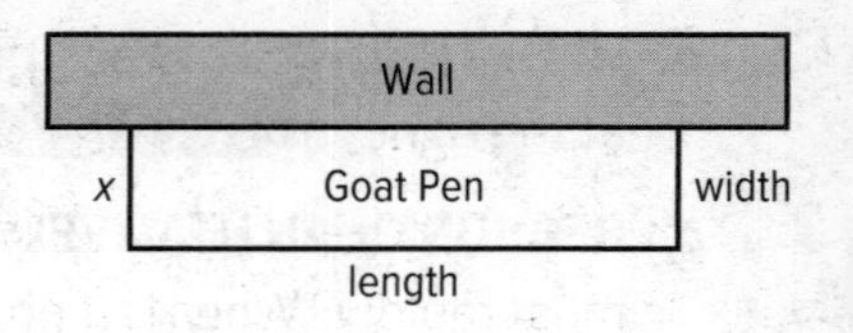

a. **USE A MODEL** Let x represent the width of the goat pen. Write a function for the length L of the pen in terms of x. CCSS F.BF.1a, SMP 4

b. **USE A MODEL** Write a quadratic function that models the area A of the goat pen as a function of x. Write the function as a product of two factors and tell what each factor represents. CCSS F.BF.1a, A.SSE.1b, SMP 4

c. **USE STRUCTURE** Graph the function on the coordinate plane at the right. CCSS A.CED.2, SMP 7

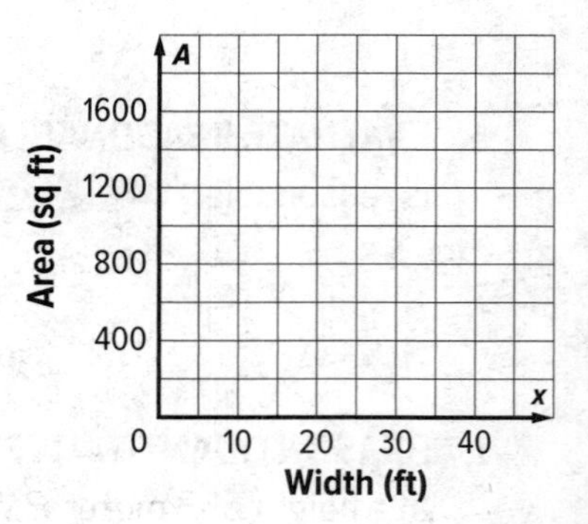

d. **REASON QUANTITATIVELY** Suppose Casey wants to make the goat pen with the greatest possible area. What length and width should she use? What will be the area of the goat pen in this case? CCSS F.IF.4, SMP 2

e. **REASON QUANTITATIVELY** Casey decides that the goat pen should have an area of exactly 1000 ft^2. Show how to write and solve an equation to find the width of the goat pen in this case. Explain how you know your answer is reasonable CCSS A.CED.1, SMP 2

3. Yoshiro hits a golf ball. The table shows data about the height of the golf ball at various times.

Time Since Golf Ball Is Hit (s)	Height of Golf Ball (ft)
0.5	28
1.0	48
1.5	60

a. **USE TOOLS** Use your calculator to find a quadratic function that models the motion of the golf ball. Tell what each variable represents. CCSS N.Q.2, A.CED.2, SMP 5

b. **USE A MODEL** Does the quadratic function you wrote in **part a** have a constant term? What does this tell you, and why does this make sense in the context of the problem? CCSS A.SSE.1a, SMP 4

c. **USE A MODEL** What do the coefficients of the quadratic function tell you? CCSS A.SSE.1a, SMP 4

d. **USE STRUCTURE** Graph the function on the coordinate plane at the right. Be sure to mark the scale on the axes and label the axes appropriately. CCSS A.CED.2, SMP 7

e. **REASON QUANTITATIVELY** Write and solve an equation to find the times when the golf ball was at a height of exactly 55 feet. Explain your steps. CCSS A.CED.1, SMP 2

h

t

0

f. **REASON QUANTITATIVELY** What is the maximum height that the golf ball reached? At what time did this occur? Explain how you found your answer. CCSS A.CED.2, SMP 2

4. Mr. Marks has enough fencing material to enclose a garden with a perimeter of 100 feet. He wants his garden to be rectangular, and he wants to enclose the greatest area possible. CCSS A.CED.2

a. **USE A MODEL** If x represents the width of the rectangular garden, what is an expression for the length, in terms of x? Explain. CCSS F.BF.1a, SMP 4

b. **USE A MODEL** Write a quadratic function that models the area A of the garden as a function of x. Write the function as a product of two factors and tell what each factor represents. CCSS F.BF.1a, A.SSE.1b, SMP 4

c. **USE STRUCTURE** Graph the function on the coordinate plane at the right. Be sure to mark the scale on the axes and label the axes appropriately. CCSS A.CED.2, SMP 7

d. **REASON QUANTITATIVELY** What are the dimensions of the garden with the maximum area? What is the maximum area that can be enclosed? CCSS A.CED.2, SMP 2

y

x

10.4 Solving Linear-Quadratic Systems

Objectives

- Solve a system consisting of a linear equation and a quadratic equation graphically.
- Solve a system consisting of a linear equation and a quadratic equation algebraically.

Content: A.REI.7
Practices: 1, 2, 4, 5, 6, 7, 8
Use with Extend 9-3

Recall that the solution for a system of two linear equations in two variables can be found either geometrically by graphing and noting where the graphs intersect or algebraically by using substitution or elimination. The following problem will have us examine ways to find the solution of a system including a linear equation and a quadratic equation.

EXAMPLE 1 Solve a System Graphically CCSS A.REI.7

EXPLORE An architect is using a coordinate plane to design a new city park that will have two paths. The equations of the paths are $y = 2x - 2$ and $y = x^2 + 2x - 3$. The architect plans to construct a fountain at each point of intersection of the paths. The architect wants to know how many fountains will be needed and the map coordinates at which they will be placed.

a. PLAN A SOLUTION What can you conclude about each of the equations? What can you conclude about the shapes of the paths of each of their graphs? Explain. CCSS SMP 7

b. USE STRUCTURE Graph each of the equations on the coordinate plane at the right. CCSS SMP 5

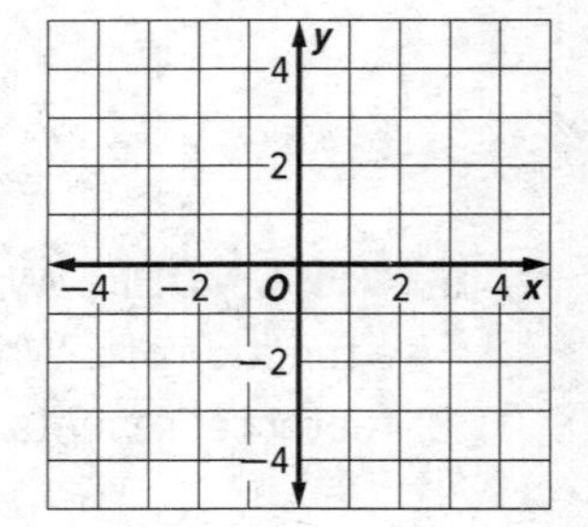

c. USE A MODEL Based on your graph, how many fountains are needed, and where should they be located? Explain. CCSS SMP 2

d. EVALUATE REASONABLENESS Show how you can check that the locations you identified in **part c** are correct. CCSS SMP 3

e. DESCRIBE A METHOD Describe a general method for solving a linear-quadratic system graphically. CCSS SMP 1

f. COMMUNICATE PRECISELY What must be true about the equations in a linear-quadratic system in order for you to able to come up with an exact answer to the system graphically? CCSS SMP 6

g. REASON ABSTRACTLY The architect is working on other projects in which one path is given by a linear function and the other is given by a quadratic function. She wants to know how many points of intersection the paths might have. Describe all of the possibilities and make sketches in the space provided to illustrate your answer. CCSS SMP 2

EXAMPLE 2 Use a Calculator to Solve a System Graphically CCSS A.REI.7

In this example you will use the intersection function on your calculator to find the solution for a system of equations. Follow these steps to solve the system shown below.

$y = x^2 - 5x + 1$

$y = -\frac{1}{2}x + 3$

a. USE TOOLS Enter the two equations in your calculator as Y_1 and Y_2, as shown. CCSS SMP 5

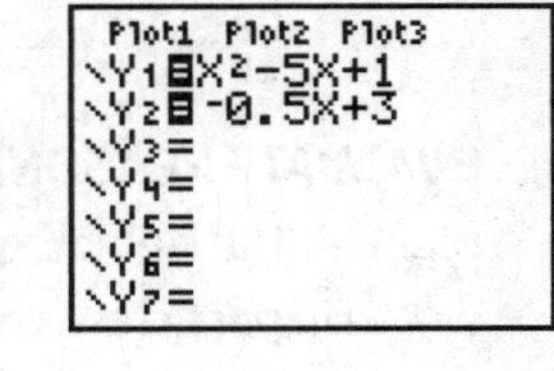

b. USE TOOLS Graph the system. Choose a viewing window that shows all points of intersection. CCSS SMP 5

c. USE TOOLS Use the calculator's Intersect tool to find the points of intersection. Round the coordinates to the nearest tenth. CCSS SMP 5

d. REASON ABSTRACTLY Explain how you can replace the linear equation with one that will create a system that has exactly one solution. Provide the new system and the new solution and explain how you know there is only one solution. CCSS SMP 2

Just as with linear systems of equations, other systems of equations can be solved algebraically using substitution or elimination. The substitution method is most often used for solving linear-quadratic systems.

EXAMPLE 3 Solve a System Algebraically CCSS A.REI.7

Follow these steps to solve the system shown at the right.

$$y = -x^2 - 2x + 1$$
$$y = 2x + 3$$

a. PLAN A SOLUTION Explain how you can use substitution to write a single quadratic equation that involves only the variable x. CCSS SMP 1

b. CALCULATE ACCURATELY Describe the steps you can use to solve the quadratic equation you wrote in **part a**. Leave the solutions in radical form. CCSS SMP 6

c. CALCULATE ACCURATELY Explain how you can find the corresponding value of y for each value of x you identified in **part b**. CCSS SMP 6

d. USE TOOLS Write the solutions to the system as ordered pairs in radical form. Then use your calculator to round the coordinates to the nearest tenth. CCSS A.CED.2, SMP 5

e. EVALUATE REASONABLENESS Graph the system of equations on the coordinate plane at the right. Explain how to use the graph to check that the solutions you wrote in **part d** are reasonable. CCSS SMP 8

Not all quadratic equations have graphs that are parabolas. Not all quadratic equations are functions. For example, a quadratic equation in which both x and y are squared may be the equation of a circle. As shown in the figure, a circle that is centered at the origin and that has radius r has the equation $x^2 + y^2 = r^2$. For example, the equation $x^2 + y^2 = 64$ is the equation of a circle centered at the origin with radius 8, since $8^2 = 64$. Systems of equations using these types of quadratic equations can still be solved by graphing or algebraically by substitution or elimination.

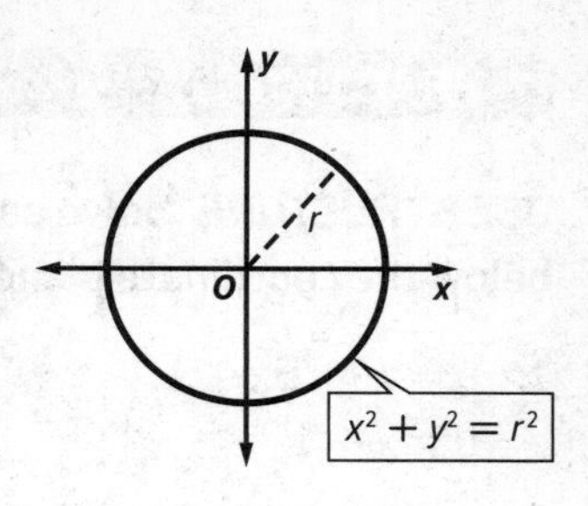

EXAMPLE 4 Solve a System Involving a Line and a Circle CCSS A.REI.7

Follow these steps to solve the system shown at the right.

$$x^2 + y^2 = 9$$
$$y = 2x$$

a. USE STRUCTURE Graph each of the equations on the coordinate plane at the right. CCSS SMP 7

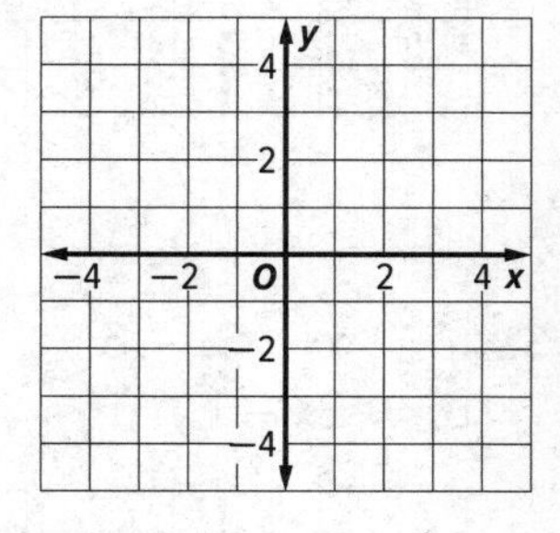

b. COMMUNICATE PRECISELY Explain how to use the graphs of the equations to estimate the solutions of the system. CCSS SMP 6

c. CALCULATE ACCURATELY Solve the system algebraically and show the steps you used to solve the equation. CCSS SMP 6

d. CALCULATE ACCURATELY Explain how you can find the corresponding value of y for each value of x you identified in **part c**. CCSS SMP 6

e. EVALUATE REASONABLENESS Write the solution to the system as ordered pairs in radical form. Then explain how you know your solution is reasonable. CCSS SMP 8

f. REASON ABSTRACTLY Examine your graph for **part a**. What do you think are the possible numbers of solutions to a system involving a line and a circle? Consider the number of solutions for the system if the equation of the circle is the same as the one given and the line is parallel to the line given. CCSS SMP 2

PRACTICE

USE STRUCTURE **Solve each system graphically. Write the solution on the line provided below the coordinate plane.** CCSS A.REI.7, SMP 7

1. $y = x^2 + 3x + 1$
$y = x + 1$

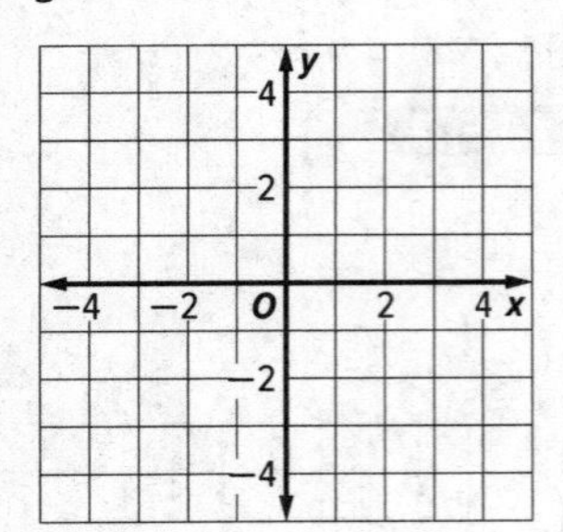

2. $y = -x^2 + 4x - 4$
$y = 2x - 3$

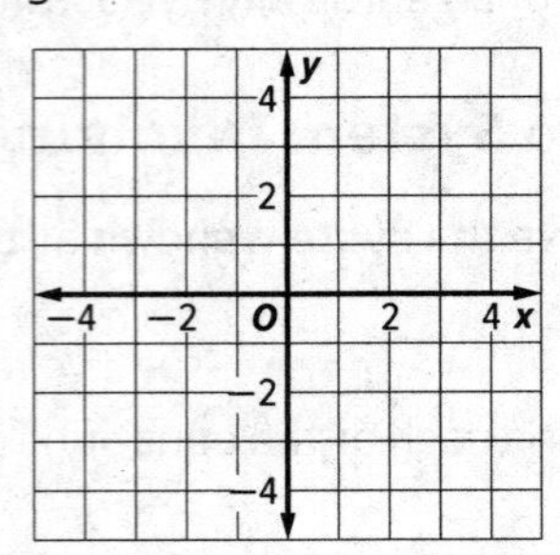

3. $y = -x^2 - 5x - 6$
$y = -3x - 1$

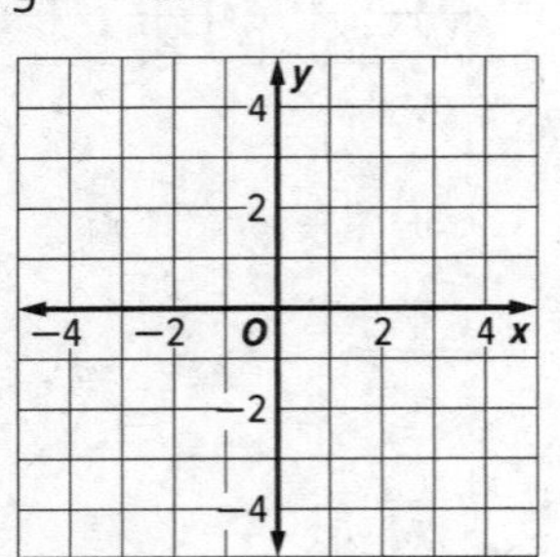

4. $y = 2x^2 - 4$
$y = 2x$

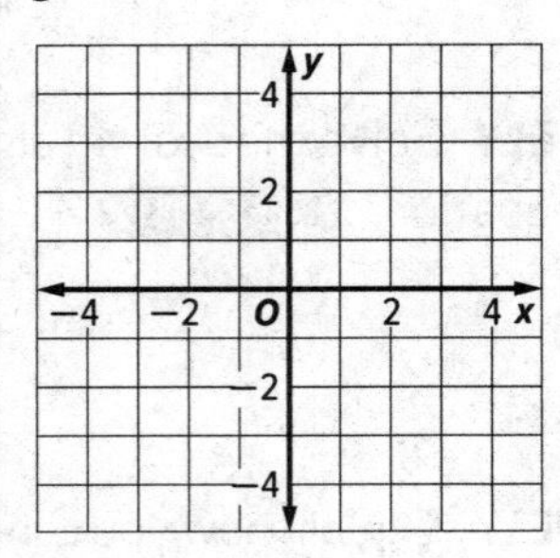

5. $x^2 + y^2 = 16$
$x + y = 4$

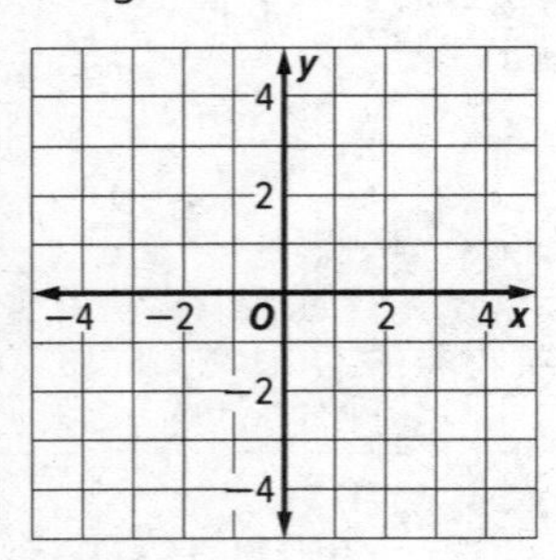

6. $x^2 + y^2 = 4$
$y = x + 3$

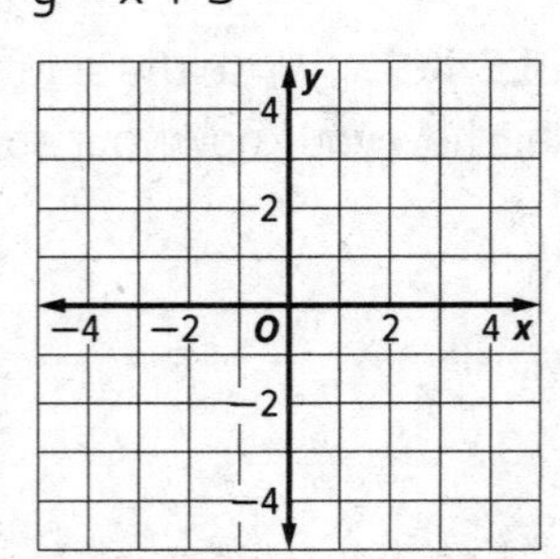

7. REASON ABSTRACTLY For what value of k does the system $y = x^2 + x$ and $y = -2x + k$ have exactly one solution? Explain your reasoning. CCSS A.REI.7, SMP 2

CALCULATE ACCURATELY **Solve each system algebraically. Write the solutions as ordered pairs in radical form when necessary.** CCSS A.REI.7, SMP 6

8. $y = x^2 - 3x + 1$
$y = x + 1$

9. $y = x^2 - 4x + 6$
$y = 2x - 3$

10. $y = -x^2 + 2x + 3$
$y = x + 2$

11. $y = x^2 + 4x - 1$
$y = 3x$

12. $x^2 + y^2 = 10$
$y = -3x$

13. $x^2 + y^2 = 15$
$-2x + y = 0$

14. In a video game, players shoot virtual rubber bands at a small target that moves around the screen. The rubber bands are all launched from the point $(-4, 0)$ and follow a path along the line $y = \frac{1}{2}x + 2$. The target moves around the screen following a parabolic path represented by the equation $y = -x^2 - 2x + 4$.

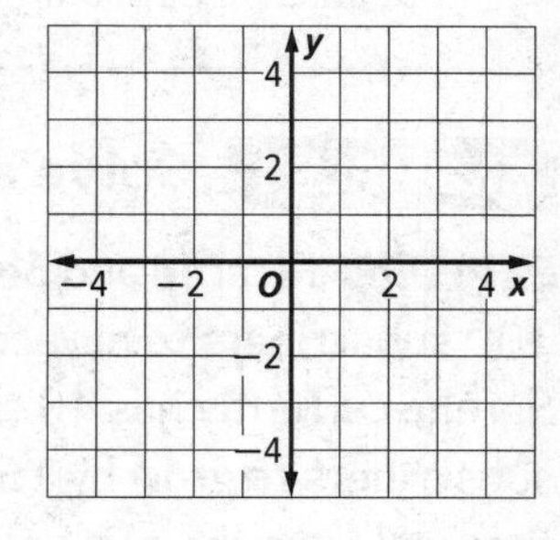

a. **USE STRUCTURE** Graph the path of the rubber bands and the path of the target on the coordinate plane at the right. CCSS A.REI.7, SMP 7

b. **USE A MODEL** If a player hits the target, at approximately what point or points on the plane will this take place? Explain. CCSS A.REI.7, SMP 4

c. **USE TOOLS** Use your calculator to find the coordinates of the point or points at which the rubber bands may hit the target. Round to the nearest tenth. How do your results compare to your answer in **part b**? CCSS A.REI.7, SMP 5

15. **REASON ABSTRACTLY** For what value of k does the system $y = x^2 + 2$ and $y = 3x + k$ have exactly one solution? Explain your reasoning. CCSS A.REI.7, SMP 2

16. **CALCULATE ACCURATELY** Verify your solutions to **Exercises 3** and **6** algebraically. CCSS A.REI.7, SMP 6

a. $y = -x^2 - 5x - 6$
$y = -3x - 1$

b. $x^2 + y^2 = 4$
$y = x + 3$

10.5 Function Intersections

Objectives

- Solve the equation $f(x) = g(x)$ by finding points where the graphs of $y = f(x)$ and $y = g(x)$ intersect.
- Find approximate solutions of the equation $f(x) = g(x)$ by using technology to graph the functions, by making tables, and by using successive approximations.

Content: A.REI.11
Practices: 2, 4, 5, 6, 8
Use with Extend 7-5

EXAMPLE 1 Solve an Equation by Graphing Functions CCSS A.REI.11

EXPLORE Ricardo and Serena each have a photo-sharing account. Ricardo currently has 100 subscribers to his account and he expects to add 10 new subscribers each month. Serena currently has 40 subscribers to her account and she expects the number of subscribers to grow by 15% each month. Ricardo and Serena would like to know when they will have the same number of subscribers.

a. USE A MODEL Write a function $f(x)$ that models the number of subscribers Ricardo will have in x months. Write a function $g(x)$ that models the number of subscribers Serena will have in x months. CCSS SMP 4

b. REASON ABSTRACTLY What equation should Ricardo and Serena solve in order to determine the number of months it will take until they have the same number of subscribers? CCSS SMP 2

c. REASON ABSTRACTLY Explain how you can use graphing to solve the equation in **part b**.

d. USE TOOLS Use your calculator to graph the two functions you wrote in **part a.** Sketch the graphs on the coordinate plane at the right. Be sure to label and write a scale on each axis. CCSS SMP 5

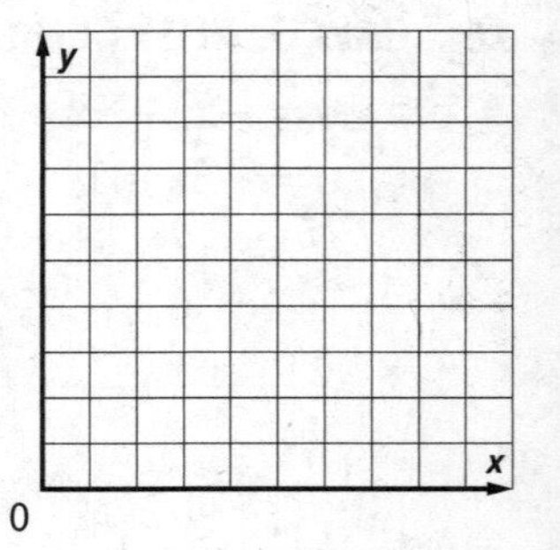

e. USE TOOLS Use your calculator's Intersect tool to find the point of intersection of the graphs. CCSS SMP 5

f. REASON QUANTITATIVELY Describe the meaning of the intersection in the context of the situation. CCSS SMP 2

EXAMPLE 2 Solve an Equation by Using a Table CCSS A.REI.11

A car company makes two different models of cars, the Regent and the Electro. In 2014, the company sold 50,000 Regents and 20,000 Electros. The company expects the sales of the Regent to drop 5% each year after 2014. They expect the sales of the Electro to be modeled by the function $g(x) = 250x^2 + 250x + 20{,}000$, where x is the number of years after 2014. The company's president wants to know the year in which sales of the two models will be approximately equal.

a. USE A MODEL Write a function $f(x)$ that models the number of Regents the company can expect to sell x years after 2014. CCSS SMP 4

b. REASON ABSTRACTLY What equation should the president of the company solve in order to determine the number of years it will take until the sales of the two models will be approximately equal? CCSS SMP 2

c. USE TOOLS Use your calculator to make a table that shows whole-number values of x, the value of $f(x)$, and the value of $g(x)$. Scroll down the table until you find two consecutive rows of the table that are useful for solving the problem. Write the values from these rows in the table at the right. CCSS SMP 5

x	$f(x)$	$g(x)$

d. COMMUNICATE PRECISELY Explain why the two rows of the table that you wrote in **part c** are important. CCSS SMP 6

e. REASON QUANTITATIVELY Explain how to use your answer to **parts c** and **d** to solve the problem. CCSS SMP 2

f. USE TOOLS Use your calculator to graph the two functions. Sketch the graphs on the coordinate plane at the right. Be sure to label and write a scale on each axis. CCSS SMP 5

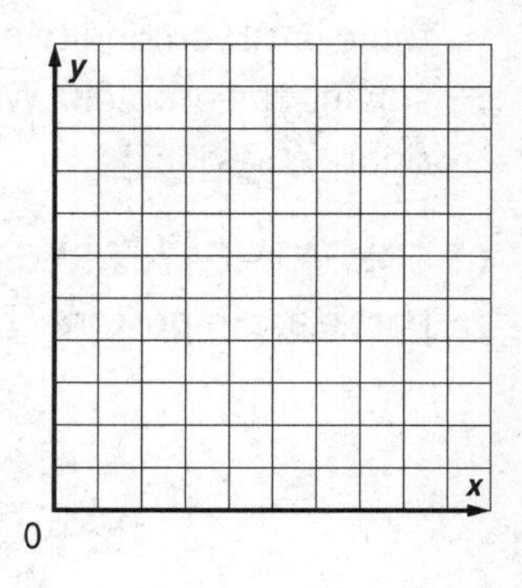

g. EVALUATE REASONABLENESS Explain how to use the graphs that you drew in **part f** to check that your solution to the problem is reasonable. CCSS SMP 8

h. COMMUNICATE PRECISELY Why do you think it would be useful to the president of the company to know when the sales of the Electro will exceed the sales of the Regent? CCSS SMP 6

EXAMPLE 3 Solve an Equation Using Successive Approximation 

A chemist has a beaker of water and a beaker of a saline solution. She measures the temperature of each liquid, in degrees Celsius, and finds that the temperature of the water is 12°C and is increasing 8% each minute and the temperature of the saline solution is decreasing according to the function $g(x) = -0.3x^2 - 0.2x + 40$, where x is the number of minutes since she started measuring the temperature. The chemist wants to know when the temperatures of the liquids will be equal.

a. USE A MODEL Write a function $f(x)$ that models the temperature of the water x minutes after the chemist started measuring the temperature. CCSS SMP 4

b. USE A MODEL What equation should the chemist solve in order to determine the number of minutes it will take for the temperatures to be equal? CCSS SMP 4

c. USE TOOLS Use your calculator to make a table that shows whole-number values of x, the value of $f(x)$, and the value of $g(x)$. Scroll down the table until you find two consecutive rows of the table that are useful for solving the problem. Write the values from these rows in the table at the right. CCSS SMP 5

x	$f(x)$	$g(x)$

d. COMMUNICATE PRECISELY Explain use why the two rows of the table that you wrote in **part c** are important. CCSS SMP 6

e. USE TOOLS Now use your calculator to make a table that shows values of x in increments of 0.1, the value of $f(x)$, and the value of $g(x)$. Scroll down the table until you find two consecutive rows of the table that are useful for solving the problem. Write the values from these rows in the table at the right. CCSS SMP 5

x	$f(x)$	$g(x)$

f. COMMUNICATE PRECISELY Explain why the two rows of the table that you wrote in **part e** are important. CCSS SMP 6

g. DESCRIBE A METHOD Suppose the chemist wants to know the number of minutes it will take for the temperatures to be equal to the nearest tenth of a minute. Explain the next steps and give the answer to the problem. CCSS SMP 8

PRACTICE

USE TOOLS Solve each equation using any of the methods from this lesson. Round solutions to the nearest tenth. CCSS A.REI.11, SMP 5

1. $2.2x^2 + x + 1 = 1.8^x + 5$

2. $-0.8x + 3.1 = x^2 - 5.4x - 2.1$

3. $14(1.9)^x = 5.5x + 33$

4. $-0.5x^2 + 3x + 21 = 2^x$

5. $8(0.8)^x = 4.6x + 60$

6. $x^2 - 24x + 144 = 60(0.9)^x$

7. Jessica is comparing the prices of two stocks. The price of Stock A is currently \$45 and is expected to decrease by 4% each month. The price of Stock B is currently \$22 and is expected to increase by \$2.25 per month. Jessica wants to know how many months it will take until the stocks have the same price. CCSS A.REI.11

a. USE A MODEL Write a function $f(x)$ that models the price of Stock A in x months and a function $g(x)$ that models the price of Stock B in x months. CCSS SMP 4

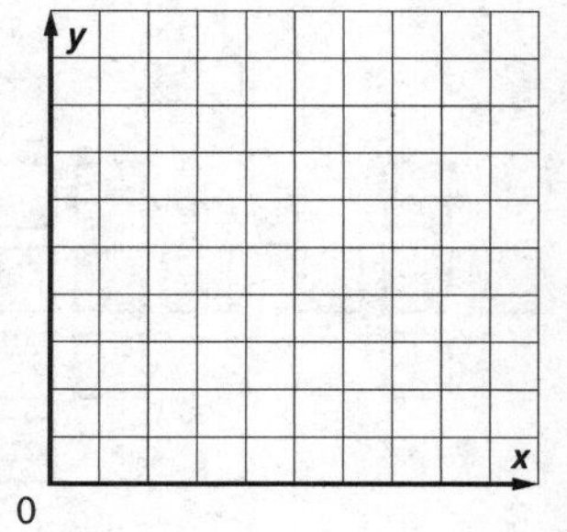

b. USE TOOLS Use your calculator to graph the two functions. Sketch the graphs on the coordinate plane at the right. Then tell how many months it will take until the stocks have the same price. To the nearest dollar, what will this price be? CCSS SMP 5

8. USE TOOLS In 2010, the population of Cedar City was 72,000. Since then, the population has increased by 2% per year. In 2010, the population of Allenville was 84,500 and the population since then has increased according to the function $g(x) = 0.5x^2 + 84{,}500$, where x is the number of years since 2010. Based on these models, in what year will the population of Cedar City be greater than the population of Allenville for the first time? Explain how you found your answer. CCSS A.REI.11, SMP 5

9. COMMUNICATE PRECISELY The amount of water, in gallons, in Tank A is modeled by $f(x) = 22(1.1)^x$, where x is the number of minutes since water began being added to the tank. The amount of water in Tank B is modeled by $g(x) = 1.3x + 35$. Use a calculator's table feature and successive approximation to find the number of minutes it will take, to the nearest tenth, until the tanks have the same amount of water. Justify your answer. CCSS A.REI.11, SMP 6

10.6 Combining Functions

Objectives

- Combine standard function types using arithmetic operations.

Content: F.IF.1a, F.BF.1b
Practices: 1, 2, 3, 4, 5, 6, 7, 8
Use with Lessons 9–3

Functions can be combined using arithmetic operations to make new functions. The examples and exercises that follow provide an introduction to and practice with combining functions.

EXAMPLE 1 Combine Linear and Quadratic Functions

EXPLORE Mia is using blue and gray square tiles to make a mosaic. She experiments with a sequence of patterns as shown in the figure below.

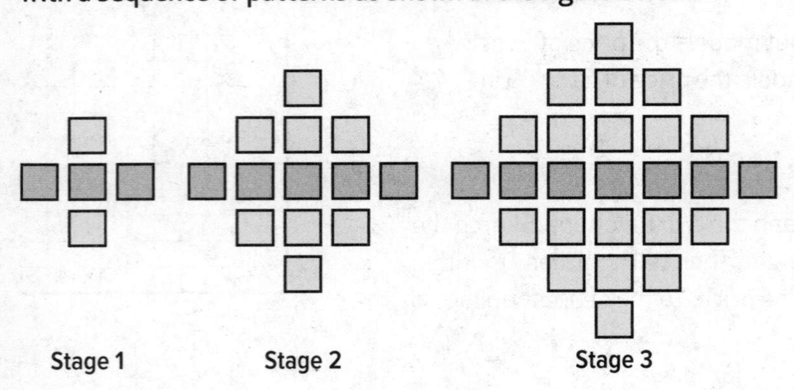

a. **FIND A PATTERN** Complete the table. CCSS SMP 7

Stage	1	2	3	4	5
Number of Blue Tiles					
Number of Gray Tiles					

b. **REASON ABSTRACTLY** Write a function $f(n)$ that gives the number of blue tiles that are needed to make stage n of the mosaic pattern. CCSS SMP 2

c. **REASON ABSTRACTLY** Write a function $g(n)$ that gives the number of gray tiles that are needed to make stage n of the mosaic pattern. CCSS SMP 2

d. **COMMUNICATE PRECISELY** Write a function $h(n)$ that gives the total number of tiles that are needed to make stage n of the pattern. Explain how $h(n)$ is related to $f(n)$ and $g(n)$. CCSS SMP 6

e. **INTERPRET PROBLEMS** As Mia builds the mosaics, she finds that it is sometimes helpful to use the function $k(n) = f(n) - g(n)$. Write a rule for $k(n)$ and explain what it represents. CCSS SMP 1

As you saw in the exploration, you can build new functions by performing arithmetic operations on existing functions.

EXAMPLE 2 Combine Linear and Exponential Functions CCSS F.BF.1b

Liam makes a cup of herbal tea and measures the temperature of the tea at 5-minute intervals, as shown in the table below. While he takes the measurements, room temperature is 75°F. He would like to develop a model that gives the temperature of the tea, in degrees Fahrenheit, at any time since the tea starts cooling.

a. USE A MODEL When a liquid cools, its temperature approaches the temperature of its surroundings (in this case, room temperature). Because an exponential model decays to 0, it is more accurate to develop an exponential model for the temperature of the tea above room temperature. Complete the third row of the table by writing the temperature of the tea above room temperature. CCSS SMP 4

Time (minutes)	0	5	10	15	20	25
Temperature of Tea (°F)	170	160	151	142	134	126
Temperature of Tea Above Room Temperature (°F)						

b. USE TOOLS Enter the data from the first and third rows of the table in two lists in your calculator. Use the calculator's exponential regression tool to write a function $f(x)$ that models the temperature of the tea above room temperature after x minutes. CCSS SMP 5

c. USE A MODEL Let $g(x)$ be the constant function that represents room temperature. Explain how to write a function $h(x)$ that models the temperature of the tea after x minutes. CCSS SMP 4

d. USE A MODEL Use your model to predict the temperature of the tea after 35 minutes. CCSS SMP 4

e. EVALUATE REASONABLENESS Sketch and label the graphs of $f(x)$, $g(x)$, and $h(x)$ on the coordinate plane at the right. Use the graphs to explain how you know your model for the temperature of the tea in **part c** is reasonable. CCSS SMP 8

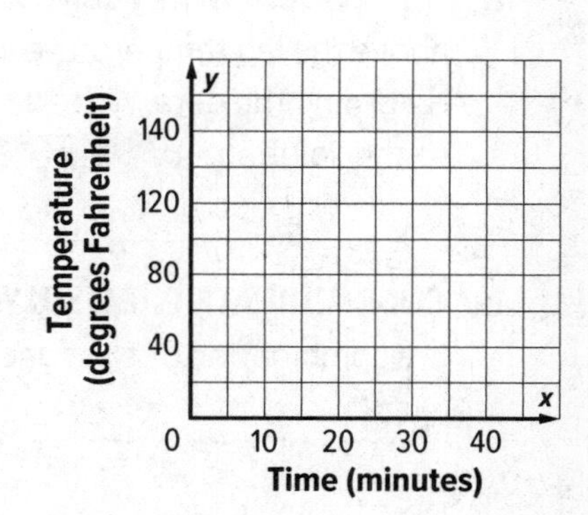

PRACTICE

1. Nathanael makes a pattern, as shown below, using gray tiles and white tiles.

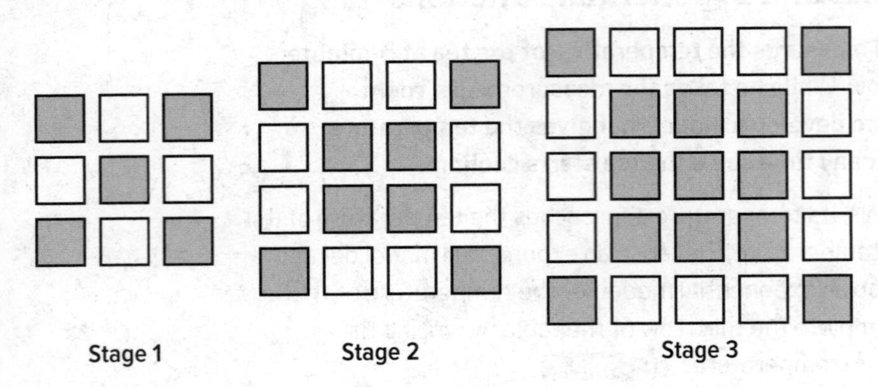

   a. **FIND A PATTERN** Write a function $f(n)$ that gives the number of gray tiles needed to make stage n of the pattern and a function $g(n)$ that gives the number of white tiles needed to make stage n of the pattern. CCSS F.BF.1a, SMP 7

   b. **USE A MODEL** Explain how to write a function $h(n)$ that gives the total number of tiles needed to make stage n of the pattern. Use the function to the find the total number of tiles needed to make stage 10. CCSS F.BF.1a, SMP 4

   c. **EVALUATE REASONABLENESS** Explain how you know your answer to **part b** is correct. CCSS F.BF.1a, SMP 8

2. A chemist works in a lab that maintains a constant temperature of 22°C. She heats a saline solution and then lets the solution cool. As the solution cools, she records its temperature every 3 minutes. Her data is shown in the table.

Time (minutes)	0	3	6	9	12	15
Temperature of Solution (°C)	82.0	79.1	76.3	73.6	70.9	68.4
Temperature Above the Lab Temperature	60.0	57.1	54.3	51.6	48.9	46.4

   a. **USE TOOLS** Write a function $P(x)$ that models the temperature of the solution above the lab temperature in degrees Celsius x minutes after the chemist starts recording the data. Write a second function, $S(x)$ that models the temperature of the solution. CCSS F.BF.1a, SMP 5

   b. **COMMUNICATE PRECISELY** Explain how the function you wrote for $S(x)$ in **part a** is a combination of two functions. CCSS F.BF.1a, SMP 6

c. **USE TOOLS** The chemist wants to know approximately how long it will take until the solution cools to a temperature of 45°C. Explain how to use your calculator to estimate this time to the nearest minute. CCSS F.BF.1a, SMP 5

3. Alba sells homemade granola at a farmers market. When she sets the price at $4 per bag, she can sell 185 bags. She finds that for every increase of $0.25 in the price of a bag of granola, she sells 10 fewer bags.

a. **USE A MODEL** Let x represent the number of $0.25 increases in the price. Write a function $P(x)$ that gives the price of the granola as a function of x. Write a function $S(x)$ that gives the number of bags of granola Alba sells as a function of x. CCSS F.BF.1a, SMP 7

b. **COMMUNICATE PRECISELY** Explain how to combine functions to write a function $R(x)$ that gives Alba's revenue from selling granola as a function of x. CCSS F.BF.1a, SMP 6

c. **REASON QUANTITATIVELY** Alba is considering raising the price of the granola to $4.75. Use your model to explain whether or not this is a good idea. CCSS F.BF.1a, SMP 2

4. A rectangular flower bed in a garden is 12 feet long and 8 feet wide. Kenji plans to add a gravel border around the flower bed so that the border is twice as wide along the 8-foot sides of the flower bed as it is along 12-foot sides.

a. **USE STRUCTURE** Let x be the width of the gravel border along the 12-foot sides of the flower bed, as shown. Write a function $A(x)$ that gives the area of the gravel border in square feet. Explain how you can write this function as a combination of simpler functions. CCSS F.BF.1a, SMP 7

b. **CRITIQUE REASONING** Kenji said that the function $A(x)$ is a quadratic function. Therefore, as x increases, the area of the gravel border will increase up to a point, reach a maximum value, and then start to decrease. Do you agree? Explain. CCSS F.BF.1a, SMP 3

10.7 Analyzing Functions with Successive Differences

Objectives

- Use successive differences to distinguish among situations that can be modeled with linear functions, quadratic functions, and exponential functions.
- Construct linear, quadratic, and exponential functions.

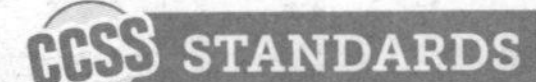

CCSS STANDARDS

Content: F.IF.6, F.IF.7a, F.IF.7e, F.LE.1b, F.LE.1c, F.LE.2, F.LE.3
Practices: 2, 3, 4, 6, 7
Use with Lesson 9-6

When the x-values in a table increase by the same amount, you can use **successive differences** of the y-values to analyze the function. The differences of successive y-values are called **first differences**. The differences of successive first differences are called **second differences**. You can also consider **successive ratios**, which are the ratios of successive y-values.

EXAMPLE 1 Investigate Successive Differences and Successive Ratios CCSS F.LE.1b

EXPLORE As you complete the following, look for connections between different function types and their successive differences and successive ratios.

a. CALCULATE ACCURATELY Write a simple linear function in the space provided. Complete the table of values for the function. Then calculate the first differences. CCSS SMP 6

Linear function: ______________________

x	0	1	2	3	4
y					
First Differences					

b. MAKE A CONJECTURE Compare your results with those of other students. Then make a conjecture about linear functions. CCSS F.LE.1b, SMP 3

____

c. CALCULATE ACCURATELY Write a simple quadratic function in the space provided. Complete the table of values for the function. Then calculate the first differences and second differences. CCSS SMP 6

Quadratic function: ______________________

x	0	1	2	3	4
y					
First Differences					
Second Differences					

d. MAKE A CONJECTURE Compare your results with those of other students. Then make a conjecture about quadratic functions. CCSS SMP 3

e. CALCULATE ACCURATELY Write a simple exponential function in the space provided. Complete the table of values for the function. Then calculate the successive ratios. CCSS SMP 6

Linear function: ____________

x	0	1	2	3	4
y					
Successive Ratios					

f. MAKE A CONJECTURE Compare your results with those of other students. Then make a conjecture about exponential functions. CCSS F.LE.1c, SMP 3

KEY CONCEPT Successive Differences and Successive Ratios

Use your findings from the previous exploration to help you complete the following.

In a table with x-values that increase by the same amount...

- if first differences of the y-values are all equal, then the data represent ____________
- if first differences of the y-values are not equal, but second differences are all equal, then the data represent ____________
- if successive ratios of the y-values are all equal (and this ratio is not equal to 1), then the data represent ____________

EXAMPLE 2 Identify and Write a Function

The table shows the time Malia has been driving and her distance from Sacramento.

Time (h), x	0	1	2	3	4
Distance (mi), y	37	92	147	202	257

a. USE STRUCTURE What type of function best models the data in the table? Why? CCSS F.LE.1b, SMP 7

b. REASON ABSTRACTLY What is the rate of change of the function over the interval from $x = 1$ to $x = 4$? How is this related to the first differences? CCSS F.IF.6, SMP 2

c. USE A MODEL Write a function that models Malia's distance from Sacramento as a function of time. What do you notice about the coefficient of x in the function? CCSS F.LE.2, SMP 4

EXAMPLE 3 Identify and Write a Function

The table shows the price of a stock at the end of several years.

Number of Years Since 2010	Price of Stock ($)
x	y
1	30.00
2	36.00
3	43.20
4	51.84

a. USE STRUCTURE What type of function best models the data in the table? Why? CCSS F.LE.1c, SMP 7

b. USE A MODEL Write a function that models the price of the stock as a function of the number of years since 2010. CCSS F.LE.2, SMP 4

c. REASON ABSTRACTLY Explain how the function you wrote in **part b** is related to the successive differences or successive ratios you observed in the table of data. CCSS F.LE.1b, SMP 2

d. CALCULATE ACCURATELY Use the function you wrote in **part b** to determine the price of the stock in the year 2020. Explain. CCSS F.LE.1c, SMP 6

e. COMMUNICATE PRECISELY Is your answer to **part d** reasonable? Explain. CCSS F.LE.1c, SMP 6

EXAMPLE 4 Write and Compare Models

The table shows the number of subscribers to three different online newspapers from 2010 to 2013.

Year	Number of Subscribers		
	InfoWorld	NewsCo	MegaTimes
2010	30,000	20,000	36,000
2011	30,150	22,000	36,360
2012	30,600	24,200	36,720
2013	31,350	26,620	37,080

a. USE STRUCTURE For each newspaper, write a function that models the number of subscribers x years after 2010. CCSS F.LE.2, SMP 7

b. USE A MODEL Sketch the graphs of the functions you wrote in **part a** on the coordinate plane at the right. Be sure to label each graph with the name of the newspaper it represents. CCSS F.IF.7a, F.IF.7e, SMP 4

c. REASON ABSTRACTLY As time goes on, what can you conclude about the number of subscribers to the three newspapers? Explain how this is connected to the types of functions you used for the models. CCSS F.LE.3, SMP 2

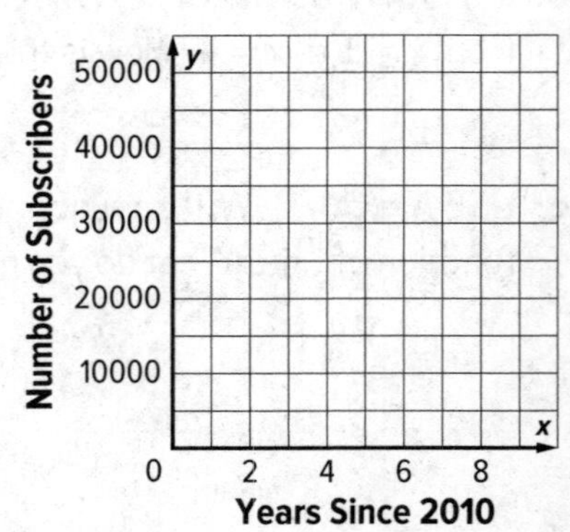

PRACTICE

USE STRUCTURE Tell what type of function best models the data in each table. Then write the function. CCSS F.LE.2, SMP 7

1.

x	2	3	4	5	6
y	12	27	48	75	108

2.

x	0	1	2	3	4
y	80	73	66	59	52

3.

x	1	2	3	4	5
y	10	20	40	80	160

4.

x	4	6	8	10	12
y	16	22	28	34	40

5. The table shows the height of an elevator above ground level at various times.

Time (s), x	0	1	2	3	4
Height (ft), y	142	124	106	88	70

a. **USE STRUCTURE** What type of function best models the data in the table? Why? CCSS F.LE.1b, SMP 7

b. **REASON ABSTRACTLY** What is the rate of change of the function over the interval $x = 0$ to $x = 4$? What does this tell you about the motion of the elevator? CCSS F.IF.6, SMP 2

c. **USE A MODEL** Write a function that models the height of the elevator as a function of time. How is the coefficient of x related to your answers to **parts a** and **b**? CCSS F.LE.2, SMP 4

6. The table shows the populations of two towns from 2010 to 2013.

Year	Population	
	Dixon	Midville
2010	80,000	96,000
2011	84,000	96,070
2012	88,200	96,280
2013	92,610	96,630

a. **USE A MODEL** For each town, write a function that models the town's population x years after 2010. CCSS F.LE.2, SMP 4

b. **REASON ABSTRACTLY** Based on the function types you used in **part a**, what can you conclude about the populations of the towns as time goes on? Explain. CCSS F.LE.3, SMP 2

10.8 Special Functions

Objectives

- Graph piecewise-defined functions, including step functions and absolute value functions.
- Interpret features of the graphs of piecewise-defined functions.

CCSS STANDARDS

Content: F.IF.4, F.IF.7b, F.BF.3, F.IF.4a, F.IF.7a
Practices: 2, 3, 4, 6, 7, 8
Use with Lesson 9-7

EXAMPLE 1 Analyze a Piecewise-Defined Function CCSS F.IF.7b

EXPLORE Malik owns a bakery. Each week, he orders walnuts from a supplier. The supplier offers a discount to customers who buy a large quantity of walnuts, as shown in the table.

Walnuts: Price List	
\$4 per pound	Up to 3 pounds
\$1.50 per pound	For every pound over 3 pounds

a. USE A MODEL Write a function $g(x)$ that gives the total cost of x pounds of walnuts, assuming Malik orders 3 pounds or less. Explain. CCSS SMP 4

b. USE A MODEL Suppose Malik orders more than 3 pounds of walnuts. For this order, he receives the discounted price only on the amount over 3 pounds. Write a function $h(x)$ that gives the total cost of x pounds of walnuts in this case. Explain. CCSS SMP 4

c. USE A MODEL You can write a single function $f(x)$ that gives the total cost of x pounds of walnuts by using different rules for different parts of the domain. Use the functions you wrote above to complete the following. CCSS SMP 4

$$f(x) = \begin{cases} \_\_\_\_\_\_\_\_\_\_ & 0 \leq x \leq 3 \\ \_\_\_\_\_\_\_\_\_\_ & x > 3 \end{cases}$$

d. USE STRUCTURE Graph $f(x)$ on the coordinate plane at the right. CCSS SMP 7

e. COMMUNICATE PRECISELY Describe the graph of $f(x)$. Include any intercepts and any intervals on which the function is increasing or decreasing. CCSS F.IF.4a, SMP 6

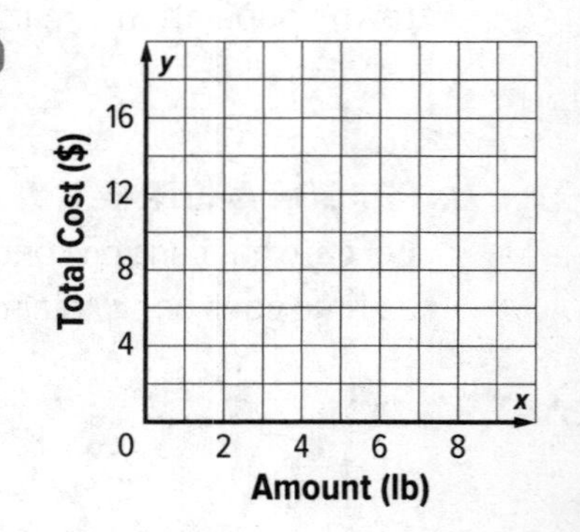

f. **EVALUATE REASONABLENESS** Malik orders 7 pounds of walnuts. Find the total cost of the walnuts and explain how you know your answer is reasonable. CCSS SMP 8

The function $f(x)$ in the previous exploration is an example of a piecewise-defined function. A **piecewise-defined function** has different rules for different intervals of the domain.
A step function is one type of piecewise-defined function. A **step function** has a graph that is a series of horizontal line segments. The figure shows the graph of the step function $s(x)$. Notice how open circles and closed circles are used at the endpoints of the intervals in the domain.

$$s(x) = \begin{cases} -3 & x \leq 0 \\ 3 & x > 0 \end{cases}$$

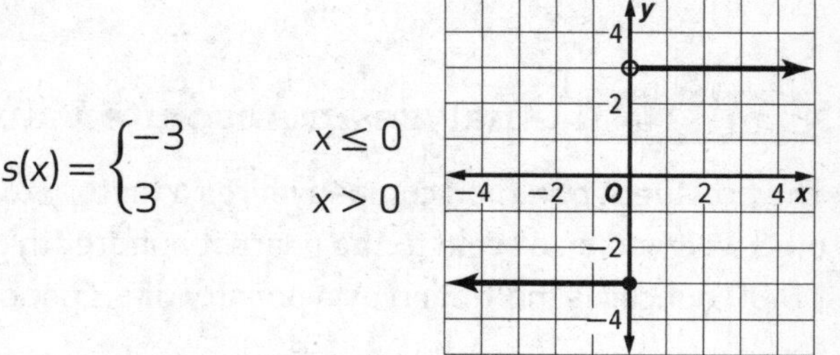

EXAMPLE 2 Analyze a Step Function CCSS F.IF.7b

A shipping company charges customers based on the weight of the package. The company charges $3 per pound for shipping, but for packages whose weights are not a whole number of pounds, the company rounds weights up to the next pound.

a. **REASON QUANTITATIVELY** Complete the table by finding the shipping costs for packages with the given weights. CCSS SMP 2

Weight (lb)	0.2	0.9	1	1.1	1.8	5.3	9.4
Cost ($)							

b. **USE A MODEL** Write a piecewise-defined function $C(x)$ that gives the shipping cost for packages that weigh up to 5 pounds, where x is the weight of the package in pounds. Write the function in the space at the right. CCSS SMP 4

$$C(x) = \begin{cases} 3 & 0 < x \leq 1 \\ 6 & 1 < x \leq 2 \\ 9 & 2 < x \leq 3 \\ 12 & 3 < x \leq 4 \\ 15 & 4 < x \leq 5 \end{cases}$$

c. **USE STRUCTURE** Graph $C(x)$ on the coordinate plane at the right. CCSS SMP 7

d. **COMMUNICATE PRECISELY** Explain how you determined where to use open circles and closed circles in the graph of $C(x)$. CCSS SMP 6

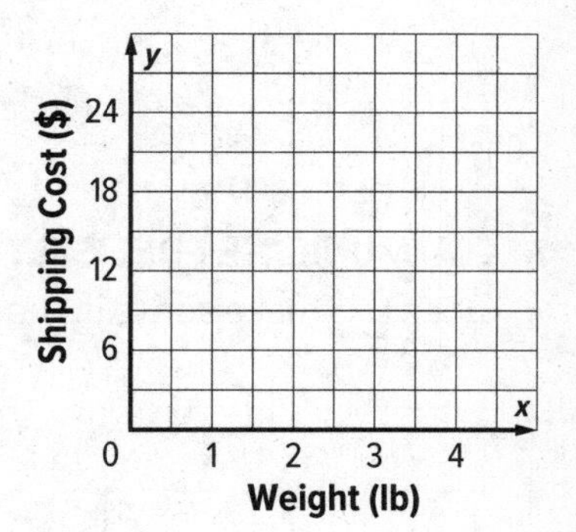

e. **COMMUNICATE PRECISELY** Describe the graph of $C(x)$. Include any intercepts and any intervals on which the function is increasing or decreasing. CCSS F.IF.4a, SMP 6

An **absolute value function** is a function whose rule contains an absolute value expression. The parent absolute value function is $f(x) = |x|$. The rule for this function may be written as a piecewise-defined function, as shown below.

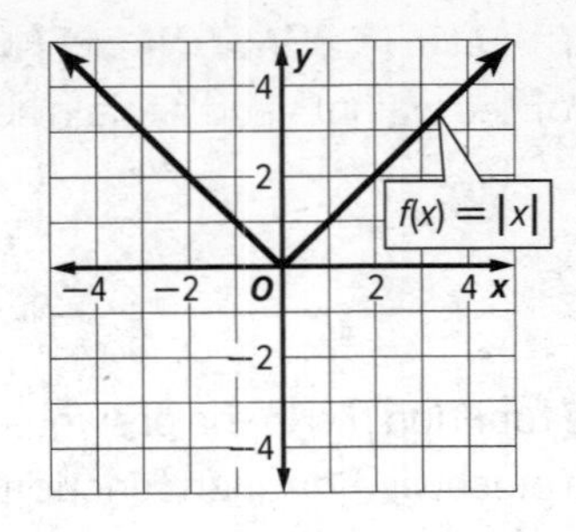

$$f(x) = \begin{cases} x & x > 0 \\ 0 & x = 0 \\ -x & x < 0 \end{cases}$$

EXAMPLE 3 Analyze an Absolute Value Function

The president of a food company hires a factory to bottle olive oil. Each bottle of olive oil is weighed on a scale to the nearest hundredth of an ounce. Any deviation from 10.00 ounces is multiplied by a penalty of $2 per ounce.

a. **USE A MODEL** Let x be the weight of a bottle of olive oil. Write an expression that tells you how many ounces the bottle is from the target weight of 10.00 ounces. Then use this expression to help you write a function $P(x)$ that gives the penalty in dollars for each bottle of olive oil. CCSS SMP 4

b. **USE STRUCTURE** Use transformations to explain how the graph of $P(x)$ is related to the graph of the parent absolute value function. CCSS F.BF.3, SMP 7

c. **USE STRUCTURE** Graph $P(x)$ on the coordinate plane provided below. CCSS F.IF.7b, SMP 7

d. **REASON QUANTITATIVELY** Find the minimum of the function $P(x)$. Does this make sense in the context of the problem? Explain. CCSS F.IF. 4, SMP 2

e. **REASON QUANTITATIVELY** Does the graph have any symmetry? If so, explain why this makes sense in the context of the problem. CCSS F.IF. 4, SMP 2

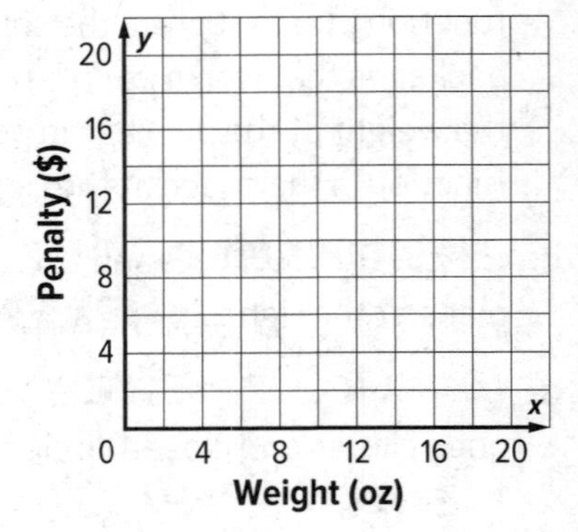

f. **COMMUNICATE PRECISELY** What is the end behavior of the graph of $P(x)$? Does this make sense in the context of the problem? CCSS F.IF. 4, SMP 6

PRACTICE

USE STRUCTURE **Graph each piecewise-defined function.** CCSS F.IF.7b, SMP 7

1. $f(x) = \begin{cases} x+2 & x < -1 \\ x^2 & -1 \leq x \leq 3 \\ 9 & x > 3 \end{cases}$

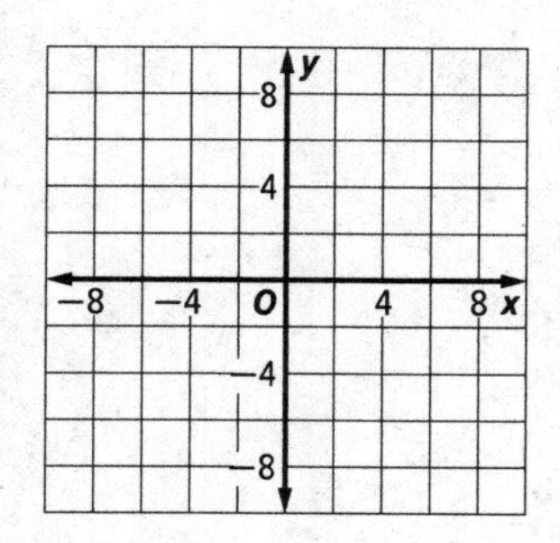

2. $g(x) = \begin{cases} -x & x < -4 \\ 2x+1 & -4 \leq x \leq 2 \\ x-3 & x > 2 \end{cases}$

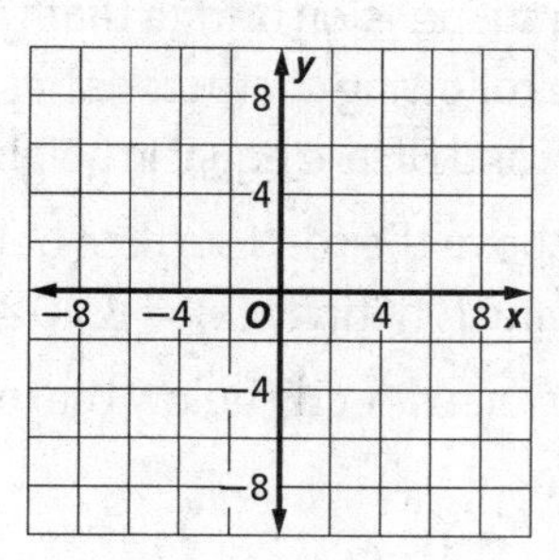

3. **COMMUNICATE PRECISELY** Use transformations to describe how the graph of $h(x) = -|x+2| - 3$ is related to the graph of the parent absolute value function. CCSS F.BF.3, SMP 6

---

4. On a straight highway, the town of Garvey is located at mile marker 200. A car is located at mile marker x and is traveling at an average speed of 50 miles per hour.

a. **USE A MODEL** Write a function $T(x)$ that gives the time, in hours, it will take the car to reach Garvey. Then graph the function on the coordinate plane at the right. CCSS F.IF.7b, SMP 4

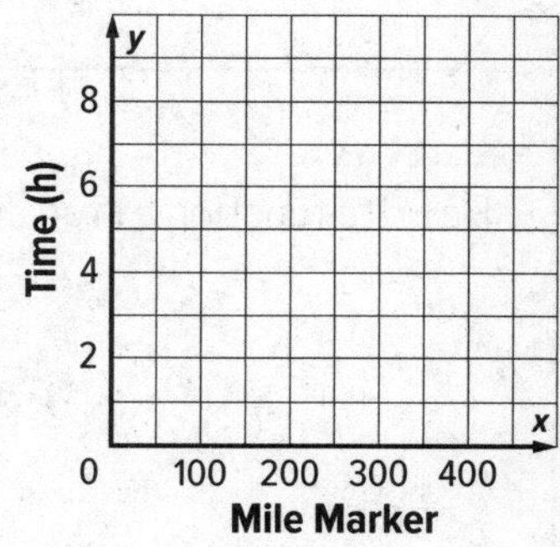

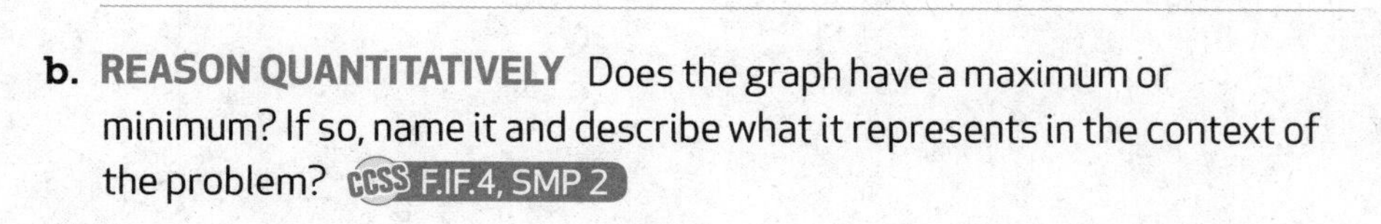

b. **REASON QUANTITATIVELY** Does the graph have a maximum or minimum? If so, name it and describe what it represents in the context of the problem? CCSS F.IF.4, SMP 2

---

5. **CRITIQUE REASONING** Amy graphed a function that gives the height of a car on a roller coaster as a function of time. She said her graph is the graph of a step function. Is this possible? Explain why or why not. CCSS F.IF.4, SMP 3

---

6. **REASON ABSTRACTLY** The function $f(x)$ is decreasing on the interval $(-\infty, -2)$, constant on the interval $[-2, 2]$, and increasing on the interval $(2, \infty)$. $f(-4) = 4$, $f(0) = 1$, and $f(4) = 2$. Sketch a graph of the function. CCSS F.IF.7a, SMP 2

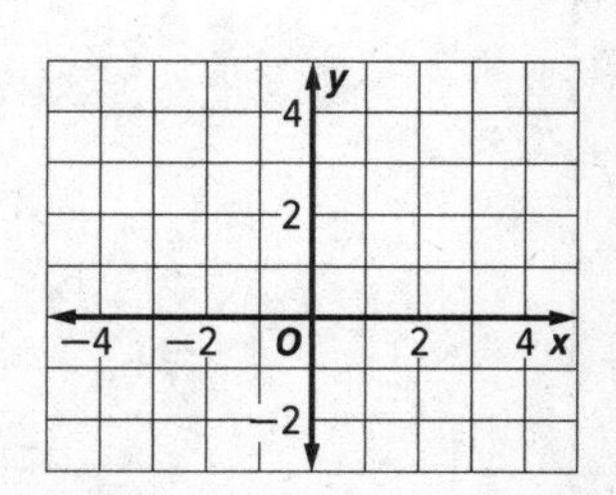

Performance Task

Engineering a Bridge

Provide a clear solution to the problem. Be sure to show all of your work, include all relevant drawings, and justify your answers.

Erica and Josh are constructing a model of a bridge for their class project. The illustration below shows a suspension bridge that they want to model. The suspension bridge has the following characteristics:

- The vertical supports $\overline{OE}$ and $\overline{BF}$ are equal in height.
- The height of the cable above the road surface between the vertical supports is modeled by the quadratic function $h(x) = 0.002x^2 - 0.8x + 86$.
- Point P is on the road surface directly below the lowest point on the cable.

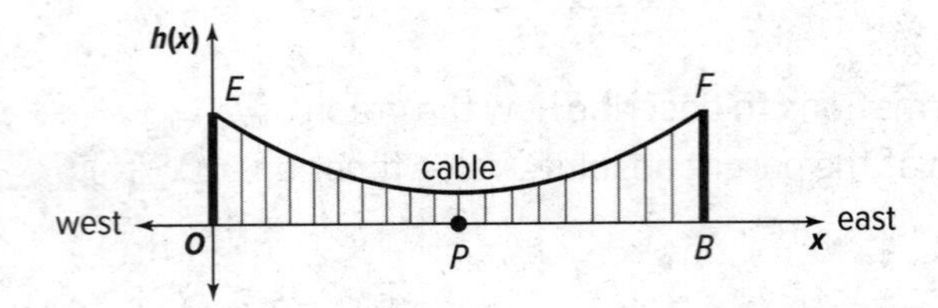

Part A

Rewrite function h in an equivalent form by completing the square.

Part B

Use your equation from **Part A** to determine the following measurements.

- the distance between the cable and point P
- the distance between point O and point P
- the heights of the vertical supports $\overline{OE}$ and $\overline{BF}$

Part C

Erica and Josh are given the diagram with its data below. Point W is on the road surface directly below the lowest point on the cable.

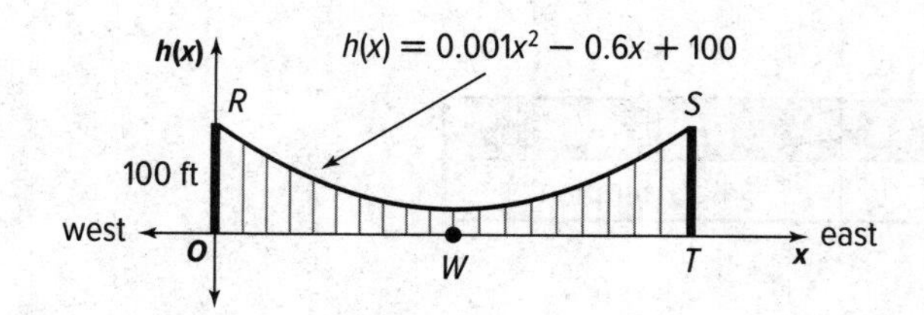

Their job is to construct a scale model of the bridge. The scale will be 1 in. : 5 ft. Determine a set of scale-model measurements that they can use to make an accurate model.

Performance Task

Arranging Plants in a Landscape

Provide a clear solution to the problem. Be sure to show all of your work, include all relevant drawings, and justify your answers.

Natalie is creating a garden, as shown. What Natalie has planted already is illustrated by the black circles in this diagram. She wants to extend the planting as suggested by the white circles in the diagram. When completed, she will have a large rectangular array of plants. (All rows of plants have the same number of plants and all the plants are spaced apart equally.)

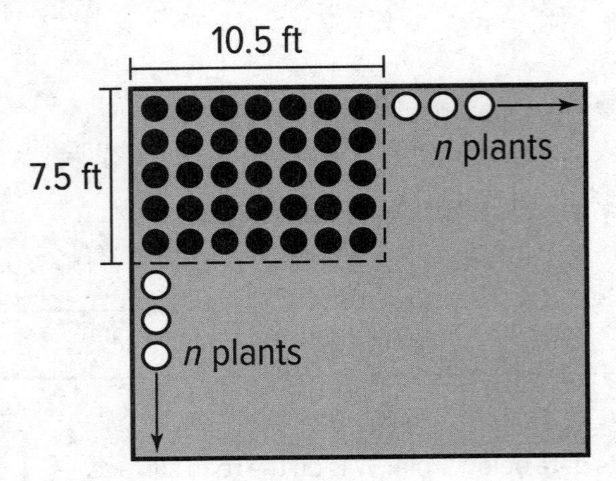

This table shows the prices of two types of plants she needs to buy for the expansion. Half of the plants will be marigolds and half of the plants will be begonias.

Marigolds	Begonias
$3.00 per plant	$4.00 per plant

Part A

Write and simplify a rule for each.

- t, the function that relates the total number of plants she needs to buy to n
- p, the function that relates the cost in dollars of marigolds she needs to buy to n
- q, the function that relates the cost in dollars of begonias she needs to buy to n
- c, the function that relates the cost in dollars of buying all the plants to n

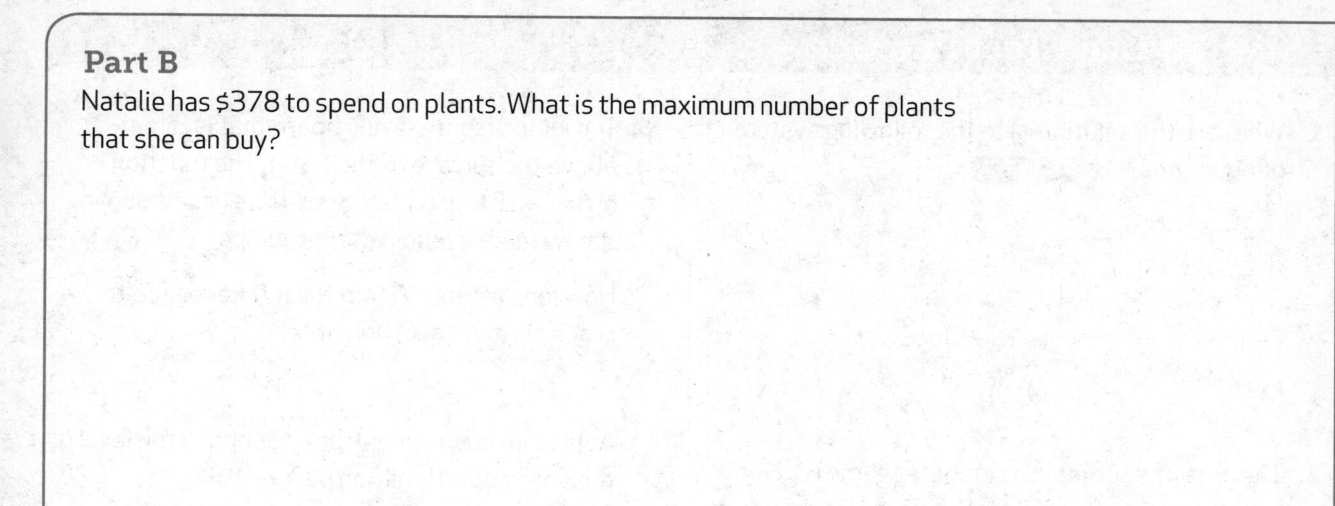

Part B

Natalie has $378 to spend on plants. What is the maximum number of plants that she can buy?

Part C

Natalie has asked Ernesto to help her with the planting. It takes each of them 6 minutes to plant one plant. They expect to begin work at 7:30 A.M. and must finish at or before noon. Will they be able to complete the task on time? If not, how much more time will they need?

Part D

Now Natalie wants to write a function for the area of the entire plot, consisting of plants already planted and plants to be bought. Write a rule for a, the function that relates the area of the entire rectangle to n. Explain your reasoning.

Standardized Test Practice

1. What are the solution(s) to the following system of equations? CCSS A.REI.7

$$\begin{cases} y = x^2 - 5x + 13 \\ y = 2x + 1 \end{cases}$$

(−4, −7) (−3, −5)

(3, 7) (4, 9)

2. The area of a target in a computer game is given by the function $y = \pi r^2 - 18$, where r is the radius of the circle in centimeters. The radius of the circle is given by the function $r = 2t - 1$, where t is the number of seconds since the game started. What function gives the area of the target as a function of the number of seconds since the game started? CCSS F.BF.1b

3. Consider the ordered pairs shown in the following tables. For each table, determine the type of function that best describes the data. CCSS F.LE.1

x	3	5	7	9	11
y	−1.5	−3	−4.5	−6	−7.5

Type of function: linear quadratic

x	6	7	8	9	10
y	2	5	9	14	20

Type of function: linear quadratic

4. What are the point(s) of intersection of $g(x) = 5x^2 + 3x - 1$ and $h(x) = 2x^2 - 4x - 3$? CCSS A.REI.7

5. For the graphs of two quadratic functions, how many points of intersection are possible? CCSS A.REI.7

0 2

1 3

6. Teo jumps from a diving board that is 8 feet above the surface of the water. The function $h(t) = -16t^2 + 8t + 8$ gives Teo's height above the water t seconds after he jumps. CCSS F.IF.4

How long after the jump will it take for Teo to reach his maximum height?

At his maximum height, how far above the level of the diving board will Teo be?

How long will it take Teo to reach the water?

Graph the function.

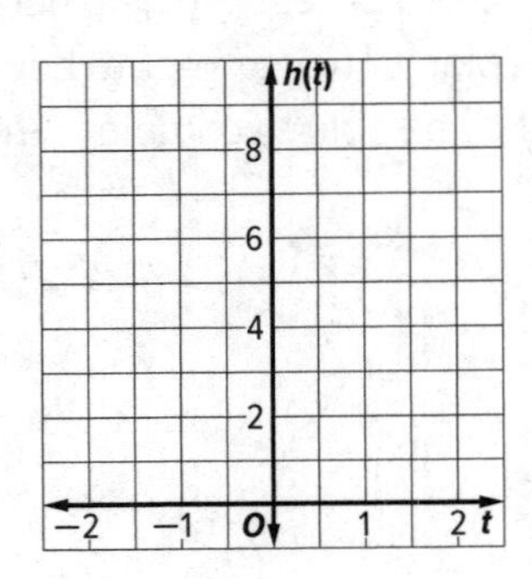

7. What are the solution(s) to the following system of equations? CCSS A.REI.7

$$\begin{cases} y = -2x^2 + 5x + 1 \\ y = 3x + 19 \end{cases}$$

8. Graph the piecewise-defined function. CCSS F.IF.7b

$$f(x) = \begin{cases} -x - 3 & x \leq -2 \\ -x^2 + 3 & -2 < x < 2 \\ x - 3 & x \geq 2 \end{cases}$$

9. Consider each function shown in the following table. Determine whether the graph opens up or down, the axis of symmetry, and the vertex. Complete the row for each function. CCSS F.IF.4

Function	Direction Graph Opens	Axis of Symmetry	Vertex
$f(x) = 3x^2 - 2x + 7$			
$f(x) = 2x^2 - 5 - 4(x^2 + 2x)$			
$f(x) = \frac{1}{4}x^2 - 5$			

10. Consider the function $f(x) = -3x^2 + 2$. CCSS F.BF.3

a. Describe the function as a transformation of $f(x) = x^2$.

b. Graph the function.

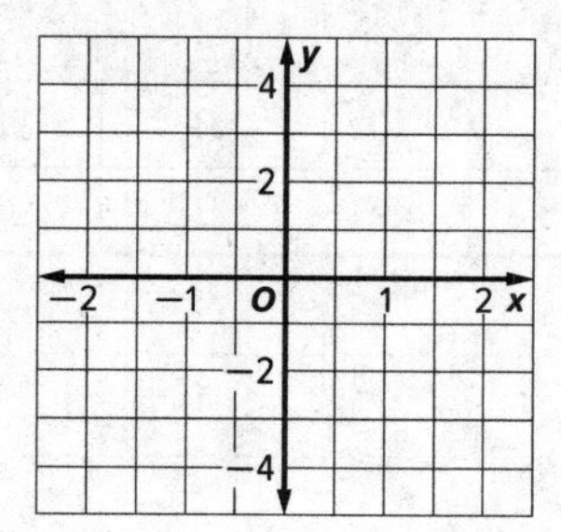

11. Suppose archaeologists found the partial remains of an ancient burial mound. A cross section of the mound is shown below.

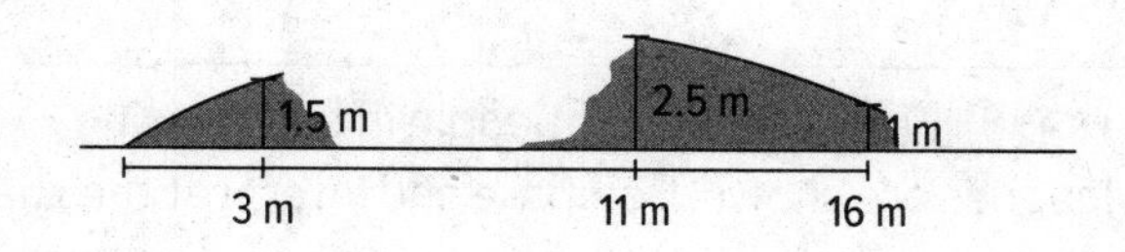

a. What quadratic function models the shape of the mound? Include the meaning and units of the variables. CCSS N.Q.2

b. Using the function you found, what do you estimate was the width of the mound? CCSS A.CED.1

c. Using your function, what do you estimate was the depth of the mound at its deepest point? CCSS A.CED.1

11 Statistics

CHAPTER FOCUS Learn about some of the Common Core State Standards that you will explore in this chapter. Answer the preview questions. As you complete each lesson, return to these pages to check your work.

What You Will Learn	Preview Question
Lesson 11.1: Statistics and Parameters	
CCSS S.ID.2 Use statistics appropriate to the shape of the data distribution to compare center (median, mean) and spread (interquartile range, standard deviation) of two or more different data sets. CCSS S.ID.3 Interpret differences in shape, center, and spread in the context of the data sets, accounting for possible effects of extreme data points (outliers).	CCSS SMP 3 A teacher recorded test scores for students who studied less than two hours with students who studied two or more hours. Compare the means and the standard deviations of each data set.

Studied Less Than Two Hours					
72	76	81	90	65	65
73	80	81	76	62	65

Studied More Than Two Hours					
76	85	92	78	85	90
93	88	75	80	88	98

What You Will Learn	Preview Question
Lesson 11.2: Distributions of Data	
CCSS S.ID.3 Interpret differences in shape, center, and spread in the context of the data sets, accounting for possible effects of extreme data points (outliers). CCSS S.ID.2 Use statistics appropriate to the shape of the data distribution to compare center (median, mean) and spread (interquartile range, standard deviation) of two or more different data sets. CCSS S.ID.1 Represent data with plots on the real number line (dot plots, histograms, and box plots). CCSS N.Q.1 Use units as a way to understand problems and to guide the solution of multi-step problems; choose and interpret units consistently in formulas; choose and interpret the scale and the origin in graphs and data displays.	CCSS SMP 3 The ages of people in two bowling leagues are shown. Compare and interpret the shape, mean, and standard deviation of the two sets.

Set 1	18	22	28	20	21	25	31	28	20	62	21
Set 2	22	18	24	20	25	32	33	21	35	24	25

What You Will Learn	Preview Question
Lesson 11.3: Comparing Sets of Data	
CCSS S.ID.2 Use statistics appropriate to the shape of the data distribution to compare center (median, mean) and spread (interquartile range, standard deviation) of two or more different data sets. CCSS S.ID.3 Interpret differences in shape, center, and spread in the context of the data sets, accounting for possible effects of extreme data points (outliers). CCSS S.ID.1 Represent data with plots on the real number line (dot plots, histograms, and box plots).	CCSS SMP 6 Describe how the mean and range are affected if each number in the data set is halved. 18, 28, 26, 10, 25, 25, 64, 62, 78, 28
Lesson 11.4: The Normal Distribution	
CCSS S.ID.2 Use statistics appropriate to the shape of the data distribution to compare center (median, mean) and spread (interquartile range, standard deviation) of two or more different data sets. CCSS S.ID.1 Represent data with plots on the real number line (dot plots, histograms, and box plots). CCSS S.ID.3 Interpret differences in shape, center, and spread in the context of the data sets, accounting for possible effects of extreme data points (outliers). **Also Addresses:** S.ID.4	CCSS SMP 1 The distribution curve for a national standardized math test is shown below. One school had a mean score of 71 and a standard deviation of 6.2 points. How does the probability that a student in this school will score 76 or less compare to the national average? Explain. 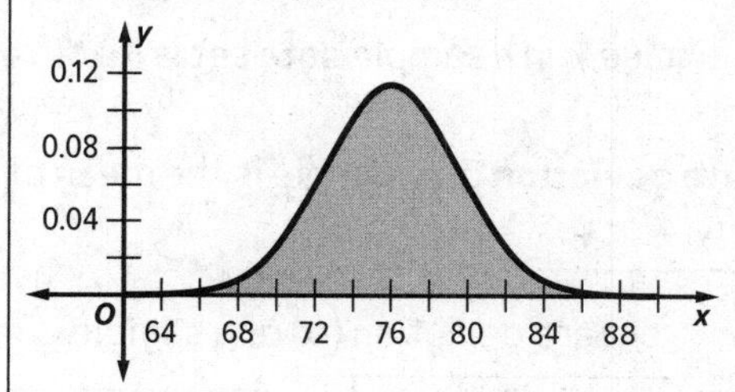
Lesson 11.5: Two-Way Frequency Tables	
CCSS S.ID.5 Summarize categorical data for two categories in two-way frequency tables. Interpret relative frequencies in the context of the data (including joint, marginal, and conditional relative frequencies). Recognize possible associations and trends in the data.	CCSS SMP 4 The table shows the results of a sample survey of two toothpaste brands. What is the probability that a surveyed woman prefers Brand A?

	Brand A	Brand B
Men	89	96
Women	72	108

11.1 Statistics and Parameters

Content: S.ID.2, S.ID.3
Practices: 1, 2, 3, 5, 6, 7
Use with Lesson 12-2

Objectives

- Compare center and spread of different data sets.
- Use statistics from data samples to estimate parameters of populations.
- Explore the effects of outliers in samples.

A **sample** is a subset of a population chosen to represent the population. A **statistic** is a measure calculated from the sample. Mean, mode, range, and interquartile range are all examples of a statistic. The equivalent measure for the population is called a **parameter** of the population. Parameters are fixed values that can be determined by the entire population but are typically estimated based on the statistics of a carefully chosen random sample.

Statistical inference is the process of drawing conclusions about a population based on information from a sample. For example, the mean height of all 18-year-old women in Texas can be inferred from a random sample of 1000 women in Texas who are 18 years old.

Making reasonable inferences about parameters of a population requires a different sort of sample statistic than the range or interquartile range.

The **deviation** of a value x_i in a sample data set is its difference from the sample mean $\overline{x}$, or $x_i - \overline{x}$.
The **mean absolute deviation** of a sample is the mean of the absolute deviations:

$\text{MAD} = \frac{|x_1 - \overline{x}| + |x_2 - \overline{x}| + \cdots + |x_n - \overline{x}|}{n} = \frac{1}{n}\sum|x_i - \overline{x}|$, where the sigma Σ means "sum over all values of *i*." The standard deviation (σ) of a sample is given by

$$\sigma = \sqrt{\frac{|x_1 - \overline{x}|^2 + |x_2 - \overline{x}|^2 + \cdots + |x_n - \overline{x}|^2}{n}} = \sqrt{\frac{1}{n}\sum|x_i - \overline{x}|^2}.$$

EXAMPLE 1 Compare Measures of Spread (CCSS S.ID.3)

The following is a table of long jump distances for students in a classroom.

Long Jump Distances (m)		
6.1	5.9	6.3
6.3	6.1	5.8

a. USE TOOLS Use a calculator to find the mean and standard deviation of the set. (CCSS SMP 5)

b. CALCULATE ACCURATELY Find the mean absolute deviation of the jumps. (CCSS SMP 6)

c. COMMUNICATE PRECISELY If the data is simply a sample, what might the population be for this example? Use the mean absolute deviation to judge how well the mean represents the data. (CCSS SMP 6)

d. REASON ABSTRACTLY A different sample of jumps from a second classroom is given in the following table. Calculate the mean and standard deviation using a calculator, and discuss how this set compares to the first set. CCSS SMP 2

Long Jump Distances (m)		
6.1	5.9	6.5
6.5	6.1	5.4

EXAMPLE 2 Identify and Allow for Outliers CCSS S.ID.2, S.ID.3

Using a GPS device, Nicholas and his son Jeff recorded the length of all their tee shots hit with a driver during a recent round of golf. The results are shown in the table.

	Drives (yds)											
Nicholas	195	203	217	198	193	188	207	200	215	205	210	207
Jeff	198	210	225	196	198	14	240	192	221	215	201	210

a. USE TOOLS Use technology to find the mean, median, interquartile range, and standard deviation of each sample to the nearest tenth. CCSS SMP 5

b. INTERPRET PROBLEMS Use the statistics to show in what aspects of driving Nicholas is better. CCSS SMP 1

c. INTERPRET PROBLEMS Use the statistics to show in what aspects of driving Jeff is better. CCSS SMP 1

d. INTERPRET PROBLEMS On one shot, Jeff loses his grip and drives the ball only 14 yards. Write what the definition for an outlier is and then use that information to decide if the 14-yard drive is an outlier for Jeff's sample drives. If it is, then use technology to recalculate the mean and standard deviation of Jeff's sample drives without the outlier and describe how this affects the statistics of the sample. CCSS SMP 1

PRACTICE

1. Two different samples on the shell diameter of a species of snail are shown. CCSS S.ID.2

Sample A (mm)		
45	35	37
40	42	40
28	38	31

Sample B (mm)		
26	44	40
27	35	28
26	39	31

   a. **INTERPRET PROBLEMS** Use the median and interquartile range to compare the samples. CCSS SMP 1

   b. **USE STRUCTURE** Based on your findings and on the data points in each sample, which sample appears to be more representative? Explain your reasons. CCSS SMP 7

2. **INTERPRET PROBLEMS** Height data samples of 17-year-old male and female students are shown. Use the mean and standard deviation to compare the samples. CCSS S.ID.2, SMP 1

Heights of Male Students (inches)		
71	69	67
68	69	70
72	74	68
71	69	72

Heights of Female Students (inches)		
67	62	69
65	71	66
63	65	68
66	63	70

3. Miranda is looking at her high jump results from the last two seasons' competitions. She feels that her performance has been skewed by unusual results. Below are her results from the last two seasons. CCSS S.ID. 2, S.ID.3

Season 1:	Results (m):		1.69	1.72
1.63	1.66	1.72	1.78	1.69
1.45	1.66	1.57	1.66	1.69

Season 2:	Results (m)		1.70	1.68
1.63	1.62	1.71	1.54	1.73
1.59	1.76	1.57	1.61	1.60

   a. **REASON QUANTITATIVELY** Consider the two sets of data. Determine whether either contains an outlier. Find the new mean and standard deviations once all outliers are removed from the sets. CCSS SMP 3

b. REASON QUANTITATIVELY Determine which was Miranda's better season after the outliers were removed. Explain your answer. CCSS SMP 2

4. Nicholas is preparing to go golfing again. He is looking for a partner. He wants to pick the best player, so he contacts two friends and obtains their totals on the last six rounds of golf. The data is given below. CCSS S.ID. 2, S.ID.3

Melinda	45	47	46	47	44	49
Wanda	49	49	47	47	48	47

a. CALCULATE ACCURATELY For each player, find the mean and standard deviation. CCSS SMP 6

b. REASON QUANTITATIVELY What would be the advantages or disadvantages to Nicholas selecting one player over the other? CCSS SMP 2

5. Consider the following set of package weights randomly selected by the Post Office during the Christmas season. CCSS S.ID.3

Package Weights		
0.56 lb	1.21 lb	1.03 lb
0.78 lb	0.88 lb	0.96 lb

a. CALCULATE ACCURATELY Find the mean and the mean absolute deviation of the weights. CCSS SMP 6

b. COMMUNICATE PRECISELY Use the mean absolute deviation to judge how well the mean represents the data. CCSS SMP 6

c. COMMUNICATE PRECISELY What could be the population for this sample? Comment on why this data set may or may not be representative of the population? CCSS SMP 6

11.2 Distributions of Data

Objectives

- Interpret differences in shape, center, and spread for data distributions.
- Select the appropriate statistics to describe samples based on their distributions.

CCSS STANDARDS

Content: S.ID.1, S.ID.2, S.ID.3, N.Q.1
Practices: 1, 4, 5, 6, 7, 8
Use with Lesson 12–3

The **distribution** of a sample or population of data describes the observed or theoretical frequencies of the data values, often in a visual way, allowing the data to be assessed at a glance.

EXAMPLE 1 Use Statistics to Create a Visual Display of a Distribution CCSS S.ID.1

EXPLORE Fifteen expectant women took part in a study of the duration of pregnancy.

a. COMMUNICATE PRECISELY Find the *five-number summary* of the data, defined as: minimum value, first quartile, median, third quartile, and maximum value. Use these statistics and the scale provided to create a box-and-whisker plot of the data. CCSS SMP 6

Pregnancy Duration (weeks)		
39	41	38
32	39	40
35	39	40
40	41	37
31	40	40

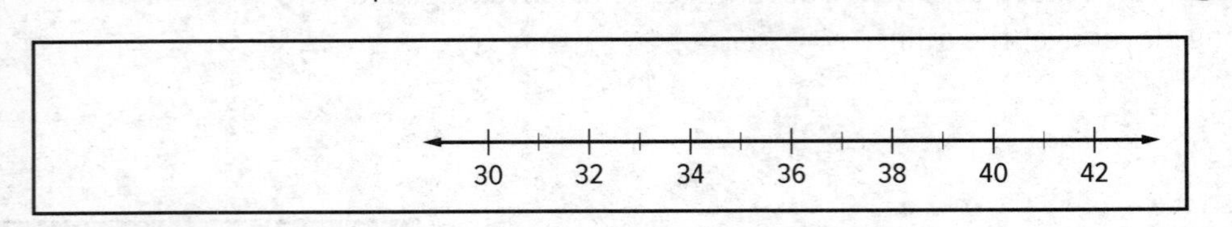

b. USE STRUCTURE Based on the box-and-whisker plot, where is the data most concentrated within the overall range of the sample? Explain your answer. CCSS SMP 7

c. INTERPRET PROBLEMS The histogram represents the same data as **part a**. How is the position of the histogram's peak related to the shape of your box-and-whisker plot? CCSS SMP 1

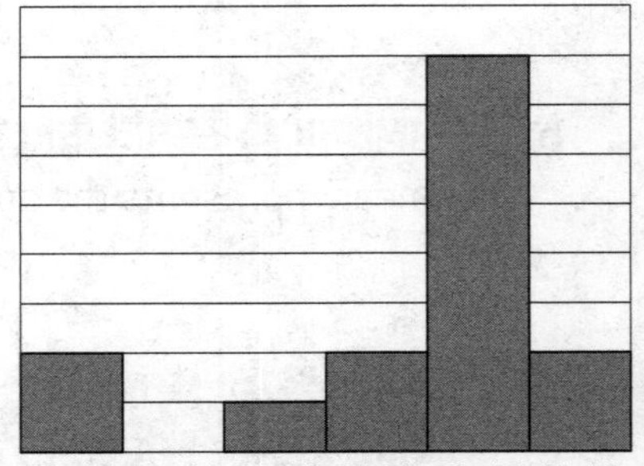

d. EVALUATE REASONABLENESS Does this data seem reasonable? Why or why not? CCSS SMP 8

A distribution is **negatively skewed** if it has a peak that is to the right within the distribution, and **positively skewed** if it has a peak to the left. An even distribution around a central peak is **symmetric**.

EXAMPLE 2 Describe the Distribution of a Data Set

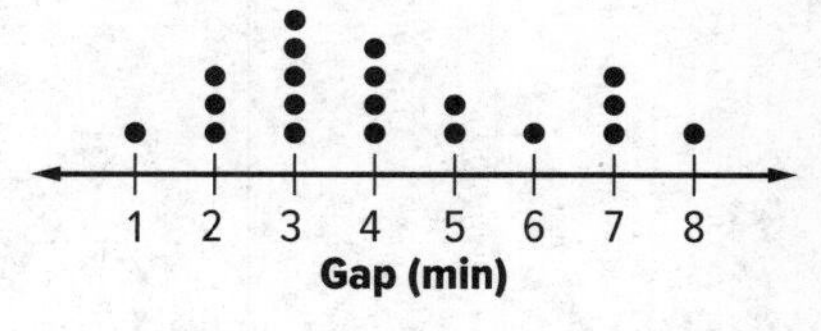

The dot plot displays data on the gaps between phone calls received at a company switchboard.

a. **USE TOOLS** Using the dot plot, enter the data into a graphing calculator and create a box-and-whisker plot. Make a sketch of the box-and-whisker plot labeled with the five-number summary data points. CCSS S.ID.1, SMP 5

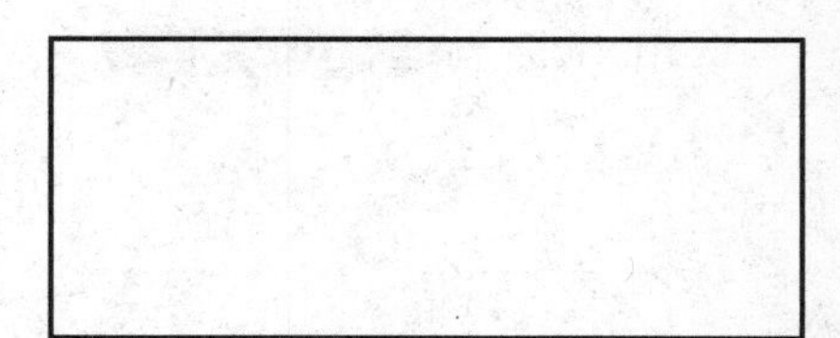

b. **COMMUNICATE PRECISELY** Describe the shape and any other significant features of the distribution. CCSS S.ID.3, SMP 6

c. **USE STRUCTURE** A second company records the gaps between phone calls in a chart. Create a histogram from the data provided. CCSS SMP 7

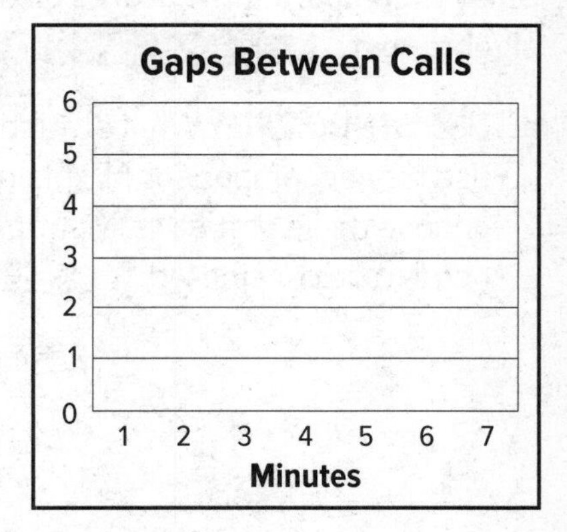

Gaps between calls, in minutes: 4, 3, 5, 4, 1, 2, 4, 4, 7, 6, 5, 3, 4, 5, 5, 3, 6, 1, 2, 6, 3, 2, 7

d. **COMMUNICATE PRECISELY** Contrast the shape of the second data set with the first. What does the difference tell you about the data? CCSS SMP 6

The shape of a sample's distribution affects which statistics best describe the data. Symmetric data sets are well described—or compared if there are two samples—by the mean and standard deviation. For skewed distributions, the five-number summary is a better description or comparison.

EXAMPLE 3 Choose Statistics to Describe a Data Set CCSS S.ID.1

During practice a coach has his kicker attempt field goals from each yard line between 17 and 40 yards. The coach keeps track of each attempt. The kicker's successful field goal distances for the practice are shown at the right.

Field Goal Distances (yd)		
17	20	22
23	33	29
28	18	36
24	26	35
32	25	19
38	30	21

a. **COMMUNICATE PRECISELY** Choose, create, and sketch a visual representation of the data. Describe its distribution. CCSS SMP 6

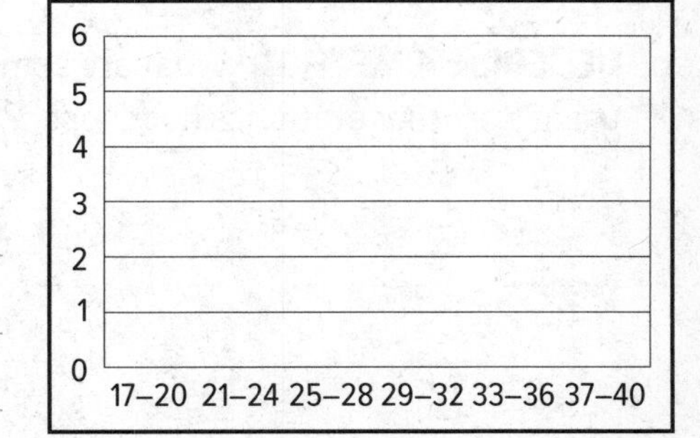

b. INTERPRET PROBLEMS Choose and calculate representative statistics for this data set. Justify your choice. CCSS SMP 1

c. FIND A PATTERN Discuss the data's distribution based on your sketch and chosen statistics. CCSS S.ID.3, SMP 7

EXAMPLE 4 Choose Statistics to Compare Data Sets

Daisy is comparing the scores of two different classes on the same algebra test. CCSS S.ID.2, S.ID.3, N.Q.1

a. USE STRUCTURE What statistics should Daisy use for her comparison? Explain why she should choose these statistics. Mention any outliers that need to be considered. CCSS SMP 7

Class A			
68	68	32	45
78	98	63	82
90	52	47	67
71	80	69	66
86	75	57	71
60	67	64	64
62	74	77	77

Class B			
60	68	88	84
53	59	70	73
62	68	73	82
82	78	54	93
79	78	73	80
85	64	90	72
72	71	67	76

b. INTERPRET PROBLEMS Calculate your chosen statistics, taking account of outliers. Then, use these statistics to suggest issues for Daisy to discuss in analyzing her comparison. CCSS SMP 1

c. DESCRIBE A METHOD What are some advantages of comparing the data in **part b** using the *other* statistics not chosen? CCSS SMP 8

1. The histograms show the weight of sample boxes of two brands of pasta. CCSS S.ID.2, S.ID.3

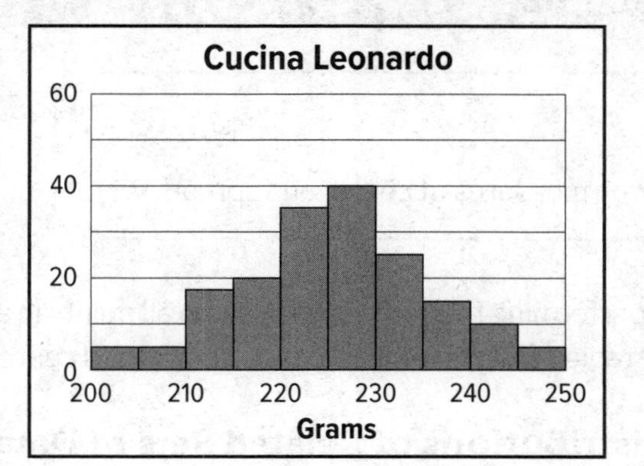

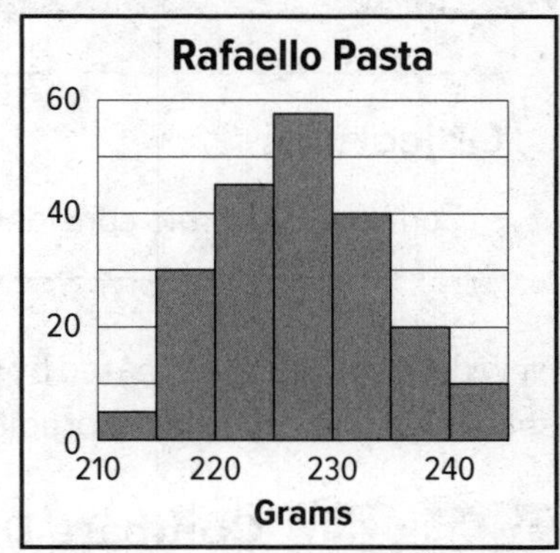

   a. **USE A MODEL** Do the two packages of pasta likely have the same advertised weight? Which manufacturer's quantity control appears better? Explain your answers based on the distributions. CCSS SMP 4

   b. **FIND A PATTERN** Infer the two population distribution shapes by sketching smooth curves across the tops of the histograms. Describe the shapes you have sketched. CCSS SMP 7

2. **CALCULATE ACCURATELY** Consider the data from **Example 3** regarding a kicker's successful field goal attempts. Find the mean and standard deviation for the data and explain what the statistics tell the coach about his kicker. CCSS S.ID.3, SMP 6

3. **INTERPRET PROBLEMS** The United States has been sending astronauts up in the Space Shuttle since 1981. The table below provides data regarding the duration of Space Shuttle flights from 1981 to 1985 and then from 2005 to 2011.

Length of Flights from 1981–1985 (days)

Days: 2, 2, 8, 7, 5, 5, 6, 6, 10, 8, 7, 6, 8, 8, 3, 7, 7, 7, 8, 7, 4, 7, 7

Length of Flights from 2005–2011 (days)

Days: 14, 13, 12, 13, 14, 13, 15, 13, 16, 14, 15, 13, 13, 16, 14, 11, 14, 15, 12, 13, 16, 13

Choose and calculate the statistics appropriate for the distribution of the data sets. Use the statistics to compare the two sets. CCSS S.ID.2, S.ID.3, N.Q.1, SMP 1

11.3 Comparing Sets of Data

Objectives

- Compare data sets considering measures of center and spread.

CCSS STANDARDS

Content: S.ID.1, S.ID.2, S.ID.3
Practices: 1, 2, 3, 5, 6, 7, 8
Use with Lesson 12-4

New data sets can be created by performing the same operation on all points in an original set. These operations have predictable effects on measures of spread and center.

EXAMPLE 1 Compare Distributions of Related Sets of Data

EXPLORE The installation times for Easiflow central air systems are shown.

a. INTERPRET PROBLEMS Easiflow develops a new system that will cut installation time on any project by 1.5 hours. The table below shows installation times for a series of projects before the new system was put in place. Fill in the predicted installation times to create a new set of data that would represent the times if the new system were used. Next, compare the two data sets. How do the mean, median, standard deviation, and interquartile range (IQR) change, and why? Verify your answer by finding these measures before and after the reduction. CCSS S.ID.2, SMP 1

Installation Times (hrs)														
12.0	12.0	9.5	10.0	11.5	10.0	17.5	6.5	9.0	10.5	11.0	13.5	8.0	9.5	15.0

b. USE TOOLS Predict the effect of implementing the new system on the box-and-whisker plot for the data set. Check your prediction by creating "before" and "after" plots. CCSS S.ID.1, S.1D.2, SMP 5

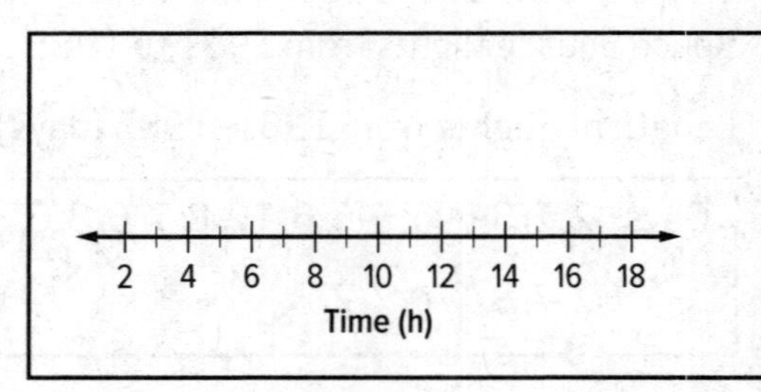

c. CONSTRUCT ARGUMENTS Explain what will happen to a histogram of the data as a result of the implementing the new system. CCSS S.ID.1, SMP 3

EXAMPLE 2 Compare Data Sets with Symmetric Distributions CCSS S.ID.2

A sports reporter wants to compare two basketball players over the first 20 games of the season. To do so, she collected the points scored by each player for each of the games. The results are shown below.

Player A			
24	32	25	29
29	19	27	36
22	35	24	32
13	21	20	17
27	23	11	25

Player B			
32	17	26	39
24	36	25	11
19	31	9	27
24	22	20	21
13	42	28	25

a. USE STRUCTURE Choose and sketch a visual representation of the data for each player and describe the shape of each distribution. CCSS SMP 7

Player A

8 6 4 2 0

10–14 15–19 20–24 25–29 30–34 35–39

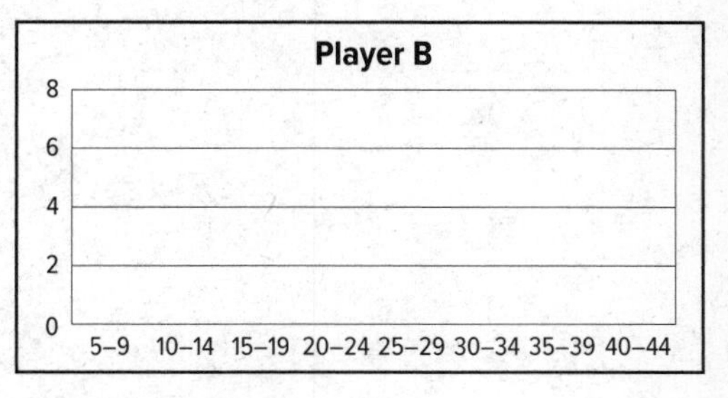

b. INTERPRET PROBLEMS Use the appropriate statistics to compare the center and spread of the data for each player. What do they tell you about the scoring tendencies of the two players? CCSS SMP 1

EXAMPLE 3 Compare Distributions in Terms of Shape

The twelve fastest times for two skiing events—a downhill race and a slalom competition—are shown. CCSS S.ID.1, S.ID.2, S.ID.3

Downhill Times (s)	
114.6	118.7
115.4	113.7
114.0	113.4
113.6	112.5
116.9	114.2
115.1	120.4

Slalom Times (s)	
91.7	90.9
93.5	87.3
96.8	99.1
90.1	94.9
87.7	89.4
88.5	89.1

a. USE STRUCTURE Compare the data sets by constructing box-and-whisker plots and using appropriate statistics. Explain why you chose these statistics. CCSS SMP 7

b. DESCRIBE A METHOD How could you transform the two data sets so that you could compare the distributions purely in terms of shape, without differences in center and spread being a factor? CCSS SMP 8

c. COMMUNICATE PRECISELY Using your method, transform both data sets and construct box-and-whisker plots of the transformed data. What would you add to your comparison in **part a**? CCSS SMP 6

PRACTICE

1. Saeed owns an electronics store. He is revising his pricing for phone accessories. His current prices for an assortment of accessories is listed at the right. He has also determined that the mean price for the same assortment of accessories at a rival store is $10.99. CCSS S.ID.3

Saeed's Price Data ($)		
14.99	4.49	9.99
18.49	12.99	6.99
8.49	21.99	13.49
13.99	9.99	10.99
12.49	4.49	12.99

   a. REASON QUANTITATIVELY Saeed wants to match his rival's prices. Use the table below to list the new prices. Explain. CCSS SMP 2

New Prices				

   b. CALCULATE ACCURATELY Compare the mean and standard deviation of the current prices to the new prices. CCSS SMP 6

2. At the end of the year Josey, has saved $5000 from her wages. She wants to invest the money in an investment vehicle that might earn more than the interest she can make in the bank. She gets information regarding the performance of two investments over the last ten years. The annual growth rates by percent for each investment are shown below. CCSS S.ID.2

Investment A	
3.51	3.54
3.52	3.57
3.50	3.51
3.53	3.47
3.52	3.52

Investment B	
3.97	1.67
3.53	5.81
2.31	3.42
3.32	5.14
3.89	3.81

a. INTERPRET PROBLEMS Determine the shape of each distribution and use the appropriate statistics to find the center and spread for each set of data. CCSS SMP 1

b. REASON QUANTITATIVELY What do the measures of center and spread tell you about the two different investment vehicles? CCSS SMP 2

c. REASON QUANTITATIVELY Josey is planning to keep her investment in whichever vehicle she chooses for one year. What advice would you give her? CCSS SMP 2

3. Samantha is planning a two-week vacation to one of two islands and wants to base her decision on the weather history for the same dates as her vacation. She has collected the number of days that it has rained during this two-week period for each island over the past 10 years. The results are shown to the right. CCSS S.ID.2

Island A	
5	0
7	6
5	6
6	6
3	2

Island B	
4	4
6	5
3	7
4	3
5	4

a. INTERPRET PROBLEMS Determine the shape of each distribution and use the appropriate statistics to find the center and spread for each set of data. CCSS SMP 1

b. REASON QUANTITATIVELY What advice would you give Samantha regarding which island to visit for her vacation? Explain your reasoning. CCSS SMP 2

11.4 The Normal Distribution

Objectives

- Use properties of the normal distribution to make predictions about data.
- Compare data sets using transformations of normal distributions.
- Know how to calculate a *z*-score, and understand its significance for comparing normal distributions.

CCSS STANDARDS

Content: S.ID.1, S.ID.2, S.ID.3, S.ID.4
Practices: 1, 2, 4, 6, 7, 8
Use with Extend 12-8

The frequency distribution of a data set with a large number of values that tends to center around the mean in a symmetric distribution is known as the **normal distribution**. The normal distribution can be completely described by its mean, μ, and standard deviation, σ. The center of the distribution is always at μ, and its width is defined by σ, as shown in the diagram at the right.

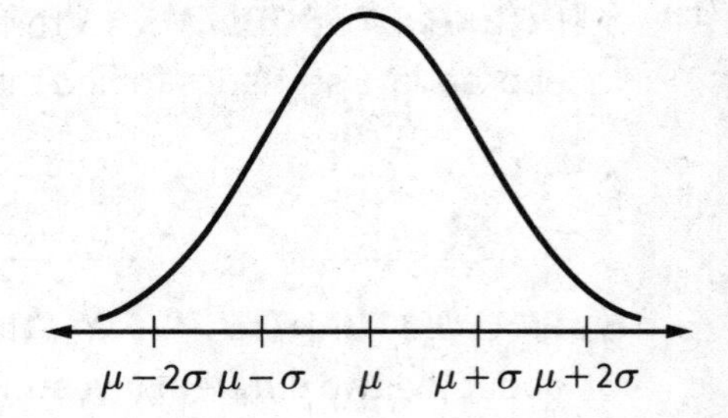

EXAMPLE 1 Recognize the Shape of Normally Distributed Data

EXPLORE Latesha has been playing a coin game with a friend for quite a while. She flips the coin, and if it comes up heads, Latesha gets to pick what they do next. The coin always comes from a particular box. One day, Latesha takes the box and flips all the coins at once. Out of the 100 coins, 35 wind up on heads.

a COMMUNICATE PRECISELY Latesha is concerned. She puts the coins away and gets 100 coins of her own. She flips them multiple times. The results are shown below. Arrange the results in a histogram. CCSS S.ID.1, SMP 6

Number of Heads out of 100 Coins:
52, 46, 50, 49, 47, 53, 51, 54, 43, 47, 55, 59, 51, 51, 49, 50, 50, 43, 52, 54, 39, 50, 51, 49, 48, 51, 41, 45, 47, 47, 50, 57

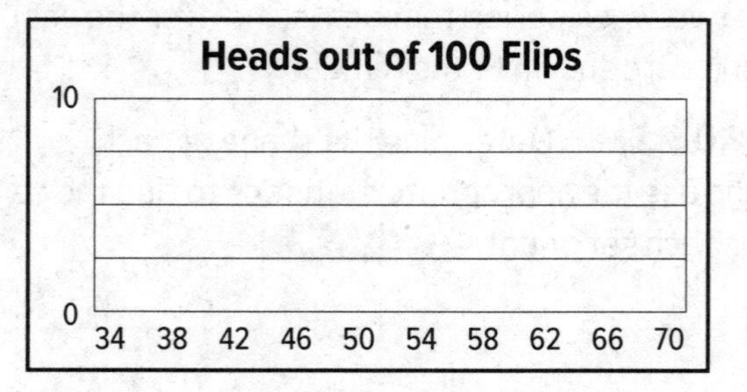

b. REASON ABSTRACTLY What can you say about the amount of data points within 1 standard deviation of the mean? Within 2 standard deviations of the mean? Beyond 3 standard deviations of the mean? CCSS S.ID.3, SMP 2

All normal distributions have the same percentage of data within one standard deviation of the mean, about 68%. Approximately 95% of data lie within two standard deviations and approximately 99.7% of data lie within three standard deviations of the mean for any normally distributed data. A normal curve with more precise percentages is shown below.

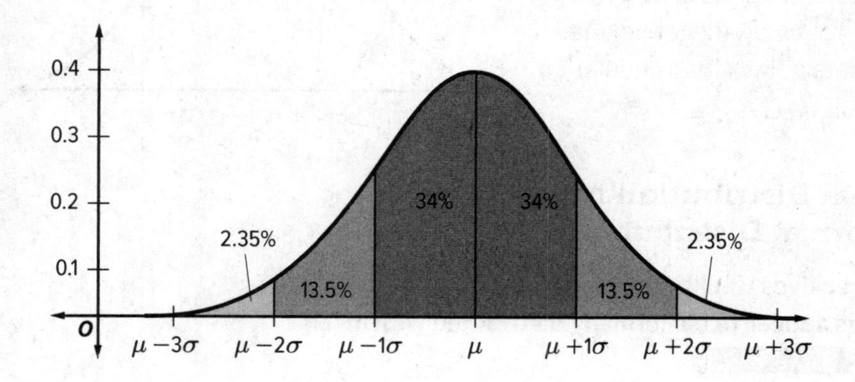

EXAMPLE 2 Find Proportions of Normally Distributed Data Within Given Limits CCSS S.ID.2

Pablo is investigating the heights of "tubes"—waves large enough to surf along the inside—and believes they are normally distributed with mean $\mu = 2.35$ m and $\sigma = 0.42$ m.

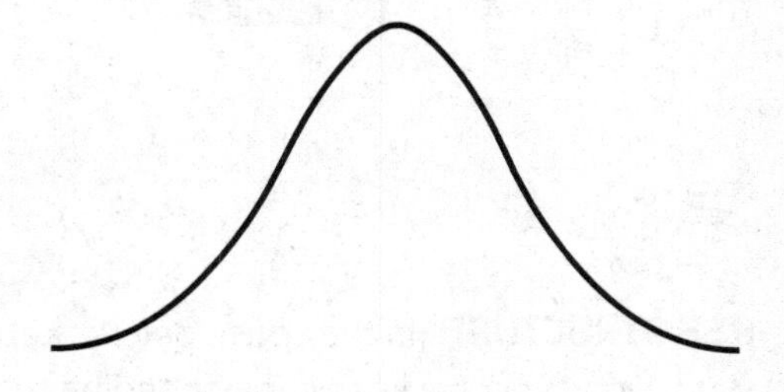

a. USE STRUCTURE Use the mean and standard deviation to label the normal curve provided. CCSS SMP 7

b. INTERPRET PROBLEMS Pablo believes that the larger the tube is, the easier it is to "catch" or successfully surf. What percent of tubes are greater than 2.77 meters in height? Explain your reasoning. CCSS SMP 1

c. USE STRUCTURE What percent of the waves have heights that are between 1.93 meters and 3.19 meters? Explain your reasoning. CCSS SMP 7

d. USE STRUCTURE Pablo believes that small waves are not worth surfing. What percent of waves are less than 1.51 meters? Explain your reasoning. CCSS SMP 7

The **standard normal distribution** is a normal distribution with a mean 0 and a standard deviation 1. A normal curve allows us to visualize the idea of data falling one, two, or three standard deviations away from the mean.

A ***z*-score** for a data point restates the value in terms of how many standard deviations it is above (positive) or below (negative) the mean. The formula for the ***z*-score** of a data point x for a population or sample with mean μ and standard deviation σ is $z = \frac{x - \mu}{\sigma}$.

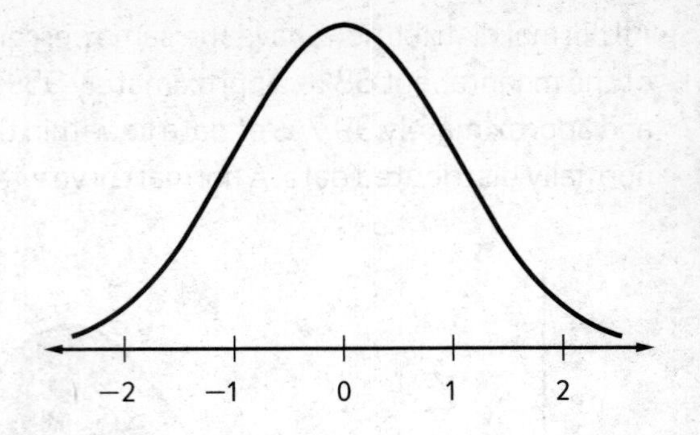

EXAMPLE 3 Convert a Normal Distribution to the Standard Normal Distribution

Andrew is a sales representative who receives weekly bonuses based on a percent of the sales he makes. His weekly earnings appear to be normally distributed with mean \$715 and standard deviation \$57. CCSS S.ID.2, S.ID.3

a. REASON QUANTITATIVELY One week Andrew makes \$829. A second week he makes \$658. What are the *z*-scores associated with these data points?

b. REASON QUANTITATIVELY What do the *z*-scores reveal in regard to how good or bad these weeks were? CCSS SMP 2

c. USE STRUCTURE In the space below sketch the normal curve suggested by Andrew's data. Draw vertical lines corresponding to 1 and 2 standard deviations above and below the mean and label the values. Plot the dollar values from Andrew's two weeks on the graph. CCSS SMP 7

The percentage of a population in any interval of the data can be found using the normalcdf function on the DISTR key of a graphing calculator. This function requires values entered for the form normalcdf(a, b, μ, σ) where a and b are the ends of the interval, μ is the population mean, σ is the standard deviation, and $a < b$. Use -1×10^{99} or 1×10^{99} when a or b refer to the least or greatest data value of the population.

EXAMPLE 4 Make Statistical Inferences about Normal Populations

The heights of male and female students in a Grade 9 math class are shown. This class is representative of the normally distributed heights of all Grade 9 male and female students.

Heights of Male Students (in.)		
65	66	66
69	71	66
63	67	70
68	62	65
68	64	73

Heights of Female Students (in.)		
64	62	63
61	66	65
69	64	64
61	59	63
67	65	62

a. USE A MODEL Find the mean and standard deviation for each population. CCSS S.ID.2, SMP 4

b. USE STRUCTURE Using the means and standard deviations, draw and label normal curves that represents the population of all Grade 9 male and female students.

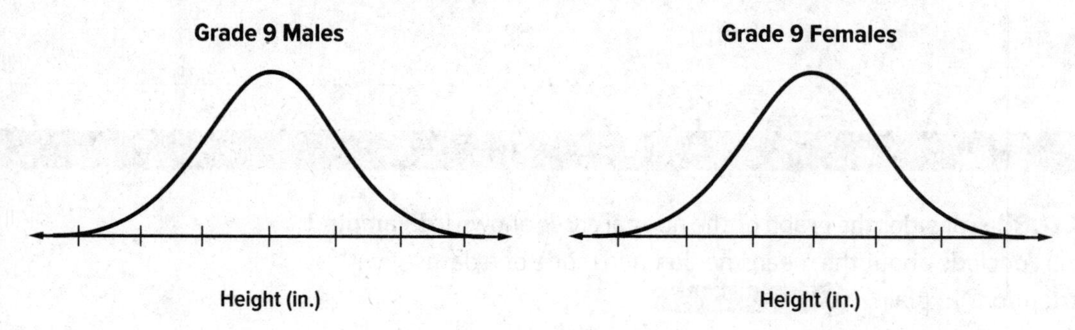

c. REASON QUANTITATIVELY Aleesha is 67 inches tall and Issac is 65 inches tall. Using the statistics from part a and a graphing calculator, what percent of the population of Grade 9 females is Aleesha taller than and what percent of the Grade 9 males is Issac shorter than? Explain your answers. CCSS SMP 2

EXAMPLE 5 Compare Data from Different Normal Distributions

SAT reading and math scores are each normally distributed on a 200–800 point scale. The mean reading score is 496, with standard deviation 115. Math scores have mean 514 and standard deviation 118. CCSS S.ID.2, S.ID.3, S.ID.4

a. **DESCRIBE A METHOD** Describe how to sketch both distributions on the same appropriately scaled horizontal axis. Then create this sketch. CCSS SMP 8

b. **REASON QUANTITATIVELY** Tamsin has scored 660 on reading and 670 on math. Choose a method to compare Tamsin's scores and use it to decide which score is stronger. CCSS SMP 2

c. **INTERPRET PROBLEMS** An SAT score always has a percentile rank associated with it. This tells what percent of college-bound seniors scored lower than this particular score. Determine the percentile ranks for Tamsin's scores and compare this to the method of comparison you used in **part b**. CCSS SMP 1

PRACTICE

1. a. **USE STRUCTURE** Consider the graph of the normal curve shown in **Example 1**. What can you conclude about the mean, median, and mode of a dataset with a normal distribution? Explain. CCSS S.ID.4, SMP 7

   b. **COMMUNICATE PRECISELY** A study of over 400,000 births in Sweden showed that the lengths of pregnancy for human beings are normally distributed with a mean of 281 days and a standard deviation of 13 days. A doctor has a patient who has been pregnant for 295 days and is getting concerned. What can the doctor tell her patient regarding the length of her pregnancy? CCSS S.ID.4, SMP 6

2. **INTERPRET PROBLEMS** Kerry is comparing the winning scores for her high school's football and soccer teams. Both samples represent normal populations. Which winning score is more unusual: 36 or more points, or 4 or more goals? CCSS S.ID.2, S.ID.3, SMP 1

Football Winning Scores (points)		
42	24	45
28	38	20
35	31	21
21	27	28
34	17	14

Soccer Winning Scores (goals)		
1	4	1
1	2	2
3	2	3
3	2	1
6	1	4

3. Considering how many standard deviations a data point is from the mean when data is distributed normally can help identify data points that may need to be reconsidered or reviewed. CCSS S.ID.2, S.ID.3

   a. **EVALUATE REASONABLENESS** For each SAT score distribution in **Example 5,** identify scores that are 2 and 2.5 standard deviations above and below the mean. Describe how you could use these scores to identify unusual scores and how unusual those scores might be. CCSS SMP 8

   b. **USE A MODEL** A certain high school has a senior class with 200 students. Academic abilities of the class are similar to the abilities of the students who take the SAT throughout the country. Is it likely that at least one student would have a score of 750 or above in math? 10 students? Justify your answers. CCSS SMP 4

4. **DESCRIBE A METHOD** Describe how to find data points corresponding to given distance away from the mean value for a normal distribution with parameters μ and σ. Based on **Example 4**, what heights (to the nearest inch) are within 1.25 standard deviations of the mean for Grade 9 male and female students? CCSS S.ID.2, SMP 8

11.5 Two-Way Frequency Tables

Objectives

- Use two-way frequency tables to summarize data, and recognize trends.
- Interpret relative frequencies and conditional, joint, and marginal relative frequencies as probabilities.

Content: S.ID.5
Practices: 1, 2, 3, 4, 6, 7
Use with Extend 12–7

A **two-way frequency table**, or contingency table, is a tool for showing frequencies of data classified in two categories. The rows in the table indicate one category and the columns indicate the other.

The values for every combination of subcategories in a two-way frequency table are called **joint frequencies**. The totals for each row or column are termed **marginal frequencies**.

EXAMPLE 1 Create and Analyze a Two-Way Frequency Table CCSS S.ID.5

In a poll for senior class president, 68 of the 145 male students said they planned to vote for Santiago. Out of 139 female students 89 planned to vote for his opponent, Measha.

a. REASON QUANTITATIVELY Enter the given poll data into the appropriate cells of the table. Then fill in the remaining cells in the table, showing your reasoning below. CCSS SMP 2

Gender	Santiago	Measha	Totals
Female			
Male			
Totals			

Joint Frequencies

Grand Total

Marginal Frequencies

b. COMMUNICATE PRECISLEY What do the joint frequencies represent? What do the marginal frequencies represent? CCSS SMP 6

c. USE STRUCTURE Express the grand total value in the bottom right cell as the sum of three different logical combinations of values, in both words and numbers. CCSS SMP 7

A **relative frequency** is the ratio of the number of observations in a category to the total number of observations. Relative frequencies are also probabilities. To create a two-way relative frequency table divide each of the values by the total number of observations and replace them with their corresponding decimals or percents.

EXAMPLE 2 Create and Analyze a Two-Way Relative Frequency Table CCSS S.ID.5

A two-way relative frequency table can be created for the poll data from Example 1.

a. CALCULATE ACCURATELY Convert the joint and marginal frequencies for the poll data into relative frequencies, rounded to the nearest percent. CCSS SMP 6

Gender	Santiago	Measha	Totals
Female			
Male			
Totals			

b. USE STRUCTURE Fill in the bottom right entry. Explain why the answer is reasonable. CCSS SMP 7

c. USE A MODEL What is the probability that a student chosen at random is female and intends to vote for Santiago? Explain. CCSS SMP 4

d. USE A MODEL Assume that the poll has been conducted without bias. What is the probability that a student chosen at random plans to vote for Measha? CCSS SMP 4

e. REASON QUANTITATIVELY Suppose a third candidate, Adil, joins the race. 10 of the females and 8 of the males who were going to vote for Santiago are going to vote for Adil. 18 of the males and 16 of the females who were going to vote for Measha are going to vote for Adil. Make a relative frequency table that illustrates this data? Assuming these numbers provide an accurate representation, who will win the election with Adil as a third candidate? Why? CCSS SMP 2

Gender	Santiago	Measha	Adil	Totals

A **conditional relative frequency** is the ratio of a joint frequency to one of the marginal frequency.

EXAMPLE 3 Use Conditional Relative Frequencies to Analyze Categorical Data CCSS S.ID.5

Instead of finding the relative frequency for each category, Measha wants to know the conditional relative probabilities of voter preference based on gender.

a. **CALCULATE ACCURATELY** Use the table from **Example 1** to enter the conditional relative frequencies in the table. Show the ratio and the percent. CCSS SMP 6

Gender	Santiago	Measha	Totals
Female			
Male			

b. **INTERPRET PROBLEMS** What does each conditional relative frequency represent? CCSS SMP 1

c. **USE A MODEL** What is the probability that a male will vote for Santiago? How does this probability differ from the probability that a student chosen at random is male and intends to vote for Santiago? CCSS SMP 4

d. **CONSTRUCT ARGUMENTS** What advice would you give Santiago on his election strategy? Base your advice on trends in the polling data. CCSS SMP 3

PRACTICE

1. a. **CALCULATE ACCURATELY** Enter the conditional relative frequencies based on voter preference in the table. Show your calculations. CCSS S.ID.5, SMP 6

Gender	Santiago	Measha
Female		
Male		
Totals		

b. **INTERPRET PROBLEMS** What does each conditional relative frequency represent? CCSS SMP 1

c. **USE A MODEL** What is the probability that a vote for Measha will come from a female student? How is this different than the probability that a female student intends to vote for Measha? CCSS S.ID.5, SMP 4

2. a. **CALCULATE ACCURATELY** The table shows the relative frequencies of drive systems for different vehicle types in a school parking lot. If there are 215 vehicles in the lot, calculate the joint and marginal frequencies. CCSS S.ID.5, SMP 6

Vehicle Type	2WD	AWD	Totals
Hatchbacks	42%	4%	
Sedans	28%	6%	
SUVs	1%	19%	
Totals			215

b. **REASON QUANTITATIVELY** Without calculating individual frequencies, how many times greater will the conditional relative frequencies based on drive systems for AWD be than the relative frequencies for AWD, and why? CCSS S.ID.5, SMP 2

3. An online poll collected a sample of Thanksgiving pie preferences for different U.S. regions. CCSS S.ID.5

a. **CALCULATE ACCURATELY** Complete the table, and also find each relative frequency, to the nearest tenth of a percent. CCSS SMP 6

Region	Apple	Sweet Potato	Pumpkin	Totals
West	77	4	13	
Midwest	32		54	
South		63	24	
Northeast	92	2		
Total	213	75	117	

b. **USE A MODEL** Assuming the poll is representative of the whole population, what is a reasonable estimate of the probability that a family will be from the northeast and will be eating pumpkin pie on Thanksgiving? CCSS SMP 4

c. **CALCULATE ACCURATELY** Construct a table of conditional relative frequencies based on pie preference. Round each percent to the nearest tenth. Interpret the meaning of the probabilities in the context of the problem. CCSS SMP 6

Region	Apple	Sweet Potato	Pumpkin

d. **REASON ABSTRACTLY** If we had found the conditional relative frequencies by dividing by the total replies from each region, what would be the meaning of the probability in each cell? CCSS SMP 2

Performance Task

Choosing a Diver

Provide a clear solution to the problem. Be sure to show all of your work, include all relevant drawings, and justify your answers.

Kate Rodriguez is the coach of a diving team. She has been asked to choose one of the members of the team to compete at a state diving meet. Coach Rodriguez is considering two of the team's divers. To help her make the decision she collects the divers' scores from their last 20 diving meets. The scores are shown in the table.

Diver	Scores
Amani	120.1, 135.2, 75.6, 102.3, 120.6, 95.2, 99.3, 125.5, 136.1, 115.3, 124.6, 136.0, 102.3, 126.0, 104.7, 90.3, 130.6, 126.9, 92.4, 80.3
Isabelle	97.5, 115.3, 104.6, 103.9, 108.3, 97.5, 117.3, 92.6, 125.3, 106.5, 108.3, 103.3, 121.1, 124.0, 91.3, 96.6, 116.3, 111.5, 103.3, 109.3

Part A

Make box plots for the data. For each data set, describe the shape of the distribution and explain what the plots tell you about each diver.

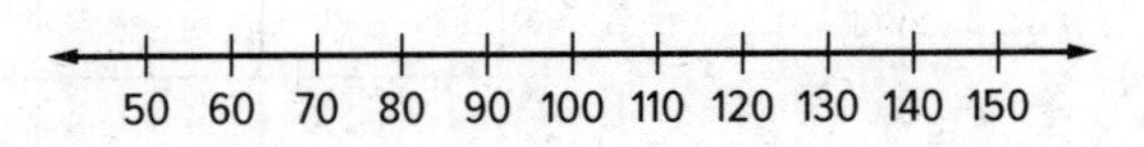

Part B

Compare the data sets using either the means and the standard deviation or the five-number summaries. Justify your choice.

Part C

A third member of the team, Tamiko, tells Coach Rodriguez that she should also be considered for the state diving meet. She tells the coach that over her last 10 diving meets her median score is 110 and her interquartile range is about the same as Isabelle's interquartile range. Give a set of 10 scores that could represent Tamiko's data and justify your answer.

Part D

Coach Rodriguez wants to choose the diver with the best scores, but she also wants to choose a diver who is consistent. Before she makes her decision, she discovers that Isabelle's scores were incorrectly reported, and each score should be 10 points greater than what is shown. Taking this into consideration, would you send Tamiko, Isabelle, or Amani to the competition? Justify your choice.

Performance Task

Handedness and Sports

Provide a clear solution to the problem. Be sure to show all of your work, include all relevant drawings, and justify your answers.

A one-on-one sport is a sport in which two people play directly against each other. Some examples of one-on-one sports are shown. Is there a connection between a person's handedness (right-handed or left-handed) and whether or not they play a one-on-one sport? Survey students in your school district. Use the data you have collected to investigate this question as follows.

Examples of One-on-One Sports

Badminton
Fencing
Judo
Ping Pong
Squash
Tennis
Wrestling

Part A

Summarize the data you collected by making a two-way frequency table that shows your results. Describe the categories and subcategories for your table.

Part B

Make a relative frequency two-way table for your data. What are the joint and marginal relative frequencies? What do they represent?

Part C

Based on your data, given that a person is left-handed, what is the probability that he or she plays a one-on-one sport? Given that a person plays a one-on-one sport, what is the probability that he or she is left-handed? Justify your answers.

Part D

It has been suggested that there is an association between being left-handed and playing a one-on-one sport. Use your data to support or refute this claim. Justify your response.

Standardized Test Practice

1. Two data sets are normally distributed, and they have the same mean. Which of the following statements are true? CCSS S.ID.2

The distribution of both data sets is symmetric about the mean.

The standard deviations of the two data sets are equal.

The medians of the two data sets are equal.

The IQR is equal to half the range for both data sets.

2. The table below shows the final score for a basketball team in the first ten games of the season. CCSS S.ID.2

43	51	39	48	60
64	56	47	52	48

Complete the following.

Mean: ☐ Median: ☐

Mode: ☐ Range: ☐

IQR: ☐

3. Forty six students took a science test. The scores were normally distributed with a mean of 52 and a standard deviation of 7. Which of the following statements are true? CCSS S.ID.2

About 95% of the students had of score of between 45 and 59.

About 23 students had a score that is above 52.

The probability that a student randomly selected from students who took the test scored lower than 31 is less than 1%.

The data is bimodal, with modes of about 45 and 59.

4. Ten students were asked how many minutes it took them to get to school. The responses from nine of the students were 15, 11, 8, 5, 6, 17, 23, 11, and 20 minutes. If the median of the data was 12 minutes, how many minutes did it take the tenth student to get to school? CCSS S.ID.2

☐ minutes

If the mean of the data was 13 minutes, how many minutes did it take the tenth student to get to school?

☐ minutes

If the range of the data was 18 minutes, which of the following could not have been the time it took the tenth student to get to school?

15 minutes 23 minutes

20 minutes 25 minutes

5. The scores for a national standardized test have a mean of 75 with a standard deviation of 9. CCSS S.ID.3

Determine the *z*-score for each of the given test scores. Round your answer to the nearest hundredth.

Test Score	*z*-score
84	☐
70	☐
90	☐
80	☐

The test score that corresponds to a *z*-score of −3.00 is ☐. The test score that corresponds to a *z*-score of 2.22 is ☐.

6. A researcher wants to estimate the average amount of time American teenagers play video games. He surveys a group of 2300 teens from all over the country and finds that the average weekly time spent playing video games is 2.3 hours. Consider each value listed in the table below. For each value, identify whether it is a parameter or a statistic. CCSS S.ID.2

Value	Parameter	Statistic
The average weekly time spent playing video games for the 2300 teens surveyed is 2.3 hours		
The average weekly time spent playing video games for all American teenagers		

7. The students at a school were asked whether they participate in sports and music. The results show that 45 students participate in sports only, 38 participate in music only, 19 participate in sports and music, and 22 participate in neither sports nor music. CCSS S.ID.5

a. Create a two-way frequency table for the data.

	Participate in Sports	Do Not Participate in Sports	Totals
Participate in Music			
Do Not Participate in Music			
Totals			

b. What percent of students surveyed participate in sports?

____

c. Of the students who participate in music, what percent do not participate in sports?

____

8. The back-to-back stem-and-leaf plot shows the ages of people who attended matinee and evening performances of a play. CCSS S.ID.5

Matinee		Evening
3556779	0	9
00349	1	6678999
38	2	001223446678
25556779	3	0134
00111246	4	56
	5	
037	6	
9	7	

a. Which data set has a higher median? Explain how you can tell.

____

____

____

b. Compare the spreads of the two data sets. Give a possible explanation for why the data sets may have different spreads.

____

____

____

c. Does either data set appear to be normally distributed? Explain how you can tell.

____

____